Physicochemical Applications of Gas Chromatography

Physicochemical Applications of Gas Chromatography

RICHARD J. LAUB
Department of Chemistry
University College of Swansea, Wales

ROBERT L. PECSOK
Department of Chemistry
University of Hawaii, Honolulu

A WILEY-INTERSCIENCE PUBLICATION

JOHN WILEY & SONS, New York • Chichester • Brisbane • Toronto

6441 — 6422

CHEMISTRY

Library of Congress Cataloging in Publication Data

Laub, Richard J.   1945-
    Physicochemical applications of gas chromatography.

    "A Wiley-Interscience publication."
    Includes bibliographical references and indexes.
    1. Gas chromatography.   I. Pecsok, Robert L.,
joint author.   II. Title.

QD79.C45L37   544.926   78-5493
ISBN 0-471-51838-7

Printed in the United States of America

10 9 8 7 6 5 4 3 2 1

To

MARIE and MARY

and

HOWARD PURNELL

**from whom we have both learned
most of what we know about
gas chromatography**

Preface

Since its inception in 1952 gas chromatography (GC) has enjoyed an explosive growth throughout the world and is now a well-known technique in all disciplines of chemistry. The reasons for its proliferation are undoubtedly many, but, chiefly, gas chromatography is an extremely powerful as well as sensitive separations method which certainly has other uses that are of compelling interest in many disciplines. These *physicochemical* (nonanalytical) applications, which are the subject of this book, require some exposition of the background physical chemistry which, however, should not be construed as irrelevant to the interests of analytical chemists. On the contrary, the topics covered here should be of value to all practicing chromatographers who wish to know more than how to inject a sample and record the results on a strip chart.

Because we hope to encourage newcomers to the field, we have tried to present the material, insofar as possible, in a "user-oriented" fashion. However, as the subject continues to expand, we must, inevitably, be found guilty of omissions. Indeed, the development of physicochemical uses of GC appears to be limited only by the number of research groups in the field and the time devoted to it; our purpose in producing this book will have been fulfilled if we stimulate or otherwise encourage new developments by workers in this area.

RICHARD J. LAUB
ROBERT L. PECSOK

Swansea, Wales
Honolulu, Hawaii
April 1978

Acknowledgments

We gratefully acknowledge the following authors, journals, and societies who have granted us permission to reproduce data and figures cited in the text:

Analytical Chemistry, © American Chemical Society for Tables 3.3–3.8, 3.11, 3.12, 4.12, 5.2A, 5.11, 6.7, 6.8, 6.11, 7.5, 7.8, 10.3, 10.5, 10.7 and Figures 7.3, 8.8, 10.4.

Aspects in Gas Chromatography, © Akademie Verlag for Table 4.3C.

Biochemical Journal, © Biochemical Society for Figure 10.2.

Canadian Journal of Chemistry, © National Research Council of Canada for Figures 7.1, 8.7.

Chromatographie, © Pergamon Press for Table 2.1.

Chromatographic Reviews, © Elsevier for Tables 10.8, 10.9.

Gas Chromatography 1962, © Butterworths for Table 4.2.

Gas Chromatography 1968, © Institute of Petroleum for Tables 4.3B, 5.12C, D, 10.6 and Figure 10.7.

Journal of American Chemical Society, © American Chemical Society for Tables 5.2B, 6.9, 6.10, 6.12–6.14, 7.2 and Figure 6.10.

Journal of the Chemical Society Faraday Transactions I, © Chemical Society for Table 5.16 and Figures 10.5, 10.6.

Journal of the Chemical Society Faraday Transactions II, © Chemical Society for Table 5.8.

Journal of Colloid and Interface Science, © Academic Press for Table 5.18.

Journal of Chromatographic Science, © Preston Technical Abstracts for Table 3.9, 5.3, 5.17, 7.4 and Figures 3.1, 3.2, 3.6.

Journal of Chromatography, © Elsevier for Tables 2.4, 3.1, 3.2, 3.10, 4.3E, F, G, 4.8, 4.9, 4.11, 5.2E, 5.5, 5.9, 5.10, 5.14, 6.15, 7.3, 7.10, 9.1A, 10.1, 10.2 and Figures 3.5, 3.7, 4.2, 7.5, 7.6, 10.1.

Journal of Physical Chemistry, © American Chemical Society for Tables 4.3A, 4.6, 5.2C, D, 5.12E, F, G, 5.13, 6.4, 6.5, 6.6, 8.1, 9.1B, 9.5, 9.6 and Figures 8.1–8.3, 8.5, 8.6.

Macromolecules, © American Chemical Society for Tables 5.15, 9.3 and Figure 7.2.

Perkin-Elmer Corp. for Tables 7.6, 7.7.

Proceedings of the Chemical Society, © Chemical Society for Figure 8.4.

Proceedings of the Royal Society, © Royal Society for Figures 3.4, 4.1.

Separation Science, © Marcel Dekker for Table 4.3D.

Transactions of the Faraday Society, © Chemical Society for Tables 4.4, 4.5, 4.10, 5.1, 5.6, 5.7, 5.12A, B, 6.1, 10.4 and Figure 3.8.

RJL thanks the Foxboro Corp. (Foxboro, Mass.) and the Science Research Council (Great Britain) for financial support. Several individuals have contributed to the preparation of the material in its final version, including, in particular, B. Edwards, E. Gleeson, L. Paul, and the text illustrator, K. Francis. Finally, we are indebted to our editor, Georgia Smith, whose thoughtfulness, kindness, and forbearance, despite innumerable delays, have continually sustained our efforts.

R.J.L.
R.L.P.

Contents

Symbols

Symbol	Definition	Page No.
a_i	Solute activity in ith phase; activity of species i	14
A	Column cross-sectional area	17
A	Eddy diffusion term in van Deemter equation	21
A_{DA}	UV charge-transfer complex absorbance	156
A_L	Liquid-phase surface area	31
A_S	Adsorbent surface area	19
A_S	Measure of peak asymmetry	28
b	Carrier fugacity coefficient divided by RT	90
B	Baseline distance from perpendicular through peak maximum to tangent to rear side of peak	28
B_o	Longitudinal diffusion term in van Deemter equation corrected to 1 atm	21
B_{11}	Solute virial coefficient	24
B_{12}	Solute-carrier virial coefficient	24
B_{22}^c	Carrier virial coefficient at column temperature	62
c	Solute concentration in mobile phase	24
c	Fraction of solute complexed in stationary phase	160
c	BET equation constant	220
C_A	Additive concentration in (binary) stationary phase	158
C_D	Donor solute concentration in (binary) stationary phase	158
C_D^0	Donor solute concentration in pure stationary phase	158
C_D^M	Donor solute concentration in mobile phase	158
C_{DA}	Complex concentration in stationary phase	158
C_L	Solute concentration in stationary phase	14
C_L	Stationary phase mass transfer non-equilibrium term in van Deemter equation	21
C_M	Solute concentration in mobile phase	14
C_M^o	Mobile-phase mass transfer non-equilibrium term in van Deemter equation corrected to 1 atm pressure	21

k'	Solute capacity factor	20
K	Packed-column specific permeability coefficient	17
K'	Defined form of equilibrium constant	160
K_1	Complex stability constant	153
K_a	Lewis relative acidity constant	139
K_{eq}	Thermodynamic equilibrium constant	160
K_R	Solute liquid/gas partition coefficient	14
$K_{R(i)}^0$	Solute partition coefficient with pure stationary phase i	159
K_S	Solute liquid/gas interfacial adsorption partition coefficient	31
l	Cell path length	156
l_A	Chart distance from point of injection to air peak maximum	48
l_R	Chart distance from point of injection to solute peak maximum	48
L	Column length	4
MW_L	Stationary-phase molecular weight	20
n	Number of carbon atoms	85
n_L	Mole number of stationary phase	20
n_1^L	Solute mole number in stationary phase	20
n_1^M	Solute mole number in mobile phase	20
N	Number of theoretical plates	4
p	Pressure	17
$\bar{p}$	Average column pressure	17
p_1^0	Solute bulk vapor pressure	19
p_{fm}	Pressure at flowmeter exit	16
p_i	Column inlet pressure	18
p_o	Column outlet pressure	17
p_w	Water-vapor pressure at flowmeter temperature	16
q	Solute concentration in stationary phase	24
r	Column tube radius	22
r	Chart speed	48
r	Ratio of solvent : solute molar volumes	119
R	Gas constant	14
R	Retention ratio	23
s	Chart speed	25
S	Adsorbent specific surface area	19
$\Delta \bar{S}_s$	Molar entropy of solution	112
t	Student factor	54
t_0	Time at which mobile phase is changed from pure carrier to solute plus carrier in FAGC	24

β	Virial parameter in expression for pressure-dependence of V_N	30
β'	Virial parameter corrected for carrier solubility effects	91
β_L	Stationary-phase mass transfer non-equilibrium term in Giddings equation	22
γ	Tortuosity factor in van Deemter equation	21
γ_1^∞	Fully-corrected solute activity coefficient	30
γ_p^∞	Uncorrected solute activity coefficient	19
δ	Solubility parameter	120
δ	NMR shift	156
Δ	NMR shift difference due to complex formation	156
ε	Packed-column porosity	17
ε_{DA}	Molar absorptivity of complex DA	156
ζ'	Virial parameter corrected for carrier solubility effects	91
η	Carrier gas viscosity	17
η	Measure of peak asymmetry	28
λ	Packing geometry term in van Deemter equation	21
λ	Carrier-stationary phase solubility term	91
μ	Chemical potential	14
ν	Reduced linear carrier velocity	22
ρ_L	Stationary-phase density	19
σ	Peak standard deviation	21
σ	Surface tension	140
σ	Molecular cross-sectional area	221
τ_c	Sorbate monolayer weight	226
ϕ	Carrier-stationary phase solubility term	91
ϕ	Volume fraction	120
χ	Interaction parameter	120
Ω	Mass transfer non-equilibrium term in Giddings equation	22

Physicochemical Applications of Gas Chromatography

PART I

Introduction

CHAPTER 1

The Chromatographic Method

All chromatographic methods have the following features in common: two mutually immiscible phases are brought into contact (possess a common interface) wherein one (the mobile phase) is made to flow over the other (the stationary phase), which remains static. The surface area of the stationary phase, which is exposed to the mobile, is generally large. When a third component, called the solute, is introduced (injected) into the system, it is partitioned between the two phases. It is also carried (eluted) through the system by the mobile (carrier) phase. While being eluted it is partitioned between the mobile and stationary phases many times, and in most situations equilibration is achieved. Because of the large number of equilibrations extant in chromatography, it is inherently more efficient than, for example placing separatory funnels in tandem.

The mobile and stationary phases can, in principle, be gases, liquids, or solids; in practice, the mobile phase is either a gas or a liquid and the stationary phase is either a liquid or a solid. Thus there are four possible techniques: gas-liquid (GLC), gas-solid (GSC), liquid-liquid (LLC), and liquid-solid (LSC) chromatography. Use of the first two (collectively abbreviated as GC) to measure physicochemical phenomena is the subject of this text.

1.1 HISTORICAL BACKGROUND

Although Day[1] and Tswett[2] are generally credited with the "discovery" of chromatography in 1903–1906, it is likely that the technique has been used in some form for centuries.[3-5] In fact, the word "chromatography" was employed as early[6] as 1731. Tswett, however, was the first to recognize that the chromatographic process consists of sequential sorption-desorption

3

interactions, as easily visualized in his original experiments in which plant pigments were separated into colored bands by elution with petroleum ether through a bed of powdered calcium carbonate.

Nevertheless, several years intervened before LSC achieved widespread recognition. Finally, in 1931; Kuhn and Lederer[7] and Kuhn, Winterstein, and Lederer[8] reported the separation of carotene and xanthophyll isomers by liquid-solid chromatography and thereby firmly established it as a new and powerful analytical method. Techniques recognizable as primitive forms of gas-solid chromatography were also in use during this period.[9, 10]

From 1931 to 1940 LSC was employed largely on an empirical basis[11, 12] for analytical separations. The next 15 years, however, saw this relatively crude approach transformed into a highly sophisticated and, by and large, well-understood family of methods, all of which are still employed today. The use of chromatography for physicochemical studies also began during this period.

Wilson,[13] in 1940, was the first to describe the chromatographic process mathematically. He assumed that solute sorption-desorption equilibria were complete. The elution process was then treated as the passage of a concentration profile through the chromatographic system. He also recognized that bandwidth phenomena were most likely dependent on column void space, diffusion, and finite rates of sorption-desorption.

In 1941 Martin and Synge[14] proposed that the chromatographic process be modeled after distillation theory[15] in which columns are divided into a number N of theoretical "plates" of equal length; the "height equivalent to a theoretical plate" H is the column length L divided by N. Equilibration of a solute between mobile and stationary phases in each plate was assumed to be complete. Plate-to-plate diffusion was said to be negligible and the ratio of solute concentrations in the phases (the partition coefficient) was taken to be constant, irrespective of the amount of solute (or solutes) present. This, of course, presupposes a linear isotherm. The model predicts that when a solute, initially deposited in the first plate, is carried down the column by the mobile phase it becomes distributed over N plates in a binomial manner due to partitioning between the mobile and stationary phases. If the number of plates exceeds 100, the distribution closely approximates a Gaussian error curve. It was shown[14] that elution times relative to some (arbitrary) standard depend on the magnitude of the solute partition coefficient and it was intuitively recognized that bandwidths are a function of N (hence H and L).

The original "plate theory" of Martin and Synge has often been criticized (e.g., see the discussion by Giddings[16]) on the grounds that it postulates a discontinuous (plate-to-plate) chromatographic distribution process. The plate model, however, does offer a simple description which, with N sufficiently large, qualitatively fits experimental data and which furthermore allowed Martin and Synge to make several important observations. They recognized, for example, that N and H depend not only on diffusion but also on flow rate, particle diameter, column pressure drop, uniformity of packing, and linearity of the solute sorption isotherm. They also demonstrated that the stationary phase support need not be granular particles and could instead consist of filter paper strips. Finally Martin and Synge recognized that since the chromatographic method consists essentially of the repeated partitioning of a solute between two phases the mobile phase could be a gas and the stationary phase, a liquid. They were also the first to measure partition coefficients in terms of the movement of solutes with respect to the solvent "front"; this constitutes the first physicochemical application of chromatography.

In 1947 Mayer and Tompkins[17] considerably expanded the original plate theory of Martin and Synge and demonstrated that the number of theoretical plates required to effect a given separation could be calculated in advance. Said[18] later verified that in the limit of large N the solute elution curve tends to a Gaussian distribution function.

In 1943 de Vault[19] offered a promising improvement to Wilson's earlier treatment by correcting several of the physical impossibilities (such as bandwidth independence of column length and flow rate) which developed from the theory. He also found that LSC band shapes could be predicted from adsorption isotherms as shown by a comparison of the theory with data for lauric acid on charcoal. de Vault's treatment was superior to that of Martin and Synge in that it was based on the continuous distribution of solute between the stationary and mobile phases along the length of the column; that is, it avoided the (hypothetical) discontinuity of the plate model. The theory proved to be intractable, however, when the elution behavior of two or more (overlapping) solutes were considered and in this respect offered no improvement. Independently, Weiss[20] arrived at substantially the same results later in 1943.

An important contribution to the continuous distribution model was made in 1944–1948 by Thomas[21] who originally examined ion exchange processes in flowing systems. He demonstrated that if the flow rate is slow enough the rate-controlled sorption-desorption column processes will approach equilibrium and can then be treated with mathematical precision. Other early contributors to this "rate" theory include Beaton and Furnas,[22] Walter,[23] Boyd and co-workers,[24, 25] and Sillen.[26]

Glueckauf[27] and Glueckauf and Coates[28, 29] made a series of prominent advances in the study of physicochemical applications of chromatography during this period; for example, they were able to solve de Vault's problem of the (hypothetical) separation of two solutes by assuming that the pair forms a pseudo-Langmuir isotherm. A graphical method was also found for solving the mathematics of two overlapping solutes when the (pseudo-)isotherm is other than Langmuir. They further demonstrated that elution curves could be calculated from adsorption isotherms and, more importantly, that adsorption and ion exchange isotherms could be abstracted from elution behavior. It was shown that localized nonequilibria can be relaxed by a suitable choice of packing particle size and flow rate and thereby a substantial increase in column efficiency could be achieved. Finally, in 1955 Glueckauf introduced the first generalized equation that quantitatively describes the elution process. Although derived for ion exchange, it was soon recognized as applicable to all forms of chromatography and indicated that much physicochemical information could be obtained from the technique.

With the advent of gas chromatography[30] in 1952 interest as well as progress in the field accelerated considerably. Development of the physicochemical aspects of the subject also divided about this time into two directions: investigations concerned with the kinetic information available from band broadening and the determination of thermodynamic quantities from sorption equilibria.

Investigations of rate-controlled kinetic processes (such as diffusion and mass transfer) were carried out by many workers. Drake[31] had considered the role of molecular diffusion in band spreading in 1949 and in 1952 Lapidus and Amundson[32] developed a mathematical model that introduced explicitly the concepts of eddy diffusion, longitudinal diffusion, and mass transfer nonequilibrium contributions. These studies formed the prelude to the well-known treatments of Klinkenberg and Sjenitzer[33] and van Deemter, Zuiderweg, and Klinkenberg[34] which appeared in 1956. Klinkenberg and co-workers simplified the mathematics of Lapidus and Amundson and as a result clarified and popularized what is now known as the rate theory of chromatography. This led in turn to substantial improvements in column efficiency which reached fruition with the work of Golay[35] on capillary columns. Thus analytical chromatography has profited considerably from rate-theory studies. Of equal importance, the investigation of diffusion and mass transfer phenomena by GC had been initiated.

Probability concepts were applied to the rate theory at this time, beginning in 1955 with Giddings and Eyring[36] and, later, Beynon and co-workers.[37] Subsequently, Giddings[38] developed a stochastic description of all sorption and diffusion processes thought to occur in chromatography; of special interest was the treatment of diffusion in terms of "random walk" statistics. A novel corollary of this work was that some rate-controlled contributions to H may, in part, be mutually dependent; that is, they are said to be "coupled." This postulate has been criticized on occasion, for it predicts that at high linear flow velocity (u) H versus u plots will asymptotically approach an upper (finite) limit. Conversely, the van Deemter treatment indicates that H tends to infinity at infinitely high values of u. Several workers, including Jones,[39] Perrett and Purnell,[40] Knox and McLaren,[41] and Littlewood,[42] have discussed this point at some length and it has on occasion provoked entertaining exchanges in the literature.[43, 44] As a result, interest in the measurement of diffusion phenomena by GC has been considerably stimulated and remains an active area of research.

On the thermodynamics side, because the partition coefficient is a (stoichiometric) concentration equilibrium constant, associated thermodynamic properties (such as the free energy and entropy) can be derived from it. Martin and Synge[14] and Consden, Gordon, and Martin[45] were the first to treat solute elution behavior as a function of the partitioning process, whereas Martin and James[30] in the first paper on GLC emphasized that much valuable physical information was available from an analysis of retention behavior. James,[46] Ray,[47] and numerous others found that plots of log (relative retention volume) versus solute carbon number were linear for a wide variety of homologous series. Hoare and Purnell[48] demonstrated in 1955–1956 that vapor pressures, boiling points, heats of vaporization, solution and mixing, and activity coefficients could all be measured by GLC. In 1955 Littlewood, Phillips, and Price[49] determined free energies and entropies by gas-liquid chromatography. Porter, Deal, and Stross,[50] in 1956, were the first to show that partition coefficients determined by GLC agreed with those measured by static techniques. Copp and Everett[51] had earlier developed semiempirical expressions for the infinite-dilution activity coefficient (γ_i^∞) of a solute in a solvent, and Herington[52] and Pierrotti and co-workers[53] used these and similar expressions in attempts to correlate GC activity coefficients with molecular structure. Purnell,[54] Bradford, Harvey, and Chalkley,[55] Barrer,[56] Kwantes and Rijnders,[57] Khan,[58] and others[59-63] contributed significantly to these efforts.

James and Phillips,[64] Griffiths and Phillips,[65] and Patton, Lewis, and Kaye[66] carried out important thermodynamic studies of GLC and GSC by frontal and displacement techniques in the years 1953–1954. Much of the GSC work at this time was based on the frontal, elution, and displacement investigations reported by Eucken and Knick,[67] Hesse and Tschanchotin,[68] Damköhler and Theile,[69] Tiselius,[70] Roth, Ohme, and Nikish,[71] Claesson,[72] Turner,[73] Phillips,[74] Turkeltaub,[75] Cremer and co-workers,[76, 77] Zhukhovitskii and co-workers,[78-80] Janak and co-workers,[81-84] and others. In several of these studies band shapes were correlated with sorption isotherms and subsequently sorption isotherms were determined for other solutes from their elution behavior.

By 1960 the thermodynamics of chromatographic partitioning was, generally speaking, well understood and advances made beyond that time were concerned mainly with refinements to the theory and with various physicochemical applications. A notable exception is the series of papers by Conder, Purnell, and co-workers[85-88] who presented a definitive treatment in 1968–1969 that embraces all known finite (as well as infinite) dilution GC techniques.

Thus, historically,[89-91] investigations commencing with liquid chromatography at the turn of this century ensured that when the postulate of a gaseous mobile phase was realized in 1952 a substantial body of knowledge about the chromatographic process was already extant. The recognized uses of chromatography to study physicochemical phenomena were greatly expanded and diversified with the advent of gas chromatography and from 1960 onward new physicochemical uses for the method were found almost as quickly as the relevant theory was developed. Today the amount and kind of useful information routinely obtained by GC is such that, broadly speaking, virtually every branch of science has found at some time or other ways to employ it to considerable advantage.

1.2 COMPARISON OF GC AND STATIC METHODS: ADVANTAGES AND LIMITATIONS

There are several advantages to the study of physicochemical phenomena by gas chromatography (as opposed to static techniques). First, GC instrumentation is generally easy to build (or, in the case of commercial instruments, inexpensive to buy), operate, and maintain.[92-97] Demands on experimental skills are therefore minimal compared with other (nonchromatographic) techniques. Second, in the normal elution mode solutes are injected into a column in such small quantities that they are effectively at infinite dilution in the solvent.[52, 53] In addition, solutes are usually separated from impurities when chromatographed in the elution mode so that only very small quantities of moderately pure material need be used. The simultaneous purification of reagents and direct measurement of their infinite-dilution properties is impossible by static methods. Further, although extrapolation to infinite dilution of finite-concentration data is in some cases possible,[98-102] in GC such extrapolation is unnecessary.

Yet gas chromatography is not limited solely to infinite-dilution studies. Solution and adsorption data can in fact be obtained over widely-extended mole fraction ranges and finite-dilution GC is today commonplace.[103-106]

As a result of these advantages little time is lost researching the instrumentation and purity of the reagents, and because data can be obtained quickly the chemistry of interest can be investigated more comprehensively by GC in a given period than it might be if other methods were used.

The accuracy of GC measurements has on numerous occasions been established to be as good as (if not superior to) static methods; this point is elaborated in detail in each chapter and illustrated by comparisons of data whenever possible.

Gas chromatography, however, has certain limitations that may necessitate the choice of other techniques. First, it is by and large limited to the study of interactions that occur on solids, in liquids, in the mobile (gas) phase, and at their interfaces. Nevertheless, it can in certain cases be used to acquire intrinsic information (such as boiling points,[107, 108] ionization potentials,[109] and molecular weights[110]) when the property of interest can be correlated with retention behavior.

The limitation of solute volatility is at times troublesome in GC. The study of underivatized steroids, for example, is generally limited to high ($>200°C$) temperatures and all-glass systems. Often some form of solute derivatization must be employed which may mask the property of interest. In this regard modern ("high-performance") LC is of obvious value,[111–113] and there is no doubt that liquid chromatography is useful for the direct measurement of liquid-liquid and liquid-solid interactions in areas in which the GC method may be only indirectly (if at all) applicable.[112]

More or less associated with the problem of solute volatility is the temperature limit of the stationary phase in the normal elution mode of GC. Liquid phases are limited normally to operation at temperatures at which their vapor pressure is less than about 0.01 torr if one is to ensure that the amount of stationary phase in a column remains virtually constant over a reasonable period. The use of volatile materials is not entirely precluded,[57, 114–118] but if there is a finite pressure drop across a column such phases must gradually be stripped off, regardless of the use of saturators and/or precolumns. This can be partially offset by employing coarse packings, small pressure gradients, and internal standards; these, however, are only limited remedies.

In general, gas chromatography can be used to study a wide variety of physicochemical phenomena quickly, simply, and accurately and in many cases is the only method available of practical value.[119–134] Experimentally, GC is limited mainly by the volatility of the solutes and stationary phases of interest and in this regard modern liquid chromatography holds much promise.[112, 132]

1.3 REFERENCES

1. D. T. Day, *Proc. Am. Phil. Soc.*, **36**, 112 (1897); *Science*, **17**, 1007 (1903).
2. M. Tswett, *Ber. Dtsch. Bot. Ges.*, **24**, 316, 384 (1906).
3. K. C. Baily, *The Elder Pliny's Chapters on Chemical Subjects*, Arnold, London, 1929.
4. H. Weil and T. I. Williams, *Nature*, **166**, 1000 (1950).
5. J. Farradane, *Nature*, **167**, 120 (1951).
6. T. I. Williams and H. Weil, *Nature*, **170**, 503 (1952).
7. R. Kuhn and E. Lederer, *Chem. Ber.*, **64**, 1349 (1931).
8. R. Kuhn, A. Winterstein, and E. Lederer, *Hoppe-Seyler's Z. Physiol. Chem.*, **197**, 141 (1931).
9. H. B. Hass, *Natl. Pet. News*, **19**, 251 (1927).
10. P. Schuftan, *Gasanalyse in der Technik*, Hirzel, Leipzig, 1931.
11. L. Zechmeister and L. V. Cholnoky, *Monatsh.*, **68**, 68 (1936); *Principles and Practice of Chromatography*, Wiley, New York, 1943.
12. T. I. Williams, *An Introduction to Chromatography*, Blackie and Son, London, 1946.
13. J. N. Wilson, *J. Am. Chem. Soc.*, **62**, 1583 (1940).
14. A. J. P. Martin and R. L. M. Synge, *Biochem. J.*, **35**, 1358 (1941).
15. K. Peters, *Ind. Eng. Chem.*, **14**, 476 (1922).
16. J. C. Giddings, *Dynamics of Chromatography*, Marcel Dekker, New York, 1965, pp. 20–26.
17. S. W. Mayer and E. R. Tompkins, *J. Am. Chem. Soc.*, **69**, 2866 (1947).
18. A. S. Said, *AIChE J.*, **2**, 477 (1956); **5**, 223 (1959).
19. D. de Vault, *J. Am. Chem. Soc.*, **65**, 532 (1943).
20. J. Weiss, *J. Chem. Soc.*, 297 (1943).
21. H. C. Thomas, *J. Am. Chem. Soc.*, **66**, 1664 (1944); *Ann. N.Y. Acad. Sci.*, **49**, 161 (1948).
22. R. H. Beaton and C. C. Furnas, *Ind. Eng. Chem.*, **33**, 1501 (1941).
23. J. E. Walter, *J. Chem. Phys.*, **13**, 332 (1945).
24. G. E. Boyd, A. W. Adamson, and L. S. Myers, Jr., *J. Am. Chem. Soc.*, **69**, 2836 (1947).
25. G. E. Boyd, L. S. Myers, Jr., and A. W. Adamson, *J. Am. Chem. Soc.*, **69**, 2849 (1947).
26. L. G. Sillen, *Nature*, **166**, 722 (1950).
27. E. Glueckauf, *Proc. Roy. Soc. Ser. A*, **186**, 35 (1946); *J. Chem. Soc.*, 1302, 1321 (1947); *Trans. Faraday Soc.*, **51**, 34, 1540 (1955); in *Ion Exchange and Its Applications*, Metcalfe and Cooper, London, 1955, p. 34.
28. J. I. Coates and E. Glueckauf, *J. Chem. Soc.*, 1308 (1947).
29. E. Glueckauf and J. I. Coates, *J. Chem. Soc.*, 1315 (1947).
30. A. J. P. Martin and A. T. James, *Biochem. J.*, **50**, 679 (1952).
31. B. Drake, *Anal. Chim. Acta*, **3**, 452 (1949).
32. L. Lapidus and N. R. Amundson, *J. Phys. Chem.*, **56**, 984 (1952).
33. A. Klinkenberg and F. Sjenitzer, *Chem. Eng. Sci.*, **5**, 258 (1956).
34. J. J. van Deemter, F. J. Zuiderweg, and A. Klinkenberg, *Chem. Eng. Sci.*, **5**, 271 (1956).
35. M. J. E. Golay, in *Gas Chromatography*, V. J. Coates, H. J. Noebels, and I. S. Fagerson, Eds., Academic, New York, 1958, p. 1; in *Gas Chromatography 1958*, D. H. Desty, Ed., Butterworths, London, 1958, p. 36.

36. J. C. Giddings and H. Eyring, *J. Phys. Chem.*, **59**, 416 (1955).
37. J. H. Beynon, S. Clough, D. A. Crooks, and G. R. Lester, *Trans. Faraday Soc.*, **54**, 705 (1958).
38. J. C. Giddings, *J. Chem. Phys.*, **26**, 169, 1755 (1957); *J. Chem. Educ.*, **35**, 588 (1958); *J. Chem. Phys.*, **31**, 1462 (1959); see also Ref. 16.
39. W. L. Jones, *Anal. Chem.*, **33**, 829 (1961).
40. R. H. Perrett and J. H. Purnell, *Anal. Chem.*, **35**, 430 (1963).
41. J. H. Knox and L. McLaren, *Anal. Chem.*, **35**, 449 (1963).
42. A. B. Littlewood, in *Gas Chromatography 1964*, A. Goldup, Ed., Institute of Petroleum, London, 1965, p. 77; *Anal. Chem.*, **38**, 291 (1966).
43. A. Klinkenberg, *Anal. Chem.*, **38**, 489, 491 (1966).
44. J. C. Giddings, *Anal. Chem.*, **38**, 490 (1966).
45. R. Consden, A. H. Gordon, and A. J. P. Martin, *Biochem. J.*, **38**, 224 (1944).
46. A. T. James, *Biochem. J.*, **52**, 242 (1952).
47. N. H. Ray, *J. Appl. Chem.*, **4**, 21 (1954).
48. M. R. Hoare and J. H. Purnell, *Research*, **8**, S41 (1955); *Trans. Faraday Soc.*, **52**, 222 (1956).
49. A. B. Littlewood, C. S. G. Phillips, and D. T. Price, *J. Chem. Soc.*, 1480 (1955).
50. P. E. Porter, C. H. Deal, and F. H. Stross, *J. Am. Chem. Soc.*, **78**, 2999 (1956).
51. J. L. Copp and D. H. Everett, *Discuss. Faraday Soc.*, **15**, 268 (1953).
52. E. F. G. Herington, *Analyst*, **81**, 52 (1956); in *Vapour Phase Chromatography*, D. H. Desty, Ed., Butterworths, London, 1957, p. 5.
53. G. J. Pierrotti, C. H. Deal, E. L. Derr, and P. E. Porter, *J. Am. Chem. Soc.*, **78**, 2989 (1956).
54. J. H. Purnell, in *Vapour Phase Chromatography*, D. H. Desty, Ed., Butterworths, London, 1957, p. 52; *J. Roy. Inst. Chem.*, **82**, 586 (1958).
55. B. W. Bradford, D. Harvey, and D. E. Chalkley, *J. Inst. Pet.*, **41**, 80 (1955).
56. R. M. Barrer, in *Gas Chromatography 1958*, D. H. Desty, Ed., Butterworths, London, 1958, p. 122.
57. A. Kwantes and G. W. A. Rijnders, in *Gas Chromatography 1958*, D. H. Desty, Ed., Butterworths, London, 1958, p. 125.
58. M. A. Khan, in *Gas Chromatography 1958*, D. H. Desty, Ed., Butterworths, London, 1958, p. 135; in *Gas Chromatography 1960*, R. P. W. Scott, Ed., Butterworths, London, 1960, p. 251.
59. G. Dijkstra, J. G. Keppler, and J. A. Schols, *Rec. Trav. Chim. Pays-Bas*, **74**, 805 (1955).
60. F. R. Cropper and A. Heywood, *Nature*, **172**, 1101 (1953); **174**, 1063 (1954).
61. M. Dimbat, P. E. Porter, and F. H. Stross, *Anal. Chem.*, **28**, 290 (1956).
62. E. M. Fredericks and F. R. Brooks, *Anal. Chem.*, **28**, 297 (1956).
63. F. T. Eggertsen, H. S. Knight, and S. Groennings, *Anal. Chem.*, **28**, 303 (1956).
64. D. H. James and C. S. G. Phillips, *J. Chem. Soc.*, 1600 (1953); 1066 (1954).
65. J. H. Griffiths and C. S. G. Phillips, *J. Chem. Soc.*, 3446 (1954).
66. H. W. Patton, J. S. Lewis, and W. I. Kaye, *Anal. Chem.*, **27**, 170 (1955).
67. A. Eucken and H. Knick, *Brennst.-Chem.*, **17**, 241 (1936).
68. G. Hesse and B. Tschachotin, *Naturwiss.*, **30**, 387 (1942).

69. G. Damköhler and H. Theile, *Chemie*, **56**, 353, 354 (1943).

70. A. Tiselius, *Arkiv. Kemi Mineral. Geol.*, **14B**, No. 22 (1940); **16A**, No. 18 (1943).

71. F. Roth, W. Ohme, and A. Nikish, *Oel u. Kohle*, **37**, 1133 (1942).

72. S. Claesson, *Arkiv Kemi. Mineral. Geol.*, **23A**, 133 (1946); **24A**, 7 (1946); *Discuss. Faraday Soc.*, **7**, 34 (1949).

73. W. C. Turner, *Natl. Pet. News*, **35**, 234 (1943).

74. C. S. G. Phillips, *Discuss. Faraday Soc.*, **7**, 241 (1949).

75. N. M. Turkeltaub, *Zh. Anal. Khim.*, **5**, 200 (1950); *Neft. Khoz.*, **32**, 72 (1954).

76. E. Cremer and R. Müller, *Z. Elektrochem.*, **55**, 217 (1951); *Mikrochim. Acta*, **36/37**, 553 (1951).

77. E. Cremer and F. Prior, *Z. Elektrochem.*, **55**, 66 (1951).

78. A. A. Zhukhovitskii, O. V. Zolotareva, V. A. Sokolov, and N. M. Turkeltaub, *Dokl. Akad. Nauk SSSR*, **77**, 435 (1951).

79. A. A. Zhukhovitskii, N. M. Turkeltaub, and V. A. Sokolov, *Dokl. Akad. Nauk SSSR*, **88**, 859 (1953).

80. A. A. Zhukhovitskii, N. M. Turkeltaub, and T. V. Georgievskaya, *Dokl. Akad. Nauk SSSR*, **92**, 987 (1953).

81. J. Janak, *Chem. Listy*, **47**, 464, 817, 828, 837, 1184, 1348 (1953); **49**, 1403 (1955).

82. J. Janak and M. Rusek, *Chem. Listy*, **47**, 1184, 1190 (1953); *Collect. Czech. Chem. Commun.*, **19**, 700 (1954).

83. J. Janak and I. Paralova, *Chem. Listy*, **47**, 1476 (1953); *Collect. Czech. Chem. Commun.*, **20**, 336 (1955).

84. J. Janak and K. Tesarik, *Chem. Listy*, **51**, 2048 (1957); *Z. Anal. Chem.*, **164**, 62 (1958); *Collect. Czech. Chem. Commun.*, **24**, 536 (1959).

85. J. R. Conder and J. H. Purnell, *Trans. Faraday Soc.*, **64**, 1505, 3100 (1968); **65**, 824, 839 (1969).

86. J. R. Conder, D. C. Locke, and J. H. Purnell, *J. Phys. Chem.*, **73**, 700 (1969).

87. D. F. Cadogan, J. R. Conder, D. C. Locke, and J. H. Purnell, *J. Phys. Chem.*, **73**, 708 (1969).

88. J. R. Conder, *J. Chromatogr.*, **39**, 273 (1969); *Chromatographia*, **7**, 387 (1974).

89. R. J. Magee, Ed., *Selected Readings in Chromatography*, Pergamon, New York, 1970.

90. A. J. P. Martin, in *Gas Chromatography in Biology and Medicine*, R. Porter, Ed., Churchill, London, 1969, p. 2.

91. L. S. Ettre, *Am. Lab.*, **4**(10), 10 (1972); *J. Chromatogr.*, **112**, 1 (1975).

92. D. H. Lichtenfels, S. A. Fleck, and F. H. Burow, *Anal. Chem.*, **27**, 1510 (1955).

93. D. H. Everett and C. T. H. Stoddart, *Trans. Faraday Soc.*, **57**, 746 (1961).

94. A. J. B. Cruickshank, M. L. Windsor, and C. L. Young, *Proc. Roy. Soc. Ser. A*, **295**, 271 (1966).

95. M. Goedert and G. Guiochon, *J. Chromatogr. Sci.*, **7**, 323 (1969).

96. G. Blu, F. Lazarre, and G. Guiochon, *Anal. Chem.*, **45**, 1375 (1973).

97. J. E. Oberholtzer and L. B. Rogers, *Anal. Chem.*, **41**, 1234 (1969).

98. J. A. V. Butler, D. W. Thomson, and W. H. McLennon, *J. Chem. Soc.*, 674 (1933).

99. R. J. L. Andon, J. D. Cox, and E. F. G. Herington, *J. Chem. Soc.*, 3188 (1954).

100. H. C. Van Ness, C. A. Soczek, and N. K. Kochar, *J. Chem. Eng. Data*, **12**, 346 (1967).

101. L. B. Schreiber and C. A. Eckert, *Ind. Eng. Chem. Proc. Des. Dev.*, **10**, 572 (1971).

102. D. A. Palmer and B. D. Smith, *Ind. Eng. Chem. Proc. Des. Dev.*, **11**, 114 (1972).

103. J. F. Parcher and C. L. Hussey, *Anal. Chem.*, **45**, 188 (1973).

104. V. Rezl, B. Kaplanova, and J. Janak, *Anal. Chem.*, **47**, 159 (1975).

105. P. Valentin and G. Guiochon, *Sep. Sci.*, **10**, 245, 271, 289 (1975).

106. J. R. Conder, *Anal. Chem.*, **48**, 917 (1976).

107. A. Matukuma, in *Gas Chromatography 1968*, C. L. A. Harbourn, Ed., Institute of Petroleum, London, 1969, p. 55.

108. J. J. Walraven and A. W. Ladon, in *Gas Chromatography 1970*, R. Stock, Ed., Institute of Petroleum, London, 1971, p. 358.

109. R. J. Laub and R. L. Pecsok, *Anal. Chem.*, **46**, 1214 (1974).

110. D. E. Martire and J. H. Purnell, *Trans. Faraday Soc.*, **62**, 710 (1966).

111. D. E. Martire and D. C. Locke, *Anal. Chem.*, **39**, 921 (1967).

112. D. C. Locke, *Adv. Chromatogr.*, **8**, 47 (1969); **14**, 87 (1976).

113. L. R. Snyder, *Principles of Adsorption Chromatography*, Marcel Dekker, New York, 1968.

114. J. Novak and J. Janak, *Anal. Chem.*, **38**, 265 (1966).

115. R. E. Pecsar and J. J. Martin, *Anal. Chem.*, **38**, 1661 (1966).

116. S. P. Wasik and W. Tsang, *J. Phys. Chem.*, **74**, 2970 (1970); *Anal. Chem.*, **42**, 1648 (1970).

117. P. E. Barker and A. K. Hilmi, *J. Gas Chromatogr.*, **5**, 119 (1967).

118. A. Hartkopf and B. L. Karger, *Acc. Chem. Res.*, **6**, 209 (1973).

119. C. J. Hardy and F. H. Pollard, *J. Chromatogr.*, **2**, 1 (1959).

120. R. A. Keller, G. H. Stewart, and J. C. Giddings, *Ann. Rev. Phys. Chem.*, **11**, 347 (1960).

121. J. H. Purnell, *Endeavour*, **23**, 142 (1964); *Ann. Rev. Phys. Chem.*, **18**, 81 (1967).

122. D. E. Martire and L. Z. Pollara, *Adv. Chromatogr.*, **1**, 335 (1965).

123. J. C. Giddings and K. L. Mallik, *Ind. Eng. Chem.*, **59**(4), 18 (1967).

124. R. Kobayashi and H. A. Deans, *Ind. Eng. Chem.*, **59**(5), 11 (1967).

125. R. Kobayashi, P. S. Chappelear, and H. A. Deans, *Ind. Eng. Chem.*, **59**(10), 63 (1967).

126. J. R. Conder, *Adv. Anal. Chem. Instrum.*, **6**, 209 (1968).

127. S. Trestianu, *Rev. Chim.*, **19**, 709 (1968).

128. C. L. Young, *Chromatogr. Rev.*, **10**, 129 (1968).

129. M. S. Vigdergauz and R. I. Izmailov, *The Application of Gas Chromatography in the Physico-Chemical Characterization of Substances*, Nauka, Moscow, 1970.

130. A. N. Korol, *Usp. Khim.*, **41**, 321 (1972).

131. A. V. Kiselev, A. V. Iogansen, K. I. Sakodynskii, V. M. Sakharov, Y. I. Yashin, A. P. Karnaukhov, N. Y. Buyanova, and G. A. Karkchi, *Physico-Chemical Applications of Gas Chromatography*, Khimiya, Moscow, 1973.

132. D. C. Locke, *Am. Lab.*, **7**(5), 17 (1975).

133. K. L. Mallik, *An Introduction to Nonanalytical Applications of Gas Chromatography*, Peacock Press, New Delhi, 1976.

134. J. R. Conder and C. L. Young, *Physico-Chemical Applications of Gas Chromatography*, in preparation.

CHAPTER 2

Precision Data: The Partition Coefficient

2.1 RETENTION EQUATIONS

The distribution of a solute between stationary (L, liquid, or S, solid) and mobile (M) phases at constant temperature and pressure corresponds to equilibrium when the solute free energy is at a minimum. Its chemical potential in one phase is then equal to that in the other phase (e.g., Ref. 1):

$$\mu_L = \mu_M \tag{2.1}$$

where

$$\mu_i = \mu_i^0 + RT \ln a_i \tag{2.2}$$

a_i is the solute activity in the ith phase and μ_i^0 is the solute chemical potential at some unit activity however it is defined. Equating expressions for μ_L and μ_M and making the approximation that activities can for the moment be replaced by concentrations gives

$$\mu_L^0 + RT \ln C_L = \mu_M^0 + RT \ln C_M \tag{2.3}$$

which on rearrangement yields

$$\frac{C_L}{C_M} = \exp\left(\frac{\Delta\mu^0}{RT}\right) = K_R \tag{2.4}$$

Since ideally $\Delta\mu^0$ ($= \mu_M^0 - \mu_L^0$) is a constant, the solute partition coefficient K_R is assumed accordingly to be invariant with the amount of solute and phases extant in the system.

Partition coefficients are of fundamental importance in GC since, from them, equilibrium (sorption) thermodynamic properties can be determined. The normal elution mode of GC is by far the most commonly employed technique used to measure K_R values. The time required for the center of gravity of the solute band to pass completely through the column (the solute retention time t_R) is a function of the relative amounts of time spent in the mobile and stationary phases; hence K_R.

2.1.1 Linear Ideal Chromatography: Elution at Infinite Dilution

In the simplest model of chromatography, first treated by Wilson,[2] a solute is said to undergo partitioning but not band broadening while being eluted. Solute equilibrium is assumed to be achieved instantaneously everywhere in the column and K_R is identified with the true thermodynamic partition coefficient. The sorption isotherm is taken to be linear; the solute retention time will therefore be independent of the amount injected (usually 0.1 to 5 μl). It will also be identical to the retention time if an infinitesimal amount is injected. When the solute is dissolved in (or adsorbed on) the stationary phase, it is assumed to be immobile; movement occurs only when the solute vaporizes and is carried down the column by the mobile phase. The linear rate of travel is therefore equal to the average carrier velocity $\bar{u}$ multiplied by the fraction of time the solute spends in the mobile phase:

$$\text{rate of travel} = \bar{u}\left(\frac{C_M V_M}{C_M V_M + C_L V_L}\right) \tag{2.5}$$

where V_M and V_L are the mobile-phase and stationary-phase volumes, respectively (in GSC V_L is replaced by A_S, the adsorbent surface area). Equation 2.5 is rearranged to give

$$\text{rate of travel} = \bar{u}\left(1 + \frac{C_L V_L}{C_M V_M}\right)^{-1} \tag{2.6}$$

Since C_L/C_M is K_R, the partition coefficient, the rate of travel becomes

$$\text{rate of travel} = \bar{u}\left(1 + K_R \frac{V_L}{V_M}\right)^{-1} \tag{2.7}$$

Alternatively, the solute rate of travel is given by

$$\text{rate of travel} = \frac{\text{column length } L}{\text{retention time } t_R} \tag{2.8}$$

Equating relations 2.7 and 2.8, followed by rearrangement, yields

$$t_R = \frac{L}{\bar{u}}\left(1 + K_R\frac{V_L}{V_M}\right) \tag{2.9}$$

The quantity $L/\bar{u}$ is just t_A, the ("dead") time a nonsorbed ($K_R = 0$) solute requires to pass through the column and

$$t_R = t_A\left(1 + K_R\frac{V_L}{V_M}\right) \tag{2.10}$$

Equation 2.10 is one form of the fundamental retention equation first deduced by Martin and Synge[3] and is applicable to all chromatographic techniques.

In order to convert observed retention times to gas volumes the mobile phase flow rate inside the column must be known. Generally it is measured at the column outlet, usually after the detector. The measured flow rate F must therefore be corrected to the conditions prevailing in the column; that is, F_c:

$$F_c = F\left(\frac{T}{T_{fm}}\right)\left(\frac{p_{fm} - p_w}{p_{fm}}\right) \tag{2.11}$$

T and T_{fm} are the column and flowmeter temperatures and p_{fm} and p_w are the flowmeter and water-vapor pressures at T_{fm}; the second bracketed term on the rhs of eq. 2.11 is included if a soap-bubble flowmeter is employed. The "raw" dead (V_A) and retention (V_R) volumes are now given by

$$V_A = t_A F_c \tag{2.12}$$

$$V_R = t_R F_c \tag{2.13}$$

and eq. 2.10 becomes

$$V_R = V_A\left(1 + K_R\frac{V_L}{V_M}\right) \tag{2.14}$$

The "adjusted" retention volume is defined as

$$V_R' = V_R - V_A = F_c(t_R - t_A) \tag{2.15}$$

In order for mobile phase to flow through a column a pressure gradient must, of course, exist. This necessitates the introduction of a gas compressibility correction factor, as first recognized by Martin and James[4] in 1952. Consider a carrier gas flowing through a packed column of uniform cross section A at a pressure p and velocity u. The volume throughput must be constant everywhere within the column so that

$$puA = p_o u_o A = \bar{p}\bar{u}A \qquad (2.16)$$

where $\bar{p}$, p_o, $\bar{u}$, and u_o are the average and outlet pressures and velocities, respectively. The velocity at any point is therefore given by

$$u = \frac{p_o u_o}{p} \qquad (2.17)$$

The velocity can also be related to the pressure gradient dp within a length dx along the column, the column specific permeability coefficient K, porosity ε, and the gas viscosity η, through Darcy's law

$$u = -\frac{K}{\varepsilon\eta}\frac{dp}{dx} \qquad (2.18)$$

Rearranging, substituting for u from eq. 2.17, and multiplying through by p,

$$dx = \left(-\frac{K}{\varepsilon\eta u_o p_o}\right)p\,dp \qquad (2.19a)$$

$$p\,dx = \left(-\frac{K}{\varepsilon\eta u_o p_o}\right)p^2\,dp \qquad (2.19b)$$

The average value of a continuous function $F(X)$ is

$$\bar{F}(X) = \frac{\int F(X)\,dX}{\int dX} \qquad (2.20)$$

so that

$$\bar{p} = \frac{\int(-K/\varepsilon\eta u_o p_o)p^2\,dp}{\int(-K/\varepsilon\eta u_o p_o)p\,dp} \qquad (2.21)$$

Integrating over the column pressure gradient, which is bounded by the inlet (p_i) and outlet (p_o) pressures,

$$\bar{p} = \frac{2}{3} \left[\frac{(p_i^3 - p_o^2)^{\frac{2}{3}}}{(p_i^2 - p_o^3)} \right] \tag{2.22}$$

Dividing by p_o followed by rearrangement gives the desired ratio:

$$\frac{p_o}{\bar{p}} = \frac{3}{2} \left[\frac{(p_i/p_o)^2 - 1}{(p_i/p_o)^3 - 1} \right] \tag{2.23}$$

Gas volumes measured at the column outlet can therefore be corrected to the average column pressure by multiplying by the fraction $p_o/\bar{p}$, which is commonly given the symbol j. (Everett[5] has suggested that the compressibility correction be represented, generally, as

$$J_n^m = \frac{n}{m} \left[\frac{(p_i/p_o)^m - 1}{(p_i/p_o)^n - 1} \right]$$

Equation 2.22 would thus be given as J_3^2; this is not commonly used, the symbol j being universally accepted as the usual compressibility correction factor. When m and n differ from 2 and 3, the distinction will be noted by the use of J_n^m.)

The fully corrected dead volume V_M, is jV_A so that eq. 2.14 becomes

$$jV_R = V_M \left(1 + K_R \frac{V_L}{V_M} \right) \tag{2.24}$$

The term jV_R is given the symbol V_R^0 and is referred to as the "corrected" retention volume:

$$V_R^0 = V_M \left(1 + K_R \frac{V_L}{V_M} \right) = V_M + K_R V_L \tag{2.25}$$

The product jV_R' is called the net retention volume V_N:

$$V_N = jV_R' = V_R^0 - V_M = K_R V_L \tag{2.26}$$

V_N values depend on the amount (and, of course, the type) of stationary phase. It is therefore common practice to divide V_N by the weight of liquid stationary phase w_L, or the weight of adsorbent phase w_S, and to convert the data to $0°C$, these now being called specific retention volumes:

$$V_g^0 = \frac{V_N}{w_L}\left(\frac{273.15}{T}\right) \tag{2.27}$$

The specific retention volume at the column temperature V_g^T is occasionally used:

$$V_g^T = \frac{V_N}{w_L} = \frac{K_R}{\rho_L} \tag{2.28}$$

where ρ_L is the stationary phase density. In GSC the surface area per gram of adsorbent (the specific surface area) S is employed:

$$K_R = \frac{V_g^T}{S} = \frac{V_N}{Sw_S} = \frac{V_N}{A_S} \tag{2.29}$$

(Here K_R pertains to a two-dimensional concentration of adsorbate on the adsorbent surface and has units of distance.)

These relations are summarized by the equation

$$K_R = \underbrace{(t_R - t_A)}_{t_R'} \left[\underbrace{F\left(\frac{T}{T_{fm}}\right)\left(\frac{p_{fm} - p_w}{p_{fm}}\right)}_{F_c} \right] \left\{ \underbrace{\frac{3}{2}\left[\frac{(p_i/p_o)^2 - 1}{(p_i/p_o)^3 - 1}\right]}_{j} \underbrace{\left(\frac{\rho_L}{w_L}\right)}_{V_L^{-1}} \right\} = \frac{V_g^0 T \rho_L}{273.15} \tag{2.30}$$

which relates the experimentally measured quantities of an elution GC experiment to the solute partition coefficient.

The solute partial pressure over its infinitely dilute solution (Henry's law region) in the liquid phase is

$$p_1 = \gamma_p^\infty x_1^L p_1^0 \tag{2.31}$$

Recognizing that $x_1^L \simeq n_1^L/n_L$, dividing both sides by V_L, and rearranging,

$$\frac{n_1^L}{V_L} = \frac{n_L p_1}{\gamma_p^\infty p_1^0 V_L} \tag{2.32}$$

where p_1 and p_1^0 are the solute partial and saturation vapor pressures, x_1^L is the solute mole fraction in the liquid phase, n_1^L and n_L are the solute and liquid phase mole numbers, and γ_p^∞ is the Henry's law activity coefficient. Since, for an ideal gas,

$$\frac{n_1^M}{V_M} = \frac{p_1}{RT} \tag{2.33}$$

and $K_R = (n_1^L V_M)/(n_1^M V_L)$, combination of eqs. 2.32 and 2.33 gives

$$K_R = \frac{RT}{\gamma_p^\infty p_1^0 \overline{V}_L} = \frac{w_L RT}{\gamma_p^\infty p_1^0 V_L MW_L} = \frac{\rho_L RT}{\gamma_p^\infty p_1^0 MW_L} \tag{2.34}$$

where $\overline{V}_L$ and MW_L are the molar volume and molecular weight of the stationary phase. Comparing eqs. 2.28 and 2.34,

$$V_g^T = \frac{RT}{\gamma_p^\infty p_1^0 MW_L} \tag{2.35}$$

Equations 2.34 and 2.35 are of fundamental importance, for they relate the solute retention behavior to its activity coefficient and vapor pressure.

The term $K_R V_L / V_M$ occurs so frequently in gas chromatography that it is given the symbol k' and is called the capacity factor (or the packing factor). van Deemter and co-workers[6] originally defined the inverse function k as

$$k = \frac{V_M}{K_R V_L} \left(= \frac{1}{k'} \right) \tag{2.36}$$

and some confusion has thereby resulted in the literature. The definitions of eq. 2.36 are maintained here. The phase (volume) ratio V_M / V_L is given the symbol β so that

$$k' = \frac{K_R}{\beta} = \frac{t_R - t_A}{t_A} \tag{2.37}$$

The most commonly reported retention parameter is α, the "relative retention" of two solutes:

$$\alpha = \frac{K_{R_2}}{K_{R_1}} = \frac{V_{g_2}}{V_{g_1}} = \frac{V_{N_2}}{V_{N_1}} = \frac{V'_{R_2}}{V'_{R_1}} = \frac{t'_{R_2}}{t'_{R_1}} \tag{2.38}$$

2.1.2 Linear Nonideal Chromatography: Plate Height Equations The equations used to correct the linear ideal model for band broadening originate from the plate model of Martin and Synge[3] which predicts that when stationary/mobile phase partitioning is the mechanism solely responsible for retention solute peaks are symmetric and Gaussian in shape. The mathematics of statistical distributions is therefore applicable and is summarized in Fig. 2.1. The various band-broadening processes which contribute to the peak variance σ^2 include solute transverse and longitudinal diffusion in the gas phase, finite rate of equilibration of solute between the mobile and stationary phases, diffusion in the liquid phase, and factors depending on the geometry of the column and packing.[7-12] To a first approximation the total peak variance is the sum of the individual variances and, for packed columns, the plate height H is given by the extended form of the van Deemter equation[12]:

$$H = \frac{\sigma^2}{L} = 2\lambda d_p + \frac{2\gamma D_{12}^o}{u_o} + \frac{2}{3}\frac{k'}{(1+k')^2}\frac{d_f^2}{D_{13}} + \frac{(k')^2 d_p^2 u_o}{96 D_{12}^o (1+k')^2} \quad (2.39)$$

which is abbreviated as

$$H = A + \frac{B}{u_o} + C_L \bar{u} + C_M u_o \quad (2.40)$$

The first contribution A, called eddy diffusion, results from the inhomogeneity of flow velocities and path lengths around the irregularly shaped

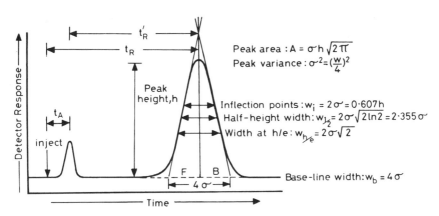

Figure 2.1 Elution chromatogram showing an "air" peak at t_A and a solute peak at t_R, the latter being treated as a Gaussian distribution.

packing particles; d_p is the particle diameter and λ is an unspecified constant which is a complex function of the packing and column geometries. The second coefficient B defines the effect of longitudinal diffusion in the gas phase where γ is a "tortuosity factor" and D_{12}^o and u_o are the solute diffusion coefficient and carrier velocity, respectively, at the column outlet. Since D_{12}^i/u_i is pressure-independent and D_{12} is inversely pressure-dependent, the appropriate diffusion coefficient is apparently that at the outlet; that is, 1 atm pressure. The third term results from solute/stationary phase resistance to mass transfer: D_{13} is the solute diffusion coefficient in the liquid phase and d_f is the thickness of the phase that is assumed to be spread on the support as an even film. Since D_{13} is pressure-independent, the appropriate velocity is $\bar{u}$. Last, radial mass transfer resistance may occur in the gas phase as given by the fourth term.

Golay[13] has derived the following for wall-coated open-tubular (WCOT) columns:

$$H = \frac{2D_{12}^o}{u_o} + \frac{2}{3}\frac{k'r^2\bar{u}}{(1+k')^2 D_{13}} + \frac{[1+6k'+11(k')^2]r^2 u_o}{24(1+k')^2 D_{12}^o} \qquad (2.41)$$

For porous-layer, open-tubular (PLOT) columns the C_L and C_M terms become

$$C_L = \left(1+\frac{2a_2}{F}\right)\left[\frac{(k')^3}{6(1+k')^2}\right]\left(\frac{r^2}{K_R^2 D_{13}}\right) \qquad (2.42)$$

$$C_M = \frac{r^2}{24D_{12}^o}\left[\frac{1+6k'+11(k')^2}{(1+k')^2} + \frac{8+32k'}{(1+k')^2}a_2 + \frac{8(k')^2}{(1+k')^2}\frac{a_1^2}{a_2}\right] \qquad (2.43)$$

where a_1 and a_2 are factors depending on the tortuosity and free gas volume, F is a liquid-phase surface area enhancement term, and r is the column radius.

Giddings[14] has rejected the notion that the total peak variance is the sum of the individual variances and has adopted the view that eddy diffusion and mass transfer are mutually dependent. Accordingly,

$$h = \frac{2\gamma}{v}(1+\beta_L) + \Omega v + \sum\left(\frac{1}{2\lambda_i}+\frac{1}{w_i v}\right)^{-1} \qquad (2.44)$$

where h is a reduced plate height ($=H/d_p$), v is a reduced gas velocity

$(= \bar{u}d_p / D_{13})$, β_L is given by

$$\beta_L = \frac{(1 - R)\gamma_L D_{13}}{R\gamma D_{12}} \qquad (2.45)$$

and Ω is defined as

$$\Omega = \frac{CD_{12}}{d_p^2} \qquad (2.46)$$

γ_L and C are stationary phase mass transfer terms that include finite rates of sorption and desorption, λ_i and w_i are configurational and mass transfer parameters that describe various modes of eddy diffusion in the mobile phase, and R, called the retention ratio, is defined by

$$R = \frac{t_A}{t_R} = \frac{V_M}{V_M + K_R V_L} = \frac{1}{1 + k'} \qquad (2.47)$$

Thus the term $2\gamma/\nu$ accounts for longitudinal diffusion in the gas phase and $2\gamma\beta_L/\nu$ reflects diffusion in the stationary phase; $\Omega\nu$ represents kinetic-controlled mass transfer effects in the stationary phase and the summation expresses the coupling of eddy diffusion and rate-controlled gas-phase mass transfer resistance.

When eq. 2.47 is abbreviated as

$$H = \frac{B}{u} + Cu + \sum \left(\frac{1}{A} + \frac{1}{C_M u} \right)^{-1} \qquad (2.48)$$

the similarity to (and disparity with) eq. 2.40 is immediately apparent.

2.1.3 Nonlinear Nonideal Chromatography: Elution at Finite Dilution

In finite-dilution GC, in contrast to the infinite-dilution mode, the solute injection profile is adjusted to resemble any one of a number of shapes of varying concentration and duration.[15] There are, however, only two methods of interest here that can be broadly classified as frontal analysis (FA) and frontal analysis by characteristic point (FACP). Although other GC techniques (such as elution on a plateau, elution by characteristic point, vacancy chromatography, and chromathermography are available,[15-49] finite-dilution modes are considered in this book solely for the purpose of determining physicochemical data which, inevitably, involve some form of the partition coefficient. With this view in mind (which, admittedly, may be a matter of some disagreement), the FA and FACP modes are considered to be the simplest and most accurate and therefore the most useful.

The theory of finite-dilution GC has with slight variations been presented elsewhere[49-57] and is mentioned here only briefly. The approach of Conder and Purnell[49] encompasses FA, FACP, and other techniques and is therefore adopted.

Frontal Analysis (FA) In the frontal analysis technique, first treated quantitatively by Glueckauf,[58] pure carrier is initially passed through the column. At some time t_0 the carrier stream at the inlet is switched to one consisting of solute plus carrier, thus introducing a finite-concentration step change in the mobile phase. The solute "breaks through" (reaches the end of the column) at time t_1 and the recorder trace rises to a plateau that corresponds to the solute concentration in the carrier at the column outlet. At some later time t_2 pure carrier is switched back to the inlet and the baseline returns to that obtained before time t_0. The front and rear boundaries of the plateau may be diffuse or self-sharpening and each may contain step changes, depending on the amount of solute used and the isotherm type. All five BET isotherm types[59] have been encountered in GC,[60] but types II, IV, and V may be treated as composites of types I and III and only these two need be considered[60]; types I and III and the respective FA chromatograms are illustrated in Figs. 2.2 and 2.3, in which the ordinate C_L is often given as the volume of sorbate per gram sorbent and the abscissa C_M is represented as a partial pressure. If the isotherm were linear, the concentration boundaries would be equivalent to the front and rear boundaries of a Gaussian curve.

Conder and Purnell showed that the generalized fundamental retention equation, applicable to all finite as well as infinite-dilution GC modes, takes the form

$$V_R^0 = V_M + V_L(1 - ajy_o)\left(\frac{\partial q}{\partial c}\right)_p \qquad (2.49)$$

where a and b are defined by

$$a = \frac{b_2^1}{b_3^2}\left[1 + \frac{2y_o P_o B_{11}}{RT}(1 - y_o J_2^1)\right] \qquad (2.50)$$

$$b_n^m = 1 + k'(1 - J_n^m y_o) \qquad (2.51)$$

in which q and c are the solute stationary- and mobile-phase concentrations, respectively, and y_o is the solute mole fraction in the carrier at the column outlet. B_{11} is the solute fugacity coefficient. For example, at infinite dilution $ajy_o \to 0$, $(\partial q/\partial c) \to K_R$, and eq. 2.49 becomes the more familiar relation (eq. 2.25). Infinite dilution is therefore the limit of eq. 2.49 at zero solute concentration in the mobile phase.

In order to employ eq. 2.49, it must first be cast in terms of q; Conder and Purnell[49] have shown that it takes the form

$$
\bar{\bar{q}} = \frac{p_o}{V_L RT_{fm}} \left\{ \left[\frac{r(\alpha+\beta)(1+k')F(o)}{ahs} \right] \left[\ln\left(1 + \frac{\bar{\bar{\psi}}}{(1+k')(1-\bar{\bar{\psi}})}\right) \right] + G \right\}
$$

$$
+ \frac{V_M}{a} \ln\left(1 - \bar{\bar{\psi}}\right) - \frac{r\beta\bar{\bar{y}}_o(1+k')F(o)}{hs\left[1 + k'\left(1 - \bar{\bar{y}}_o^*\right)\right]} \tag{2.52}
$$

where α is the area, KMNQ, in Figs. 2.2 and 2.3, β is the area $\pm$TNQ ($+$ for Fig. 2.3, $-$ for Fig. 2.2, and zero for self-sharpening boundaries), h is the plateau height, s is the chart speed, r is a units conversion factor, $F(o)$ is the flow rate of pure carrier at the column outlet, and

$$
\bar{\bar{y}}_o^* = 0.8\bar{\bar{y}}_o \tag{2.53}
$$

$$
\bar{\bar{\psi}} = aj\bar{\bar{y}}_o \tag{2.54}
$$

$$
G = \frac{3p_o B_{11}}{RT} \sum_{i=3} \left\{ \frac{\left(\bar{\bar{\psi}}\right)^i}{i} \left[\left(\frac{k'}{1+k'}\right)^{i-2} - 1 \right] \right\} \tag{2.55}
$$

The terms with double bars refer to the plateau maximum. The capacity factor is given by

$$
k' = \frac{RT_L x_1^L}{\bar{\bar{y}}_o p_o V_M (1 - X_1)} \tag{2.56}
$$

x_1^L must be measured independently or estimated. Alternatively, the infinite-dilution capacity factor (eq. 2.37) may be used without introducing appreciable ($>0.5\%$) errors in $\bar{\bar{q}}$. The plateau value of the solute mole fraction in the mobile phase at the column outlet $\bar{\bar{y}}_o$ may be calculated from the flow rate of carrier before and after passage through a solute

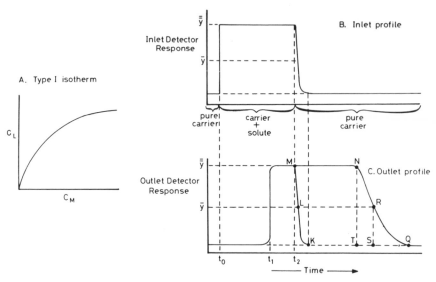

Figure 2.2 A. BET type I isotherm. B. Profile of solute at column inlet for FA and FACP GC modes. C. Resultant solute profile at column outlet.

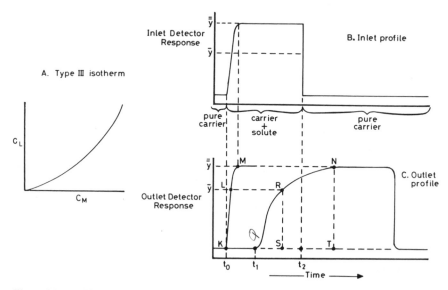

Figure 2.3 A. BET type III isotherm. B. Profile of solute at column inlet for FA and FACP GC modes. C. Resultant solute profile at column outlet.

saturator, and the solute saturation vapor pressure, followed by conversion to the flowmeter temperature and outlet pressure. A detector response curve can also be constructed such that $\bar{\bar{y}}_o$ is read directly from the chromatogram.

Once $\bar{\bar{q}}$ is known, the partition coefficient is given by the ratio $\bar{\bar{q}}/\bar{\bar{c}}$, where $\bar{\bar{c}}$ is calculated from

$$\bar{\bar{c}} = \frac{\bar{\bar{y}} j p_o}{RT + B_{11} j p_o} \tag{2.57}$$

The isotherm, which is the locus of points described by $K_R(q,c)$, is constructed over the range of interest by varying the solute concentration in the mobile phase and reducing the data as described for each FA chromatogram.

In practice, if $\bar{y}_o$ does not exceed approximately 0.2, eq. 2.52 can be simplified[55] by neglecting virial effects and expanding the logarithmic terms:

$$\bar{\bar{q}} = \frac{p_o \bar{\bar{y}}_o}{V_L RT_{fm}} \left\{ \left[\frac{r(\alpha + \beta) j F(o)}{hs\left(1 - j\bar{\bar{y}}_o\right)} \right] \left[1 - \frac{j\bar{\bar{y}}_o}{2(1 + k')\left(1 - j\bar{\bar{y}}_o\right)} \right] \right.$$

$$\left. - V_M\left(1 + \frac{j\bar{\bar{y}}_o}{2}\right) - \frac{r\beta(1 + k')F(o)}{hs\left[1 + k'\left(1 - \bar{y}_o^*\right)\right]} \right\} \tag{2.58}$$

Conder[55] has considered other simplified forms of eq. 2.52 which become progressively less satisfactory as further approximations are introduced. Equation 2.52 is, in fact, no more difficult to employ than eq. 2.58, B_{11} values being the only additional data required.

Frontal Analysis by Characteristic Point (FACP) The FACP technique simplifies the application of frontal analysis in that only a single FA chromatogram need be run. The solute concentration, which corresponds to the maximum range over which the isotherm is to be constructed, is used. Several horizontal lines, $\bar{y}$, are drawn (such as LR in Figs. 2.2 and 2.3) and the areas KLRQ (α) and SRQ (β) are employed in eq. 2.52; FACP is thus identical to FA except that data are taken from segments of the chromatogram.

Strictly, the derivation of eq. 2.52 assumed finite differentials, hence is not applicable to FACP when the boundary is self-sharpening.

Since, in FACP, only a portion of the chromatogram is used, band broadening may contribute substantially to measured areas; small (60 to 80 mesh or finer) particle sizes and optimized flow rates, which in turn may introduce significant viscosity effects, should therefore be employed. In order to ensure that band broadening is negligible, it is usually necessary to reconstruct part of the isotherm by FA or other methods. Conversely, the FA method requires measurement of the total curve area and is unaffected by band broadening. In this respect the apparent advantage of FACP over FA, that is, the determination of several points on an isotherm with a single chromatographic run, may be minimal.[49]

2.2 SOURCES OF SYSTEMATIC ERROR

2.2.1 Peak Symmetry The symmetry of chromatographic peaks, an important criterion for judging the validity of infinite-dilution GC data, has been measured in several ways. First,[75-77] η is defined as the ratio of the absolute value of the rear-to-front tangent slopes. "Trailing" peaks (sharp front, diffuse rear) give η values that are greater than unity. In another method[78] the asymmetry A_S is defined as   lower.

$$A_{\dot{s}} = \frac{B - F}{B + F} \qquad (2.59)$$

where B and F are the baseline distances measured from the rear and front tangents, respectively, to a vertical line drawn through the peak maximum. It is a simple matter to show that

$$\eta = \frac{F}{B} = \frac{1 - A_S}{1 + A_S} \qquad (2.60)$$

The first measure, namely η, seems to be preferred currently; values of 1 ± 0.2 are generally regarded as indicative of symmetric peaks. When $\eta \neq 1 \pm 0.2$, retention data must be regarded with some suspicion, since adsorption, extracolumn factors, and so forth, may be operative.

Ettre[79] has also pointed out that the ratio $w_b / w_{\frac{1}{2}}$ (cf. Fig. 2.1) should be 1.698 for Gaussian peaks, deviations from this value being an additional measure of peak asymmetry.

2.2.2 Mass Transfer Nonequilibrium Effects The gas chromatographic method is by definition a dynamic technique in which solute stationary phase/mobile phase partitioning is continuously displaced from equilibrium. Factors affecting the magnitude of the displacement are important (and controversial) areas of research and some comment is required here about the dependence of partition coefficients on the various column processes extant during elution, for these may well be considered to affect the accuracy of K_R values.

Referring back to the plate height relations (eqs. 2.39 or 2.44), it appears that the only effect that may cause asymmetric band spreading is stationary-phase mass-transfer nonequilibrium. One can imagine, for example, that at a linear carrier velocity of sufficient magnitude the resultant rate of travel of solute molecules through the column may approach the rate at which they sorb into and desorb from the stationary phase (presumably, peak asymmetry would, in such a case, be severe). Alternatively, the view may be adopted that equilibrium is not reached even at zero flow because the stationary phase is spread as a thin film on a support and is therefore not representative of bulk solution properties.

In the absence of flow, solute stationary-phase/immobile (gas)-phase equilibrium has been shown to be reached more quickly when the liquid is spread as a film on a support rather than used in bulk.[80, 81] It has also been confirmed[80, 81] that bulk and thin-film liquid/gas sorption isotherms are identical. In addition, many workers (beginning with Glueckauf[82]) have demonstrated that fully corrected retention data are independent of flow rate in dynamic systems; for example, Cruickshank and co-workers[83] found that a fivefold increase in flow rate did not produce experimentally significant deviations from $\log V_N$ versus $p_o J_3^4$ plots. Those cases in which flow-rate dependence appeared to exist (e.g., Refs. 84–86) may be attributable to the use of initial retention times or failure to correct completely for carrier gas solution in the liquid phase, interfacial adsorption, and/or virial and fugacity effects. Thus, for example, a comparison of GC versus static data shown in Table 2.8, in which retention data were extrapolated to zero flow, is no better than those cases in which finite-flow retention data were used (Table 2.9) in which the largest discrepancy is less than 0.5%.

Finally, it is a matter of experience that even in the absence of all effects other than those corresponding to bulk solution GC peaks are not symmetric, particularly when they rise from and return to the baseline. The former is slightly sharper than the latter because of the longer retention time and consequent greater broadening of the "trailing" molecules. The effect is small, however, even when baseline widths are of several minutes' duration. It may be supposed, nonetheless, that moments analysis will in these cases yield K_R values, which are more accurate than those derived from peak maxima, for, thereby, the true center of gravity of the solute band is measured. The advantage of this use of statistical moments has, however, still to be verified experimentally and appears at best to complicate unnecessarily an otherwise simple yet elegant technique.[87]

2.2.3 Fugacity and Virial Effects Derivation of the partition and activity coefficient relations, using the linear ideal model, requires the assumption of ideal gas-phase behavior. This, of course, is an approximation at best, as recognized[88–92] in many early GC studies, and for greatest accuracy fugacity and virial corrections must be applied to K_R and γ_p^∞ data. The corrections are small (ca. 0.5 to 5%) when helium, hydrogen, or nitrogen are employed as mobile phases at low ($p_i < 1.2$ atm at $p_o = 1$ atm) column pressures but may become appreciable at higher pressures and when other carrier gases are used.

When the column pressure drop is less than about 5 atm and carrier solubility in the stationary phase can be neglected, the virial correction in terms of V_N is[93, 94] (cf. Chapter 4):

$$\ln V_N = \ln V_N^0 + \beta p_o J_3^4 \qquad (2.61)$$

where

$$\beta = \frac{2B_{12} - v_1^\infty}{RT} \qquad (2.62)$$

B_{12} is the solute-carrier virial coefficient, V_N^0 is the net retention volume at zero column pressure drop, and v_1^∞ is the solute molar volume at infinite dilution in the solvent. The virial- and fugacity-corrected activity coefficient is given by

$$\ln \gamma_1^\infty = \ln \gamma_p^\infty - \frac{p_1^0 (B_{11} - v_1^0)}{RT} + \beta p_o J_3^4 \qquad (2.63)$$

where B_{11} is the solute-solute virial (fugacity) coefficient and v_1^0 is the solute bulk molar volume. Recalling that $K_R = V_N / V_L$, eqs. 2.61 and 2.63 are related by

$$\ln \gamma_1^\infty = \ln \frac{RT}{K_R^0 \bar{V}_L p_1^0} - \frac{p_1^0 (B_{11} - v_1^0)}{RT} \qquad (2.64)$$

Martire and Pollara[95] have pointed out that when helium is the carrier eq. 2.64 can be approximated by

$$\ln \gamma_1^\infty \cong \ln \gamma_p^\infty - \frac{p_1^0 B_{11}}{RT} \qquad (2.65)$$

which essentially applies only a fugacity correction to γ_p^∞.

The application of eqs. 2.61 to 2.65 is not so difficult as one might suppose: $\ln V_N$, determined in the usual manner (eq. 2.26) at several pressures, is plotted versus $p_o J_3^4$ and $\ln V_N^0$ taken from the intercept; B_{11} data are rarely available in the literature but can, generally speaking, be calculated[96-101] or measured independently.[102, 103] B_{12} values can be predicted[104-108] from the McGlashan-Potter[101] equation and the Hudson-McCoubrey combining rules[109-112] or determined experimentally.[113, 114] Substitution of v_1^0 for v_1^∞ in eq. 2.62 introduces only a small error[115]; v_1^∞ values can alternatively be measured by dilatometry.[116] Neglect of carrier solubility even for permanent gases above 2 atm inlet pressure may lead to appreciable ($>5\%$) errors[83, 105, 106, 115, 117, 118] but the effect is negligible when $p_i < 1.2$ atm.

2.2.4 Gas-Liquid Interfacial Adsorption Another source of systematic error of potentially significant importance in GLC is solute adsorption at the gas-liquid interface found in several early studies.[119-121] Martin[122] was the first to account correctly for the phenomenon in 1961, although there was subsequently considerable disagreement about its source[123-129] and importance[130-132] (these disparate interpretations have still to be rationalized[133]; cf. Section 5.2).

When gas-liquid interfacial adsorption contributes to retention, eq. 2.26 is expanded[122, 127] to

$$V_N = K_R V_L + K_S A_L \qquad (2.66)$$

where K_S is the liquid surface adsorption partition coefficient (cm) and A_L is the liquid phase surface area. The bulk partition coefficient K_R is found by plotting V_N/V_L versus $1/V_L$ and extrapolating to the ordinate. Since A_L/V_L is not a linear function of $1/V_L$, the plots will be convex to the abscissa when adsorption contributes to retention. In the absence of this effect V_N/V_L versus $1/V_L$ will be a horizontal line. To ensure, therefore, that measured partition coefficients correspond to solution and not mixed sorption, solutes must be eluted from columns containing various liquid phase/support ratios (usually over a weight-percent range of 5 to 25% w/w) and the data plotted as described above.

It is important to note that eq. 2.66 applies only to retention data obtained from symmetric peaks and as a corollary the absence of peak asymmetry does not preclude the presence of adsorption effects.[127, 132]

2.2.5 Noninert Supports It has long been recognized[123, 126, 130, 131, 134-145] that coated GLC supports (including such supposedly inert materials as Teflon[146, 147]) may contribute substantially to retention due mainly to adsorption at the liquid-solid interface. The column tube material may also cause anomalous results[148] (and in at least one case[149] the retention times of components of mixtures were at variance with those obtained when the solutes were eluted individually, although in this study the effect was more than likely caused by column overload). A number of deactivation reagents and procedures have been employed[150-163] to eliminate these effects, dimethyldichlorosilane (DMCS) and hexamethyldisilazane (HMDS) being frequently used; HMDS had originally proved efficacious as a silylating agent for phenolic solutes[164-166] and later[157] was shown to be about twice as effective as DMCS for GC support materials. This may, in part, be a consequence of the stoichiometric requirement of two adjacent surface-active sites per molecule of DMCS (silylation thus being incomplete when only one site is available), whereas HMDS reacts at only a single site.[153] Many silylating agents for solute derivatization are currently available commercially and several [such as bis(trimethylsilyl)acetamide, BSA], although considerably more expensive than HMDS, are probably equally as effective for the suppression of support adsorption.

Perrett and Purnell[157] found that for Sil-O-Cel, Celite, Chromosorb P, and acid-washed Chromosorb W the ratio of HMDS-treated/untreated support *adsorptive* surface area was 0.28, which was identical to the ratio of the saturation-point amounts of adsorbed benzene or acetone for the treated and untreated materials. Silylation therefore proved effective in reducing support adsorption either because the pores of these solids were blocked or because the surfaces were actually rendered inert. Because the uptake of nonvolatile solvents was unchanged by silylation, it was concluded that the support pores were unaffected by treatment with HMDS and that surface-active sites were in fact converted to trimethylsilyl ethers.

Although in retrospect the success of silylation due to surface deactivation seems self-evident, there remains the possibility that retention is a function of support pore structure. Cruickshank, Windsor, and Young[83] were the first to point out that when liquid loadings, for example, of 20% w/w or so squalane are coated onto Celite virtually all the liquid is taken into the approximately cylindrical (0.2 to 1 μ diameter) pores of the support. Calculations based on the Kelvin equation,[167] however, indicated that the decrease of the vapor pressure of solutes sorbed into these pools will produce an error of less than 0.002 on $\log \gamma_i^\infty$ measurements. [This effect, at times referred to as the Kelvin contribution to retention, becomes apparent only at low liquid loadings,[168] whereas in studies of bulk solution phenomena 10–20% w/w loadings are usually employed. It is therefore of little consequence here.]

At low liquid loadings adsorbents (such as graphitized carbon) have distinctive orientation effects on the first few molecular layers of stationary phase.[169–173] The porous structure of siliceous materials such as diatomites (as well as pretreatment with silylating agents) also has a marked influence on the surface distribution of liquid phases.[174–182] Gidddings[174] presented a semiempirical treatment in 1962 which indicated that the thickness of multilayers of solvents and pore cylinder radii were interrelated, but Serpinet[182] found experimentally that this was not the case. Giddings' original concept of the combination of pore filling and film formation was, however, substantiated first by Martire, Pecsok, and Purnell[125] and later by Serpinet[182] (e.g., it appears that the reputed increase in analytical column efficiency by heat-treatment of coated supports[183] is due to the redistribution of solvent with consequent formation of an evenly spread, strongly adsorbed film).

In contrast, retention behavior at high liquid loadings appears to be independent of the (treated or untreated) support even though hydrocarbon solvents such as squalane are thought to be distributed as droplets on silylated materials[182] (a consequence of the autophobic behavior[184] of these systems).

2.2.6 Retention of Nonsorbed Solutes The correction of retention data for column dead space requires the use of a nonsorbed solute, the choice of which is at times not unambiguous. Air is usually suitable for katharometry except, of course, at cryogenic temperatures. Within the limit that $K_R \cong 0$ methane is satisfactory for flame detectors, a qualification, however, that may not be met in GLC and is almost never true below 100°C in GSC. The latter may, in fact, require extended procedures (e.g., Ref. 185).

Several methods have been proposed for the measurement of dead space in flame-detection GLC[186–191] and these and others have been reviewed by Kaiser,[192] Riedmann,[193] and Ebel and Kaiser.[194] More recently Versino[195] reported that a carrier stream consisting of a few parts per million methane (pre- or postcolumn addition) in nitrogen rendered a flame detector sensitive to helium, nitrogen, and argon. Helium was eluted first in all cases. A simple variation of this technique was later described by Cramers, Luyten, and Rijks[196] who found that the trace gas was not necessary if the FID hydrogen and air flows were reduced to about 60% of those normally required for optimum detector sensitivity. Negative peaks were then seen for 0.5-ml injections of helium, oxygen, air, argon, nitrogen, and hydrogen and as shown in Table 2.1, all eluted before methane (the data $\Delta t_A = \Delta t_{CH_4} - t_{inert\ gas}$ were obtained with a 100-m squalane capillary column at 70°C). The retention time of CH_4 was 965.2 sec so that, assuming that the true (helium) dead time was 963.5 sec (where the phase ratio β was 100), the partition coefficient of methane with squalane at this temperature is 0.18. The elution of methane therefore appears to be a satisfactory method for

Table 2.1 Difference in Retention Times Between Methane (965.2 sec) and Permanent Gases[196] for a Squalane Capillary Column at 70°C

Solute	Δt_A (sec) $(= t_{CH_4} - t_{\text{inert gas}})$	Standard Deviation	Number of Observations
O_2	1.6	0.29	12
He	1.7	0.36	11
Air	1.5	0.27	11
Ar	1.5	0.18	11
H_2	1.3	0.28	8
N_2	1.5	0.28	12

the determination of dead space as long as partition coefficients are larger than 100, for it is difficult to suppose that it will be more soluble in stationary phases other than hydrocarbon solvents such as squalane.

2.2.7 Determination of V_L V_L is calculated from w_L and the density of the stationary phase at the column temperature, any one of the several variants of dilatometry[116, 197–200] or other methods[201] being used to determine ρ_L. Dilatometers are most easily calibrated with high-purity organic liquids for which the data of Orwoll and Flory[197] are of particular interest:

n-Hexane: -15 to $89°C$

$$\rho_L = 0.6773 - 0.08914 \times 10^{-2} t - 0.00086 \times 10^{-4} t^2 - 0.00652 \times 10^{-6} t^3$$

n-Hexadecane: 18 to $207°C$

$$\rho_L = 0.7871 - 0.07003 \times 10^{-2} t + 0.00185 \times 10^{-4} t^2 - 0.00134 \times 10^{-6} t^3$$

The accurate determination of w_L, however, is the most troublesome source of systematic error in gas-liquid chromatography. Wicarova and co-workers[202] soxhlet-extracted their packings, evaporated the solvent, and weighed the residue, a procedure said to be accurate to about half a milligram. Martire and Riedl[203] ashed weighed amounts of packings in a porcelain crucible with a Bunsen burner and claimed $\pm 0.5\%$ accuracy on the weight percent (use of a muffle furnace improves the accuracy appreciably[87]). Laub, Martire, and Purnell[204] found similar errors (see Table 2.2) and Aue and co-workers[183] reported that replicate ashings by an industrial laboratory were no better.

Table 2.2 Determination of Corrected (for Bare Support Weight Loss) Weight Percent of Named Liquid Phases on Chromosorb G (60 to 80-mesh, AW-DMCS) by Ashing[204]

Run No.	Wt. Packing (g)	Wt. Loss (g)	Corrected Wt. %	ΔX $(= X_i - \bar{X})$	$\Delta X / X_i$ (%)
A. *n*-Octadecane					
1	0.9973	0.0746	7.33	−0.03	−0.41
2	0.9365	0.0706	7.38	0.02	0.27
3	0.8489	0.0641	7.40	0.04	0.54
4	1.0181	0.0763	7.34	−0.02	−0.27
5	1.0308	0.0773	7.34	−0.02	−0.27
		Average:	7.36% w/w		
B. *n*-Hexatriacontane					
1	1.0822	0.1040	9.46	0.06	0.64
2	1.2776	0.1213	9.34	−0.04	−0.43
3	1.3455	0.1281	9.37	−0.01	−0.11
4	0.9591	0.0910	9.34	−0.04	−0.43
		Average:	9.38% w/w		
C. Approximately 50/50 v/v Mixture of A and B					
1	1.0665	0.0888	8.17	0.03	0.37
2	1.1930	0.0987	8.12	−0.02	−0.25
3	1.1201	0.0933	8.18	0.04	0.49
4	1.2052	0.1002	8.16	0.02	0.25
5	1.1378	0.0945	8.11	−0.03	−0.37
6	1.0835	0.0896	8.12	−0.02	−0.25
		Average:	8.14% w/w		

Table 2.3   Weight Loss[204] of Ashed Support (60 to 80-mesh Chromosorb G, AW-DMCS)

Run No.	Wt. Support (g)	Wt. Loss (g)	Wt. Loss, (g/g)
1	0.8882	0.0015	0.0017
2	0.9937	0.0018	0.0018
3	0.9499	0.0015	0.0016
4	0.9850	0.0017	0.0017
5	1.2916	0.0020	0.0016
		Average:	0.0016_8 g/g

When w_L is determined by ashing, a correction for the weight loss of bare support is required as shown in Table 2.3 for DMCS-treated Chromosorb G.

Petsev and co-workers[205, 206] in a recent comparison of the measurement of w_L by soxhlet extraction and high-temperature evaporation (below the ignition point) found that Sterchamol appeared to contain extractable inorganic material (ethanol eluent) which amounted to about 2% of the support weight even after drying at 110°C for extended periods. Boiling the support in water, 95% ethanol, and diethyl ether reduced the error to 0.14%, which indicated nonetheless that soxhlet extraction will always produce high results regardless of the use of the weight of (extracted) support or liquid residue to calculate the packing weight percent. The evaporation technique, however, did not appear to suffer from these drawbacks as long as the temperature was maintained below the combustion point of the liquid phase. Weight changes occurred with the supports tested (Sterchamol, Chromosorb P, and acid-washed Chromosorb W) but were constant and reproducible and thought to be due to strongly adsorbed water; for example, Sterchamol showed a weight loss of 0.11% at 110°C, an additional 0.24% on further heating at 200°C, 0.28% at 300°C, and 0.41% at 400°C (weight losses were not observed at a given temperature after heating at that temperature for 2 hr). The consequence of these results is that the support should be dried at the same temperature as that used for evaporation; 300°C was a suitable choice for moderately volatile phases such as squalane and di-nonyl phthalate (DNP) (400°C was required for materials somewhat less volatile such as Apiezon M[206]); the concomitant support weight losses (and percent standard deviations) at this temperature were Sterchamol, 0.28% ($\pm$0.01), Chromosorb P, 0.16% ($\pm$0.01), and acid-washed Chromosorb W, 0.20% ($\pm$0.01).

Table 2.4 Precision of Weight Percent Determination of Di-nonyl Phthalate Packing by Stationary Phase Evaporation Technique[205]

Run No.	Wt. % Found	$\Delta X\,(=X_i-\overline{X})$	$\Delta X/X_i$ (%)
1	4.77	0.04	0.84
2	4.70	−0.03	−0.64
3	4.75	0.02	0.42
4	4.69	−0.04	−0.84
5	4.71	−0.02	−0.42
6	4.77	0.04	0.84
7	4.71	−0.02	−0.42
8	4.75	0.02	0.42
9	4.69	−0.04	−0.84

Average: 4.73% w/w

Reproducibility of the stationary phase evaporation technique was tested by replicate analysis of a 5% w/w Sterchamol packing; the results are given in Table 2.4. The percent standard deviation for a single measurement was 0.03% and the probable error at the 95% confidence level was 0.07%. The relative errors are on average higher than those shown in Table 2.2 which may be due in part to the use of a silylated support in the former but also obviously because of the lower liquid loading in the latter. Therefore, and as proposed in Section 2.2.5 for different reasons, high solvent/support ratios are recommended.

2.3 ACCURACY AND REPRODUCIBILITY OF GC K_R DATA

Porter, Deal, and Stross[207] and Anderson and Napier[208] were the first to show that, within an experimental error of approximately 5%, GC and static partition coefficients are in agreement. However, interlaboratory reproducibility and GC-static data comparisons have, unfortunately, been tested primarily with squalane solvent and, based on the reported retention data, are not encouraging even when it is recognized that virial effects were not taken into account in most studies. Table 2.5 shows a comparison of K_R data for n-alkanes with squalane at 80°C which were obtained by GC and a static method, and Table 2.6 gives similar data for n-hexane at 30 °C.

Table 2.5 Comparison of GLC and Static Partition Coefficients for n-Alkanes with Squalane at 80°C

	K_R^{GLC}						K_R^{static}
	Ref. No.						
Solute	151	209	210	211	212	213	214[a]
n-Butane	26.6	23.9	25.0	24.6	25.9	25.6	27.2
n-Hexane	62.6	62.3	60.5	61.5	63.7	61.1	63.9
n-Heptane	146	145	145	144	150	142	148
n-Octane	...	329	338	333	346	329	333

[a] Gas entrainment method similar to that of Shaw and Butler.[215]

Table 2.6 Comparison of GLC and Static Partition Coefficients for n-Hexane with Squalane at 30°C

Ref. No.	K_R^{GLC}	K_R^{static}
49	306	
80	303	
77		301
93	285	
125		298
151	294	
209	298	
211	306	
216	291	
217	285	
218		306

Much of the irreproducibility of K_R data with squalane can apparently be attributed to the purity of commercial versions of this stationary phase[80, 202, 218] manufactured by hydrogenating squalene (over a nickel catalyst) and which in turn is isolated as a natural product. In addition, butylated hydroxytoluene is generally added as a preservative which may further compound the difficulty.

A more encouraging interlaboratory comparison of n-alkane data with n-alkane solvents (which are easily purified) is presented in Table 2.7 in which the largest discrepancy is 1.96%.

The most accurate static data available appear to be those reported by McGlashan and Williamson[221] for n-hexane with n-hexadecane at 20 to

Table 2.7   Comparison of GLC Partition Coefficients of n-Alkane Solutes with n-Alkane Solvents at 80°C

	Solvent						
	$n-C_{24}H_{50}$		$n-C_{30}H_{62}$			$n-C_{36}H_{74}$	
	Ref. No.						
Solute	219	220	219	220	204	213	219	220
n-Hexane	61.84	61.97	57.01	57.16	53.56	53.40	52.83	52.91
n-Heptane	146.5	145.7	134.7	134.1	123.5	125.2	125.3	123.4
n-Octane	339.0	340.8	317.5	311.8	291.6	291.2	289.5	285.6

60°C; these are compared with the fully corrected GLC K_R values of Cruickshank, Gainey, and Young[217] in Table 2.8. Similar data for n-hexane and n-heptane in n-octadecane at 30°C are presented in Table 2.9.

Finite-concentration GC data appear to be accurate[49, 55, 70] to ±3% and, as stated, may be much better than this figure, for squalane has often been used as the stationary phase when comparisons have been made. It is, of course, an important boundary condition in finite-dilution methods that the data, when extrapolated, are in agreement with those determined independently at infinite dilution. This has been observed in several studies[49, 55, 70] and has also been shown to apply to systems in which liquid surface adsorption contributes substantially to retention.[126, 131, 133, 224]

Table 2.8   Comparison of GLC[217] and Static[221] Partition Coefficients for n-Hexane with n-Hexadecane at 20 to 60°C

T (°C)	K_R^{GLC}	K_R^{static}	K_R^{GLC}/K_R^{static}
20	493.1	492.0	1.0022
40	224.9	226.4	0.9934
60	116.4	115.6	1.0069

Table 2.9   Comparison of GLC[105] and Static[222, 223] Partition Coefficients for n-Hexane and n-Heptane with n-Octadecane at 30°C

Solute	K_R^{GLC}	K_R^{static}	K_R^{GLC}/K_R^{static}
n-Hexane	275.4	274.1	1.0047
n-Heptane	898.9	898.6	1.0003

2.4 REFERENCES

1. E. A. Guggenheim, *Thermodynamics*, North-Holland, Amsterdam, 1967, pp. 32–33.

2. J. N. Wilson, *J. Am. Chem. Soc.*, **62**, 1583 (1940).

3. A. J. P. Martin and R. L. M. Synge, *Biochem. J.*, **35**, 1358 (1941).

4. A. J. P. Martin and A. T. James, *Biochem. J.*, **50**, 679 (1952).

5. D. H. Everett, *Trans. Faraday Soc.*, **61**, 1637 (1965).

6. J. J. van Deemter, F. J. Zuiderweg, and A. Klinkenberg, *Chem. Eng. Sci.*, **5**, 271 (1956).

7. D. de Vault, *J. Am. Chem. Soc.*, **65**, 532 (1943).

8. L. Lapidus and N. R. Amundson, *J. Phys. Chem.*, **56**, 984 (1952).

9. N. N. Tunitskii, *Dokl. Akad. Nauk SSSR*, **99**, 577 (1954).

10. E. Glueckauf, *Discuss. Faraday Soc.*, **7**, 12, 202 (1949); *Analyst*, **77**, 903 (1952); *Trans. Faraday Soc.*, **51**, 1540 (1955).

11. A. Klinkenberg and F. Sjenitzer, *Chem. Eng. Sci.*, **5**, 258 (1956).

12. J. J. van Deemter, Second Informal Discussion, Gas Chromatography Discussion Group, Cambridge, England, 1957.

13. M. J. E. Golay, in *Gas Chromatography*, V. J. Coates, H. J. Noebels, and I. S. Fagerson, Eds., Academic, New York, 1958, p. 316; in *Gas Chromatography 1958*, D. H. Desty, Ed., Butterworths, London, 1958, p. 36; in *Gas Chromatography 1960*, R. P. W. Scott, Ed., Butterworths, London, 1960, p. 139; *Nature*, **199**, 370 (1963); *Z. Anal. Chem.*, **236**, 38 (1968); *Anal. Chem.*, **40**, 382 (1968).

14. J. C. Giddings, *Anal. Chem.*, **35**, 439 (1963); *Dynamics of Chromatography*, Marcel Dekker, New York, 1966, Ch. 2 and references therein.

15. C. N. Reilley, G. P. Hildebrand, and J. W. Ashley, *Anal. Chem.*, **34**, 1198 (1962).

16. E. Cremer, *Monatsh.*, **92**, 112 (1961).

17. E. Cremer and H. Huber, *Angew. Chem.*, **73**, 461 (1961).

18. P. E. Eberly, Jr., *J. Phys. Chem.*, **65**, 1261 (1961).

19. L. Bachmann, E. Bechtold, and E. Cremer, *J. Catal.*, **1**, 113 (1962).

20. G. R. Fitch, M. E. Probert, and P. F. Tiley, *J. Chem. Soc.*, 4875 (1962).

21. A. Petho, P. Fejes, and J. Engelhardt, *Acta Chem. Acad. Sci. Hung.*, **30**, 63 (1962).

22. J. F. K. Huber and A. I. M. Keulemans, in *Gas Chromatography 1962*, M. van Swaay, Ed., Butterworths, London, 1962, p. 26.

23. F. Helfferich and D. L. Peterson, *Science*, **142**, 661 (1963).

24. O. V. Altschuler, O. M. Vinogradova, S. Z. Raginskii, and Y. N. Chirkov, *Proc. Acad. Sci. USSR, Phys. Chem. Sec. (Engl. transl.)*, **152**, 862 (1963).

25. F. I. Stalkup and H. A. Deans, *AIChE J.*, **9**, 106 (1963).

26. F. I. Stalkup and R. Kobayashi, *AIChE J.*, **9**, 121 (1963).

27. A. A. Zhukhovitskii, in *Gas Chromatography 1964*, A. Goldup, Ed., Institute of Petroleum, London, 1965, p. 161; *Zh. Anal. Khim.*, **27**, 971 (1972).

28. L. E. Green, L. J. Schmauch, and J. C. Worman, *Anal. Chem.*, **36**, 1513 (1964).

29. F. Steinbach, *J. Chromatogr.*, **15**, 432 (1964).

30. H. Knözinger and H. Spannheimer, *J. Chromatogr.*, **16**, 1 (1964).

31. D. R. Owens, A. G. Hamlin, and T. R. Phillips, *Nature*, **201**, 901 (1964).

32. F. Helfferich, *J. Chem. Educ.*, **41**, 410 (1964).

33. D. L. Peterson and F. Helfferich, *J. Phys. Chem.*, **69**, 1283 (1965).
34. K. T. Koonce, H. A. Deans, and R. Kobayashi, *AIChE J.*, **11**, 259 (1965).
35. C. F. Chueh and W. T. Ziegler, *AIChE J.*, **11**, 508 (1965).
36. H. B. Gilmer and R. Kobayashi, *AIChE J.*, **11**, 702 (1965).
37. P. J. Kipping and D. G. Winter, *Nature*, **205**, 1002 (1965).
38. J. F. Parcher and P. Urone, *Nature*, **211**, 628 (1966).
39. L. D. Van Horn and R. Kobayashi, *J. Chem. Eng. Data*, **12**, 294 (1967).
40. R. Kobayashi, P. S. Chappelear, and H. A. Deans, *Ind. Eng. Chem.*, **10**, 63 (1967).
41. J. R. Conder, *J. Chromatogr.*, **39**, 273 (1969).
42. A. A. Zhukhovitskii and M. L. Sazonov, *J. Chromatogr.*, **49**, 153 (1970).
43. H. E. Dubsky, *Chromatographia*, **3**, 366 (1970).
44. F. Helfferich and G. Klein, *Multicomponent Chromatography*, Marcel Dekker, New York, 1970.
45. K. Asano, T. Nakahara, and R. Kobayashi, *J. Chem. Eng. Data*, **16**, 16 (1971).
46. M. V. Sussman, K. N. Astill, R. Rombach, A. Cerullo, and S. S. Chen, *Ind. Eng. Chem. Fundam.*, **11**, 181 (1972).
47. A. N. Korol, *Russ. Chem. Rev.*, **41**, 174 (1972).
48. G. G. Arenkova and A. A. Zhukhovitskii, *Zavod. Lab.*, **40**, 145 (1974).
49. J. R. Conder and J. H. Purnell, *Trans. Faraday Soc.*, **64**, 1505, 3100 (1968); **65**, 824, 839 (1969).
50. J. Boeke, in *Gas Chromatography 1960*, R. P. W. Scott, Ed., Butterworths, London, 1960, p. 88.
51. R. S. Henly, A. Rose, and R. F. Sweeny, *Anal. Chem.*, **36**, 744 (1964).
52. G. J. Kriege and V. Pretorius, *J. Gas Chromatogr.*, **2**, 115 (1964); *Anal. Chem.*, **37**, 1186, 1191, 1202 (1965).
53. G. Guiochon and L. Jacob, *J. Chim. Phys.*, **67**, 185, 291 (1970); *Bull. Soc. Chim. Fr.*, 1224 (1970); *J. Chromatogr. Sci.*, **9**, 307 (1971).
54. G. Schay, *Usp. Khromatogr.*, 179 (1972).
55. J. R. Conder, *Adv. Anal. Chem. Instrum.*, **6**, 209 (1968); *Chromatographia*, **7**, 387 (1974).
56. L. Jacob and G. Guiochon, *J. Chromatogr. Sci.*, **13**, 18 (1975).
57. P. Valentin and G. Guiochon, *Sep. Sci.*, **10**, 245, 271, 289 (1975).
58. E. Glueckauf, *J. Chem. Soc.*, 1302 (1947).
59. S. Brunauer, P. H. Emmett, and E. Teller, *J. Am. Chem. Soc.*, **60**, 309 (1938).
60. S. J. Gregg and R. Stock, in *Gas Chromatography 1958*, D. H. Desty, Ed., Butterworths, London, 1958, p. 90.
61. C. H. Bosanquet and G. O. Morgan, in *Vapour Phase Chromatography*, D. H. Desty, Ed., Butterworths, London, 1957, p. 35.
62. C. H. Bosanquet, in *Gas Chromatography 1958*, D. H. Desty, Ed., Butterworths, London, 1958, p. 107.
63. P. Fejes, L. Czaran, and G. Schay, *Magy. Kem. Foly.*, **68**, 11 (1962).
64. P. Fejes, E. Fromm-Czaran, and G. Schay, *Acta Chim. Acad. Sci. Hung.*, **33**, 87 (1962).
65. M. J. E. Golay, *Nature*, **202**, 489 (1964).
66. P. C. Haarhoff and H. J. Van der Linde, *Anal. Chem.*, **37**, 1742 (1965).

67. N. Dyson and A. B. Littlewood, *Trans. Faraday Soc.*, **63**, 1895 (1967); *Anal. Chem.*, **39**, 638 (1967).

68. M. L. Conti and M. Lesimple, *J. Chromatogr.*, **29**, 32 (1967).

69. O. Knapp and M. Lochner, *Chromatographia*, **3**, 178 (1970).

70. C. J. Chen and J. F. Parcher, *Anal. Chem.*, **43**, 1738 (1971).

71. R. P. W. Scott, *Anal. Chem.*, **36**, 1455 (1964).

72. R. P. W. Scott, *Anal. Chem.*, **35**, 481 (1963).

73. G. R. Luckhurst, *J. Chromatogr.*, **16**, 543 (1964).

74. K. DeClerk and T. S. Buys, *J. Chromatogr.*, **84**, 1 (1973).

75. D. H. Everett and C. T. H. Stoddart, *Trans. Faraday Soc.*, **57**, 746 (1961).

76. A. J. B. Cruickshank and D. H. Everett, *J. Chromatogr.*, **11**, 289 (1963).

77. A. J. B. Cruickshank, D. H. Everett, and M. T. Westaway, *Trans. Faraday Soc.*, **61**, 235 (1965).

78. S. Dal Nogare and R. S. Juvet, Jr., *Gas-Liquid Chromatography*, Interscience, New York, 1962, p. 168.

79. L. S. Ettre, *J. Chromatogr.*, **11**, 267 (1963).

80. A. J. Ashworth and D. H. Everett, *Trans. Faraday Soc.*, **56**, 1609 (1960).

81. G. F. Freeguard and R. Stock, in *Gas Chromatography 1962*, M. van Swaay, Ed., Butterworths, London, 1962, p. 102.

82. E. Glueckauf, *Trans. Faraday Soc.*, **51**, 34 (1955).

83. A. J. B. Cruickshank, M. L. Windsor, and C. L. Young, *Proc. Roy. Soc. Ser. A*, **295**, 259 (1966).

84. C. P. Hicks, Ph.D. Dissertation, University of Bristol, England, 1970.

85. S. Wicar, J. Novak, and N. Ruseva-Rakshieva, *Anal. Chem.*, **43**, 1945 (1971).

86. J. Villermaux, *J. Chromatogr.*, **83**, 205 (1973).

87. R. J. Laub, J. H. Purnell, P. S. Williams, M. W. P. Harbison, and D. E. Martire, *J. Chromatogr.*, *in press*.

88. M. A. Khan, in *Gas Chromatography 1958*, D. H. Desty, Ed., Butterworths, London, 1958, p. 135.

89. E. R. Adlard, M. A. Khan, and B. T. Whitham, in *Gas Chromatography 1960*, R. P. W. Scott, Ed., Butterworths, London, 1960, p. 251.

90. D. H. Desty, A. Goldup, G. R. Luckhurst, and W. T. Swanton, in *Gas Chromatography 1962*, M. van Swaay, Ed., Butterworths, London, 1962, p. 67.

91. D. H. Desty, A. Goldup, and W. T. Swanton, in *Gas Chromatography*, N. Brenner, J. E. Callen, and M. D. Weiss, Eds., Academic, New York, 1962, p. 105.

92. A. Goldup, G. R. Luckhurst, and W. T. Swanton, *Nature*, **193**, 333 (1962).

93. A. J. B. Cruickshank, M. L. Windsor, and C. L. Young, *Proc. Roy. Soc. Ser. A*, **295**, 271 (1966).

94. M. L. Windsor and C. L. Young, *J. Chromatogr.*, **27**, 355 (1967).

95. D. E. Martire and L. Z. Pollara, *Adv. Chromatogr.*, **1**, 335 (1965).

96. J. A. Beattie and O. C. Bridgeman, *J. Am. Chem. Soc.*, **49**, 1665 (1927).

97. J. D. Lambert, G. A. H. Roberts, J. S. Rowlinson, and V. J. Wilkinson, *Proc. Roy. Soc. Ser. A*, **196**, 113 (1949).

98. E. A. Guggenheim and M. L. McGlashan, *Proc. Roy. Soc. Ser. A*, **206**, 448 (1951).

99. J. S. Rowlinson, *Trans. Faraday Soc.*, **45**, 974 (1949).

100. K. A. Kobe and R. E. Lynn, *Chem. Rev.*, **52**, 117 (1953).

101. M. L. McGlashan and D. J. B. Potter, *Proc. Roy. Soc. Ser. A*, **267**, 478 (1962).

102. P. Y. Feng and M. Melzer, *J. Chem. Educ.*, **49**, 375 (1972).

103. R. B. Spertell and G. T. Chang, *J. Chromatogr. Sci.*, **10**, 60 (1972).

104. J. R. Conder and S. H. Langer, *Anal. Chem.*, **39**, 1461 (1967).

105. A. J. B. Cruickshank, B. W. Gainey, and C. L. Young, *Trans. Faraday Soc.*, **64**, 337 (1968).

106. B. W. Gainey and C. L. Young, *Trans. Faraday Soc.*, **64**, 349 (1968).

107. B. W. Gainey and R. L. Pecsok, *J. Phys. Chem.*, **74**, 2548 (1970).

108. B. W. Gainey and C. P. Hicks, *J. Phys. Chem.*, **75**, 3687, 3691 (1971).

109. G. H. Hudson and J. C. McCoubrey, *Trans. Faraday Soc.*, **56**, 761 (1960).

110. E. A. Guggenheim and C. W. Wormald, *J. Chem. Phys.*, **42**, 3775 (1965).

111. A. J. B. Cruickshank, M. L. Windsor, and C. L. Young, *Trans. Faraday Soc.*, **62**, 2341 (1966).

112. R. J. Munn, *Trans. Faraday Soc.*, **57**, 187 (1961).

113. C. M. Knobler, *Rev. Sci. Instrum.*, **38**, 184 (1967).

114. C. R. Coan and A. D. King, Jr., *J. Chromatogr.*, **44**, 429 (1969).

115. E. M. Dantzler, C. M. Knobler, and M. L. Windsor, *J. Chromatogr.*, **32**, 433 (1968).

116. L. G. Hepler, J. M. Stokes, and R. H. Stokes, *Trans. Faraday Soc.*, **61**, 20 (1965).

117. D. H. Everett, B. W. Gainey, and C. L. Young, *Trans. Faraday Soc.*, **64**, 2667 (1968).

118. R. L. Pecsok and M. L. Windsor, *Anal. Chem.*, **40**, 1238 (1968).

119. F. T. Eggertsen and H. S. Knight, *Anal. Chem.*, **30**, 15 (1958).

120. T. Fukuda, *Jap. Anal.*, **8**, 627 (1959).

121. R. L. Martin and J. C. Winters, *Anal. Chem.*, **31**, 1954 (1959).

122. R. L. Martin, *Anal. Chem.*, **33**, 347 (1961); **35**, 116 (1963).

123. R. L. Pecsok, A. deYllana, and A. Abdul-Karim, *Anal. Chem.*, **36**, 452 (1964).

124. D. E. Martire, R. L. Pecsok, and J. H. Purnell, *Nature*, **203**, 1279 (1964).

125. D. E. Martire, R. L. Pecsok, and J. H. Purnell, *Trans. Faraday Soc.*, **61**, 2496 (1965).

126. R. L. Pecsok and B. H. Gump, *J. Phys. Chem.*, **71**, 2202 (1967).

127. J. R. Conder, D. C. Locke, and J. H. Purnell, *J. Phys. Chem.*, **73**, 700 (1969).

128. H.-L. Liao and D. E. Martire, *Anal. Chem.*, **44**, 498 (1972).

129. A. B. Littlewood and F. W. Willmott, *Anal. Chem.*, **38**, 1031 (1961).

130. P. Urone and J. F. Parcher, *Anal. Chem.*, **38**, 270 (1966).

131. J. F. Parcher and C. L. Hussey, *Anal. Chem.*, **45**, 188 (1973).

132. D. E. Martire, *Anal. Chem.*, **38**, 244 (1966).

133. J. R. Conder, *Anal. Chem.*, **48**, 917 (1976).

134. D. H. Desty, F. M. Godfrey, and C. L. A. Harbourn, in *Gas Chromatography 1958*, D. H. Desty, Ed., Butterworths, London, 1958, p. 200.

135. R. G. Scholz and W. W. Brandt, in *Gas Chromatography*, N. Brenner, J. E. Callen, and M. D. Weiss, Eds., Academic, New York, 1962, p. 7.

136. D. M. Ottenstein, *Adv. Chromatogr.*, **3**, 137 (1966).

137. J. F. Palframan and E. A. Walker, *Analyst*, **92**, 71 (1967).

138. M. Krejci and M. Rusek, *Collect. Czech. Chem. Commun.*, **33**, 3448 (1968).

139. N. A. Cockle and P. F. Tiley, *Chem. Ind.*, 1118 (1968).

140. W. Jequier and J. Robin, *Chromatographia*, **1**, 297 (1968).

141. C. M. Drew and E. M. Bens, in *Gas Chromatography 1968*, C. L. A. Harbourn, Ed., Institute of Petroleum, London, 1969, p. 3.

142. M. Krejci and K. Hana, in *Gas Chromatography 1968*, C. L. A. Harbourn, Ed., Institute of Petroleum, London, 1969, p. 32.

143. E. D. Smith, J. M. Oathout, and G. T. Cook, *J. Chromatogr. Sci.*, **8**, 291 (1970).

144. V. G. Berezkin, *Zh. Anal. Khim.*, **28**, 982 (1973).

145. V. G. Berezkin, C. Eon, and G. Guiochon, *Bull. Soc. Chim. Fr.*, 94 (1975).

146. J. J. Kirkland, in *Gas Chromatography*, L. Fowler, Ed., Academic, New York, 1963, p. 77; in *Gas Chromatography 1964*, A. Goldup, Ed., Institute of Petroleum, London, 1965, p. 285.

147. J. R. Conder, *Anal. Chem.*, **43**, 367 (1971).

148. R. J. Levins and D. M. Ottenstein, *J. Gas Chromatogr.*, **5**, 539 (1967).

149. K. Amaya and K. Sasaki, *Bull. Chem. Soc. Jap.*, **35**, 1507 (1962).

150. A. T. James and A. J. P. Martin, *Biochem. J.*, **50**, 679 (1952).

151. A. Kwantes and G. W. A. Rijnders, in *Gas Chromatography 1958*, D. H. Desty, Ed., Butterworths, London, 1958, p. 125.

152. E. C. Ormerod and R. P. W. Scott, *J. Chromatogr.*, **2**, 65 (1959).

153. J. Bohemen, S. H. Langer, R. H. Perrett, and J. H. Purnell, *J. Chem. Soc.*, 2444 (1960).

154. W. R. Supina, R. S. Henly, and R. F. Kruppa, *J. Am. Oil Chem. Soc.*, **42**, 459A (1965); **43**, 202A (1966).

155. B. M. Mitzner, V. J. Mancini, and T. A. Shiftan, *J. Gas Chromatogr.*, **4**, 336 (1966).

156. H. Rotzsche, in *Aspects in Gas Chromatography*, H. G. Struppe, Ed., Akademie Verlag, Berlin, 1971, p. 120.

157. R. H. Perrett and J. H. Purnell, *J. Chromatogr.*, **7**, 455 (1967).

158. I. I. Bardyshev and V. I. Kulikov, *Vestsi Akad. Nauk B., SSSR* (2), 26 (1971).

159. V. G. Berezkin, V. P. Pakhomov, and N. G. Starostina, *Neftepererab. Neftekhim.* (2), 35 (1973).

160. W. A. Aue, C. Hastings, and S. Kapila, *J. Chromatogr.*, **77**, 299 (1973).

161. R. J. DePasquale, *Am. Lab.*, **5**(6), 35 (1973).

162. Z. Suprynowicz, *Chem. Anal.*, **18**, 251, 513 (1973).

163. L. Zoccolillo and F. Salomoni, *J. Chromatogr.*, **106**, 103 (1975).

164. S. H. Langer, S. Connell, and I. Wender, *J. Org. Chem.*, **23**, 50 (1956).

165. S. H. Langer, P. Pantages, and I. Wender, *Chem. Ind.*, 1664 (1958).

166. S. H. Langer, R. A. Friedel, I. Wender, and A. G. Sharkey, Jr., *Anal. Chem.*, **30**, 1353 (1958).

167. W. Thomson, *Phil. Mag.*, **42**, 448 (1871).

168. C. Devillez, C. Eon, and G. Guiochon, *J. Colloid Interface Sci.*, **49**, 232 (1974).

169. J. H. Clint, J. S. Cunie, J. F. Goodman, and J. R. Tate, *Nature*, **223**, 51 (1969).

170. D. H. Everett and G. H. Findenegg, *Nature*, **223**, 52 (1969).

171. S. G. Ash and G. H. Findenegg, *Spec. Discuss. Faraday Soc.*, **1**, 105 (1970).

172. A. J. Groszek, *Proc. Roy. Soc. Ser. A*, **314**, 473 (1970).

173. J. Serpinet, *J. Chromatogr.*, **68**, 9 (1972); **77**, 289 (1973).

174. J. C. Giddings, *Anal. Chem.*, **34**, 458 (1962).

175. M. Krejci, *Collect. Czech. Chem. Commun.*, **32**, 1152 (1967).

176. J. Serpinet and J. Robin, *C. R. Acad. Sci., Ser. C*, **272**, 1765 (1971).

177. J. Serpinet, C. Daneyrolle, M. Trocaz, and C. Eyrand, *C. R. Acad. Sci., Ser. C*, **273**, 1290 (1971).

178. G. Untz and J. Serpinet, *Bull Soc. Chim. Fr.*, 1591 (1973).

179. G. V. Filonenko, T. I. Dovbush, and A. N. Korol, *Chromatographia*, **7**, 293 (1974).

180. V. G. Berezkin, D. Kourilova, M. Krejci, and V. M. Fateeva, *J. Chromatogr.*, **78**, 261 (1973).

181. J. Serpinet, G. Untz, C. Gachet, L. de Mourgues, and M. Perrin, *J. Chim. Phys.*, **71**, 949 (1974).

182. J. Serpinet, *J. Chromatogr. Sci.*, **12**, 832 (1974); *Chromatographia*, **8**, 18 (1975); *J. Chromatogr*, **119**, 483 (1976); *Anal. Chem.*, **48**, 2264 (1976).

183. W. A. Aue, C. R. Hastings, and S. Kapila, *J. Chromatogr.*, **77**, 299 (1973).

184. W. A. Zisman, *J. Paint Technol.*, **44**, 41 (1972).

185. S. Masukawa and R. Kobayashi, *J. Gas Chromatogr.*, **6**, 257 (1968).

186. M. L. Peterson and J. Hirsch, *J. Lipid Res.*, **1**, 132 (1959).

187. A. A. Zhukhovitskii and N. M. Turkeltaub, *Dokl. Akad. Nauk SSSR*, **143**, 646 (1961).

188. J. F. Smith, *Nature*, **193**, 679 (1962).

189. H. Gold, *Anal. Chem.*, **34**, 174 (1962).

190. A. A. Zhukhovitskii, in *Gas Chromatography 1964*, A. Goldup, Ed., Institute of Petroleum, London, 1965, p. 162.

191. G. Lai and J. A. Roth, *Chem. Eng. Sci.*, **22**, 1299 (1967).

192. R. E. Kaiser, *Chromatographia*, **2**, 215 (1969).

193. M. Riedmann, *Chromatographia*, **7**, 59 (1974).

194. S. Ebel and R. E. Kaiser, *Chromatographia*, **7**, 696 (1974).

195. B. Versino, *Chromatographia*, **3**, 231 (1970).

196. C. A. Cramers, J. A. Luyton, and J. A. Rijks, *Chromatographia*, **3**, 441 (1970).

197. R. A. Orwoll and P. J. Flory, *J. Am. Chem. Soc.*, **89**, 6814 (1967).

198. R. J. Laub and R. L. Pecsok, *Anal. Chem.*, **46**, 2251 (1974).

199. Y. A. Kanchencko, V. G. Berezkin, A. Y. Mysak, and L. P. Paskal, *Zavod. Lab.*, **40**, 1450 (1974).

200. R. J. Laub and J. H. Purnell, *J. Am. Chem. Soc.*, **98**, 30 (1976).

201. D. F. Folland, *Instrum. Control Systems*, **41**, 97 (1968).

202. O. Wicarova, J. Novak, and J. Janak, *J. Chromatogr.*, **51**, 3 (1970).

203. D. E. Martire and P. Riedl, *J. Phys. Chem.*, **72**, 3478 (1968).

204. R. J. Laub, D. E. Martire, and J. H. Purnell, *J. Chem. Soc. Faraday Trans. II*, **74**, 213 (1978).

205. N. D. Petsev, R. N. Nikolov, and A. Kostova, *J. Chromatogr.*, **93**, 369 (1974).

206. N. D. Petsev, V. H. Petkov, and Chr. Dimitrov, *J. Chromatogr.*, **114**, 204 (1975).

207. P. E. Porter, D. H. Deal, and F. H. Stross, *J. Am. Chem. Soc.*, **78**, 2999 (1956).

208. J. R. Anderson and K. H. Napier, *Austl. J. Chem.*, **10**, 250 (1957).

209. S. Evered and F. H. Pollard, *J. Chromatogr.*, **4**, 451 (1960).

210. T. Hofstee, A. Kwantes, and G. W. A. Rijnders, *Proc. Int. Symp. Dist.*, Brighton, England, 1960, p. 105.

211. D. H. Desty and W. T. Swanton, *J. Phys. Chem.*, **65**, 766 (1961).

212. A. B. Littlewood, *J. Gas Chromatogr.*, **1**(5), 6 (1963).

213. E. C. Pease and S. Thorburn, *J. Chromatogr.*, **30**, 344 (1967).

214. M. G. Burnett, *Anal. Chem.*, **35**, 1567 (1963).

215. R. Shaw and J. A. V. Butler, *Proc. Roy. Soc. Ser. A*, **129**, 519 (1930).

216. D. E. Martire, Ph.D. Dissertation, Stevens Institute of Technology, Hoboken, New Jersey, 1963.

217. A. J. B. Cruickshank, B. W. Gainey, and C. L. Young, in *Gas Chromatography 1968*, C. L. A. Harbourn, Ed., Institute of Petroleum, London, 1969, p. 76.

218. A. J. Ashworth, *J. Chem. Soc. Faraday Trans. I*, **69**, 459 (1973).

219. Y. B. Tewari, D. E. Martire, and J. P. Sheridan, *J. Phys. Chem.*, **74**, 2345 (1970).

220. J. F. Parcher, P. H. Weiner, C. L. Hussey, and T. N. Westlake, *J. Chem. Eng. Data*, **20**, 145 (1975).

221. M. L. McGlashan and A. G. Williamson, *Trans. Faraday Soc.*, **57**, 588 (1961).

222. G. M. Janini and D. E. Martire, *J. Chem. Soc. Faraday Trans. II*, **70**, 837 (1973).

223. W. L. Zielinski, Jr., Ph.D. Dissertation, Georgetown University, Washington, D.C., 1972.

224. G. V. Filonenko and A. N. Korol, *J. Chromatogr.*, **119**, 157 (1976).

CHAPTER 3

Precision
Apparatus and
Techniques

The acquisition of physicochemical data by gas chromatography requires precision apparatus and techniques that are for the most part unimportant in analytical GC. In the latter, for example, it is generally unnecessary to measure the stationary phase weight or volume. In the former, however, because the parameters of interest are most often the partition coefficient or specific retention volume, the relevant experimental variables (eq. 2.30) must be measured in a manner to ensure that the resultant K_R or V_g values are of sufficient accuracy to be useful. In this connection absolute (as opposed to relative) retention data are more difficult to determine; for example, "internal standards" cannot be relied on to negate fluctuations in one portion of the chromatographic system that may in any event not be canceled by those in another.

3.1 SOURCES OF RANDOM ERROR

A large number of workers[1-36] have considered one aspect or another of precision measurements by GC; the studies by Wicarova, Novak, and Janak[37] and Goedert and Guiochon[38] are of particular interest, for they relate the precision of retention data to experimentally measured quantities like the column inlet pressure and flow rate. To a first approximation each experimental error can be considered to be random and independent of all other error sources. The variance of the partition coefficient (or specific retention volume) can then be treated statistically as a propagation of errors in which for a generalized function, Q, such that

$$Q = a^m b^n \tag{3.1}$$

the variance of the mean of Q is given by[39]

$$\sigma_Q^2 = \left(\frac{\partial Q}{\partial a}\right)_b^2 \sigma_a^2 + \left(\frac{\partial Q}{\partial b}\right)_a^2 \sigma_b^2 \tag{3.2}$$

and the fractional standard deviation of the mean calculated from[39]

$$\frac{\sigma_Q}{Q} = \left[m^2 \left(\frac{\sigma_a}{a}\right)^2 + n^2 \left(\frac{\sigma_b}{b}\right)^2 \right]^{\frac{1}{2}} \tag{3.3}$$

Consider, for example, the determination of specific retention volumes that in practice are calculated from the relation

$$V_g^0 = \frac{(l_R - l_A) 273 F p_{fm} j}{r w_L T_{fm} p_o} \tag{3.4}$$

where l_R and l_A are the chart distances from the point of injection to the solute and air peaks, respectively, r is the chart speed, and T_{fm} and p_{fm} are the flowmeter temperature and pressure. According to eq. 3.2, the standard deviation of V_g^0 is given by[37]

$$\sigma_{V_g^0} = \left[\left(\frac{\partial V_g^0}{\partial l_R}\right)^2 \sigma_{l_R}^2 + \left(\frac{\partial V_g^0}{\partial l_A}\right)^2 \sigma_{l_A}^2 + \cdots + \left(\frac{\partial V_g^0}{\partial j}\right)^2 \sigma_j^2 \right]^{\frac{1}{2}} \tag{3.5}$$

Since j is a function of the inlet and outlet pressures,

$$\sigma_j = \left[\left(\frac{\partial j}{\partial p_i}\right)^2 \sigma_{p_i}^2 + \left(\frac{\partial j}{\partial p_o}\right)^2 \sigma_{p_o}^2 \right]^{\frac{1}{2}} \tag{3.6}$$

Table 3.1 presents the calculations of Wicarova and co-workers[37] who estimated the standard deviation of each of the variables of eq. 3.4 for the elution of hexane from a column containing squalane at 50°C and Table 3.2 shows their experimental results. The estimates in Table 3.1 appear to be confirmed by the data, although it is not clear why the fractional standard deviation should vary by a factor of 3 over the temperature

Table 3.1   Estimated[37] Standard Deviation and Percent Standard Deviation for the Variables of Eq. 3.4 for the Elution of n-Hexane from a Squalane Column at 50°C

Variable	Nominal Value	σ_x (est.)	σ_x/X (%)
Δl	437.9 mm	0.28	0.06
r	0.3339 mm/sec	2×10^{-4}	0.06
F	0.3535 ml/sec	2×10^{-4}	0.06
T_{fm}	303.15°K	0.06	0.02
$p_i \ (=p_{fm})$	1.8360 atm	2.237×10^{-4}	0.01
p_o	0.9917 atm	1.316×10^{-4}	0.01
j	0.6812	1.4×10^{-4}	0.02
w_L	3.4817 g	5.2×10^{-3}	0.15
T	323.16°K	1×10^{-3}	0.001
V_g^0	151.30 ml/g	0.284	0.19

Table 3.2   Experimental[37] Mean (15–20 Measurements) $\overline{V}_g^0$ Values, and Standard Deviations and Percent Standard Deviations for Single V_g Determinations for n-Hexane with Squalane

T (°K)	$\overline{V}_g^0$ (ml/g)	$\sigma_{V_g^0}$	$\sigma_{\overline{V}_g^0}/\overline{V}_g^0$ (%)
303.14	327.66	0.315	0.10
313.19	219.58	0.316	0.14
318.19	182.81	0.480	0.26
323.21	151.64	0.288	0.19
328.17	127.91	0.189	0.15
333.15	108.36	0.222	0.20
338.21	92.38	0.285	0.31

range, 303 to 338°K. Nevertheless, these results indicate that to a first approximation the use of eqs. 3.2 and 3.3 offers a reasonable approach to error analysis in GC.

Noteworthy is that, although the standard deviation of Δl is 0.28 whereas that for w_L is 5.2×10^{-3}, the nominal values of the respective variables are 437.9 mm and 3.4817 g. Thus, because the magnitude of the retention distance is large compared with the weight of liquid phase, inaccurate determination of the latter results in a greater error on V_g^0 than the incorrect measurement of the former. It appears therefore to be unwarranted to control, for example, the column pressure drop and flow rate to a percent standard deviation of 0.05 if another error (such as that for w_L) exceeds this value by a factor of 2 or 3. (In practice w_L is indeed generally found to be the largest source of error in the measurement of K_R or V_g values in gas-liquid chromatography.)

Goedert and Guiochon[38] have critically evaluated the effects of temperature and pressure fluctuations on the precision of retention-time measurements in a manner similar to that of Wicarova and co-workers. The retention time can be cast in the form[40]

$$t_R = \frac{4L^2}{K}\eta(1+k')\left[\frac{p_i^3 - p_o^3}{\left(p_i^2 - p_o^2\right)^2}\right] \qquad (3.7)$$

where K is the packing permeability coefficient and η is the carrier viscosity. The column length and permeability are affected only slightly by small fluctuations in temperature and pressure and so may be treated as constants.[38] Similarly, the variation of η relative to minor pressure variations (approximately 10^{-3}) can be neglected. The capacity factor, on the other hand, appears to be subject to appreciable changes due to temperature fluctuations; errors due to variations in virial effects, however, can be ignored; for example,[38] for a change of 1 mbar in the mean column pressure with carbon dioxide carrier the relative variation on k' is on the order of 10^{-5}. Thus the factors in eq. 3.7 which determine the error on t_R are the temperature variation of k' and η and fluctuations in the inlet and outlet pressures (hence the flow rate).

In order to examine each of these effects, Goedert and Guiochon[38] defined fluctuations relative to t_R as

$$\frac{dt_R}{t_R} = \frac{x}{t_R}\left(\frac{\partial t_R}{\partial x}\right)_y \frac{dx}{x} + \frac{y}{t_R}\left(\frac{\partial t_R}{\partial y}\right)_x \frac{dy}{y} + \cdots \qquad (3.8)$$

where the relative standard deviation is given by

$$\frac{\sigma_{t_R}}{t_R} = \left\{\left[\frac{x}{t_R}\left(\frac{\partial t_R}{\partial x}\right)_y\right]^2 \sigma_x^2 + \left[\frac{y}{t_R}\left(\frac{\partial t_R}{\partial y}\right)_x\right]^2 \sigma_y^2 + \cdots\right\}^{\frac{1}{2}} \qquad (3.9)$$

The terms in square brackets are called the error propagation coefficients.

Two situations arise when considering pressure variations: first, the inlet and outlet pressures may be controlled independently (thereby fixing the pressure drop), in which case,

$$\frac{dt_R}{t_R} = \frac{p_o^2(p_o^2 + 4p_i^2 + p_i p_o)}{(p_i + p_o)(p_i^3 - p_o^3)}\frac{dp_o}{p_o} - \frac{p_i^2(p_i^2 + 4p_o^2 + p_i p_o)}{(p_i + p_o)(p_i^3 - p_o^3)}\frac{dp_i}{p_i} \quad (3.10)$$

Table 3.3 gives the inlet and outlet pressure error propagation coefficients calculated by Goedert and Guiochon[38] using eq. 3.10. Second, it is possible to control the pressure drop and outlet pressure independently[41]; letting $\Delta p = p_i - p_o$, eq. 3.10 becomes

$$\frac{dt_r}{t_R} = \frac{\Delta p^2 p_o}{(2p_o + \Delta p)(3p_o^2 + 3p_o \Delta p + \Delta p^2)}\frac{dp_o}{p_o}$$

$$- \frac{(p_o + \Delta p)(6p_o^2 + 3p_o \Delta p + \Delta p^2)}{(2p_o + \Delta p)(3p_o^2 + 3p_o \Delta p + \Delta p^2)}\frac{d\Delta p}{\Delta p} \quad (3.11)$$

The resultant error propagation coefficients are given in Table 3.4.

Comparison of the data in Tables 3.3 and 3.4 indicates that the second method, namely, independent control of the pressure drop and outlet pressure, is preferrable because slight variations in one direction of the inlet and outlet pressures are approximately self-canceling. Furthermore, the magnitude of each error propagation coefficient is reduced appreciably.

Table 3.3 Error Propagation Coefficients for Inlet and Outlet Pressures Controlled Independently[38]

Pressure (bars)		Pressure Drop	Error Propagation Coefficient	
Inlet	Outlet	(bars)	p_i	p_o
1.0	1.0	0	$-\infty$	∞
1.5	1.0	0.5	-2.95	1.95
1.5	1.1	0.4	-3.70	2.70
2.0	1.1	0.9	-2.14	1.14
4.0	1.1	2.9	-1.26	0.26
4.0	1.5	2.5	-1.50	0.50
∞	1.0	∞	-1	0

Table 3.4   Error Propagation Coefficients for Pressure Drop and Outlet Pressure Controlled Independently[38]

Pressure (bars)		Pressure Drop	Error Propagation Coefficient	
Inlet	Outlet	(bars)	Δp	p_o
1.0	1.0	0	−1	0
1.5	1.0	0.5	−0.98	−0.012
1.5	1.1	0.4	−0.99	−0.013
2.1	1.1	1.0	−0.96	−0.043
4.1	1.1	3.0	−0.91	−0.085
4.5	1.5	3.0	−0.92	−0.077
∞	1.0	∞	−1	0

The variation of k' with temperature is treated by first casting it in the form[38]

$$k' = K_R \frac{V_L}{V_M} = \frac{V_L}{V_M} \exp\left(\frac{\Delta G^0}{RT}\right) = \frac{V_L}{V_M} \exp\left(-\frac{\Delta S^0}{R}\right) \exp\left(\frac{\Delta H^0}{RT}\right) \quad (3.12)$$

where

$$\frac{dk'}{k'} = -\frac{\Delta H^0}{R} \frac{dT}{T^2} \quad (3.13)$$

The carrier viscosity has been shown[40] to vary with temperature as

$$\frac{d\eta}{\eta} = 0.8 \frac{dT}{T} \quad (3.14)$$

Employing these relations in eq. 3.8, the relative variation of retention time with respect to k' and η is found to be

$$\frac{dt_R}{t_R} = \left(\frac{0.8}{T} - \frac{k'}{1+k'} \frac{\Delta H^0}{RT^2}\right) dT \quad (3.15)$$

Table 3.5 presents the calculated error propagation coefficients for k' and t_R for assumed values of ΔH^0. Clearly, the change of carrier viscosity with temperature is unimportant when compared with that of k'.

Even though the parameters of eq. 3.7 have now been treated, other instrumental factors, chief among which is the method of retention-time measurement, must be considered; that is, computer (or manual) versus

Table 3.5   Calculated Error Propagation Coefficients[38] for k' and t_R for Assumed values of ΔH^0 and k'

T ($^{\circ}$C)	$0.8/T$ ($^{\circ}\text{K}^{-1} \times 10^2$)	k'	ΔH^0 (kcal/mol)	Coefficient for k' ($\times 10^2$)	Overall Coefficient ($\times 10^2$)
77	0.23	0.5	4	-0.54	-0.31
77	0.23	1.0	4	-0.81	-0.57
77	0.23	3.0	4	-1.22	-1.00
77	0.23	10.0	4	-1.48	-1.25
127	0.20	10.0	4	-1.14	-0.94
127	0.20	10.0	10	-2.88	-2.68
127	0.20	10.0	15	-4.29	-4.09
177	0.11	10.0	4	-0.89	-0.71

integrator data acquisition. Goedert and Guiochon[38] showed that retention time, when measured from the point at which the peak signal reaches a maximum h_M, is related to the noise level i (approximately four standard deviations of the signal measurement), the plate number N, and the absolute error on the retention time Δt_R by

$$\frac{\Delta t_R}{t_R} = \left(\frac{i}{h_M} \frac{2}{N} \right)^{\frac{1}{2}} \tag{3.16}$$

which will be obeyed qualitatively both by mass-flow and concentration-dependent detectors. Conversely, when an integrator is employed, the retention time is taken from the point at which the first derivative of the peak signal becomes (ideally) zero. Ignoring systematic errors (such as improper setting of the threshold triggering level), the random error is to a first approximation given by[38]

$$\frac{\Delta t_R}{t_R} = \frac{i' t_R}{h_M N} \tag{3.17}$$

where i' is the noise level of the signal derivative. The effects of noise on these two methods of retention-time measurement are contrasted in Table 3.6. Peaks with short retention times, hence sharp maxima, are measured with greater accuracy with an integrator, which, however, becomes less reliable as the retention time lengthens, for the maximum is broadened and the rate of change of the signal's first derivative with time is decreased. This is particularly troublesome because for a mixture of solutes (some of which may be eluted quickly, whereas others remain on the column for a considerable period relative to the first) the operator can, in general,

Table 3.6   Effects of Noise on Two Methods of Retention-
Time Measurement[38]

A. Data Acquisition via Peak Height Maximum

Signal-to-Noise Ratio (h_M/i)	N	t_R (sec)	$\Delta t_R/t_R$ (%)
2.5	2×10^3		2
10^2	2×10^3		0.3
10^3	2×10^3		0.1
10^3	2×10^5		0.01

B. Data Acquisition via Peak Maximum Zero Signal Derivative

10	2×10^3	600	3
10^2	2×10^3	600	0.3
10^3	2×10^3	600	0.03
10^3	2×10^3	60	0.003
10^3	2×10^5	3×10^3	0.0015

operate an integrator only at some suitable mean triggering level that will overestimate retention times of the earliest eluting compounds and underestimate those for the last eluting materials. Consequently, one is left with the choice of chromatographing individual solutes in which the integrator response is optimized for each or using the peak maximum signal method with a sensitive detector and ensuring that the signal-to-noise ratio is kept to an absolute minimum; for example,[38] suppose that for a given flame detector $i = 10^{-13}$ A/sec and $i' = 5 \times 10^{-13}$ A/sec. If $N = 2500$ plates for a solute whose elution time is 250 sec and a 1-μg sample produces a peak maximum signal of 10^{-9} A, the signal maximum method of retention-time measurement will be precise to about 2.8×10^{-4}, whereas the signal maximum derivative method will yield a precision of 5×10^{-5}. If the sample size is reduced to 10 ng, the respective values become 2.8×10^{-3} and 5×10^{-3}.

Overall, levels of precision of the order 10^{-3} require careful control of all experimental aspects of gas chromatography. That these levels can, however, be achieved in practice has been demonstrated beyond doubt. Table 3.7, for example, presents the range of fluctuations of experimental parameters found by Goedert and Guiochon[38] for their high-precision instrument[41] and Table 3.8 shows the precision (defined as $t\sigma/t_R\sqrt{n}$, where t is the Student factor and n is the number of observations; here $t/\sqrt{n} \cong \frac{1}{3}$)

Table 3.7   Variations in Experimental Parameters[38] for a Laboratory-
Constructed Gas Chromatograph[41]

Parameter	Average Value	Fluctuations over		
		1 hr	10 hr	100 hr
Outlet pressure, atm	1.1	2.5×10^{-4}	2.5×10^{-4}	3.5×10^{-4}
Pressure drop, atm	0.4	4×10^{-4}	4×10^{-4}	7×10^{-4}
Temperature, °C	52	5×10^{-3}	1×10^{-2}	3×10^{-2}
Bridge Current, mA	190	2×10^{-4}	2×10^{-4}	5×10^{-4}

observed for the elution of four solutes from a column containing Molecular Sieve 5A. As these workers noted,[38] however, other parameters (such as
those considered by Wicarova and co-workers) must be measured in
physicochemical studies and at the present time determination of w_L limits
the inherent precision of V_g and K_R data to about 0.5%.

3.2   INSTRUMENTATION

The requirements of accurate temperature and pressure control and data
acquisition generally preclude the use of "as is" commercially available gas
chromatographs for physicochemical measurements. The vast majority of
these investigations, however, have been conducted at less than 5-atm inlet
pressure and ambient to 200°C column temperature so that in some cases
it may be feasible to employ existing instrumentation with only slight
modifications. Other situations may require fabrication of the entire GC,
as, for instance, in high-temperature/pressure studies. In any event it is
clear that the choice of the level of instrumental sophistication is never
wholly subjective in that the experimentalist may not know at the outset to
what extent precise data are important to the project at hand. Each feature
of the chromatographic system should therefore be designed and/or modified within the means available with a view to obtaining an overall percent
standard deviation consistent with an optimum calculable from eq. 3.5,
which, at worst, should be 2%.

3.2.1   Column Thermostats   A variety of devices has been used as GC
column ovens,[42-44] the earliest of which included vapor baths,[45,46] Dewar
flasks,[47] heated metal blocks,[48] and stainless steel[49] or wooden[50] boxes or
pipes[51] wrapped with Nichrome wire. Another technique involved direct
electrical heating of metal columns,[52] and a method by which thermostated
water was pumped through a column jacket in an all-glass system has been

described[53] (this enabled the injector and detector of an existing commercial unit to be used while still controlling the column temperature accurately). By far the most common type of thermostat today, however, is a liquid bath or a forced-air oven in which, generally, the latter is more useful at higher temperatures than the former, for most liquids (with the possible exception of silicone fluids) tend to decompose slowly when heated for extended periods. Several early designs of air ovens have been presented elsewhere[54, 55]; more recently Rogers and co-workers[56–58] described the modification of a Becker Model 1452 SH oven which consisted of an upright cylinder surrounded by heating coils and contained in a second concentric cylinder. A high-speed fan forced air upward past the heating coils, then rapidly down over the column. Inclusion of an impeller blade at the top of the inner cylinder was said to reduce temperature gradients to $\pm 0.02°C$ everywhere within the oven. A Melabs CTC-1A proportional temperature controller was used; the temperature was monitored with a precision platinum resistance thermometer system and the stability ($\pm 0.01°C$ over an 8-hr period), measured with a thermistor that formed one arm of a Wheatstone bridge. Parcher and co-workers[59] have reported the use of a Beckman Model GC-65, the column temperature of which was claimed to be constant to $\pm 0.1°C$. Premru and Kremser[60] described an interesting modification of a Varian 1800 gas chromatograph: a homemade oil bath was built into the oven compartment and the heaters for the original air thermostat, used to warm the liquid bath. In this manner, although a longer period (than that for air) was required to reach thermal equilibrium, a temperature stability of $\pm 0.02°C$ was achieved.

Liquid baths have of late been used more frequently than air ovens for physicochemical studies, with vigorous stirring temperature stability is more readily achieved. Thus, for example, Wicarova and co-workers[37] and Goedert and Guiochon[41] used glycerol (30–150°C) and ethylene glycol (30–130°C) thermostats, respectively. The latter system included a Melabs CTC-1A proportional temperature controller and the bath temperature was found to vary less than 0.01°C over 15 hours with gradients of less than 0.05°C. Figure 3.1 illustrates the stability obtained with this unit. Rijks and Cramers[61] found gradients of less than 0.01°C when a Tamson Model TEV 70 liquid bath was employed. Most recently, Laub, Martire, and Purnell[62] obtained a stability of at worst 0.03°C with a Grant Model LB 50 water bath, measured with a Hewlett-Packard Model 2802A platinum resistance thermometer system. Accurate temperature control (as well as extension of the range of the unit beyond the factory-set limit of 75°C) was achieved by incorporation of a 1000-ohm precision rheostat across the 10,000-ohm arm of the control bridge circuit.

High-temperature gas chromatography has been employed for some years for the direct analysis of inorganic materials and can, of course, also

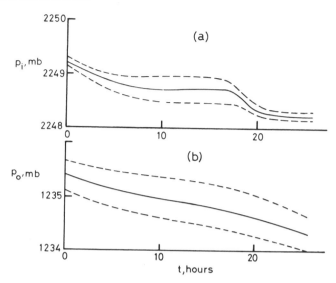

Figure 3.1 (a) Output pressure fluctuation of a Negretti-Zambra R 182 precision regulator over a 20-hr period.[41] A variation of 1 mbar corresponds to 0.04% of 2249 mbar. (b) Outlet pressure fluctuation.

be used for physicochemical studies; for example, De Boer[63] described a GC capable of the analysis of cadmium-zinc mixtures in 1960 and in 1966 Barrett[64] patented an instrument that was designed to operate from 500 to 3000°C for the production of high-purity cadmium from cadmium-mercury amalgams. Cremer and Deutscher[65] have also reported a high-temperature unit that was useful up to 1000°C. Several Russian groups have investigated high-temperature instrumentation, chief among whom were Sokolov and co-workers[66-69] who recently[69] described the direct determination of bismuth in alloys containing tin, lead, zinc, and cadmium at 1250°C.

At the other end of the temperature scale cryogenic gas chromatographs have beeen used to analyze isotopes and to examine exchange equilibria and adsorption; for example, Merritt and co-workers[70] described a programmable cryogenic GC in 1964 and Bruner, Cartoni, and Liberti[71] used a combination of liquid and air baths to achieve temperatures of -190 to 190 ± 1°C in 1966. Conti and Lesimple[72] also found their cryostat to be stable to ± 1°C. Better stability (± 0.2°C) was reported by Haubach and co-workers[73] in 1967 at temperatures between 27 and 55°K. Finally, a rather more versatile (and consequently more elaborate) system has been described by Giannovario, Gondek, and Grob[74] which was said to be capable of operation from -190 to 300°C.

In general, most modern commercial forced-air ovens are stable to at least 0.5°C (and in some cases an order of magnitude better than that) and can be improved still further by the addition of a 10-turn precision rheostat in series with the oven temperature variable resistor control as a "fine" adjustment. Gradients within a unit can usually be limited to 0.02°C at worst by inclusion of baffles or a second fan placed to create maximum turbulence. Liquid thermostats enable very accurate (±0.01°C or better) temperature control up to about 200°C and well-stirred systems appear to suffer little from the problem of gradients.

3.2.2 Pressure/Flow Controllers In addition to inadequate temperature control, most commercially available gas chromatographs are unsuited to physicochemical measurements because of imprecise pressure regulation. Many modern units are equipped with flow regulators of various types, which, however, merely exacerbates the problem because such units are generally placed next to the column oven and are not thermostated. That this practice has continued is, in fact, somewhat surprising, for it has long been recognized that even minor temperature fluctuations will generate large (> 1%) changes in the "regulated" flow rate; for example, Mikkelsen[8] reported in 1967 that a 3°C change produced a 1% change in the output of a flow controller (which subsequently resulted in a 1% change in the sensitivity of the thermal conductivity detector). Deans[75] cited an even larger variation of 1%/degree at ambient temperature and in addition pointed out that inlet (to the controller) pressure fluctuations of 1% may result in a further 1% flow-rate variation. Thus the sensitivity of such devices to inlet pressure and temperature variations renders them inadequate for precision measurements unless these variables are carefully controlled. Pressure controllers,[75] on the other hand, appear to be relatively insensitive to inlet pressure and ambient temperature variations and fluctuations in the postcolumn pressure.

Table 3.8 Reproducibility[38] of High-Precision[41] Retention Data for Named Solutes Eluted from a Molecular Sieve Column at 325°K

	Peak		Precision $(\times 10^4)$	
Solute	Height (mV)	t_R (sec)	Inlet Pressure Controlled	Pressure Drop Controlled
Ar	19	137	11	15
N_2	15	284	8	7
CH_4	11	484	4	5
CO	3.5	1064	9	5

Table 3.9 Variation[41] of the Outlet Pressure from a Negretti-Zambra R 182 Precision Regulator with Changes in the Operating Parameters

Operating Parameter	Parameter Variation	Variation in Outlet Pressure
Inlet pressure	1%	$1.5 \times 10^{-2}\%$
Reference pressure	1 mbar	1 mbar
Temperature	1°C	$2 \times 10^{-2}\%$

A variety of pressure/flow control systems has been described in the literature [76-84]; for example, Cruickshank and co-workers[11] used a lagged high-precision pressure regulator with a thermostated needle valve to achieve a variance of 0.25% for replicate retention-time measurements. A twofold improvement was obtained by thermostating both the regulator and needle valve to ±0.05°C (where the inlet pressure was measured with a Bourdon-tube gauge calibrated to ±0.01 atm and the pressure drop with a differential mercury manometer). Goedert and Guiochon[41] reported a detailed investigation of Negretti-Zambra Model R 182 precision pressure regulators in 1969. Table 3.9 shows the influence of experimental parameters on the regulated output pressure from one R 182 regulator and Fig. 3.2 depicts the long-term stability at an inlet pressure of 2500 mbar. According to these data, the regulated output pressure is most seriously affected by the reference pressure, which could be varied between vacuum and 1.5 bars in the test apparatus. To minimize all the variations a second R 182 regulator was connected in tandem with the first and for this initial study both were referenced against a vacuum of $4-5 \times 10^{-2}$ mbar. The column inlet pressure variation was thereby reduced to ±0.3 mbar, an improvement of a factor of 3 over the use of a single regulator. As Goedert and

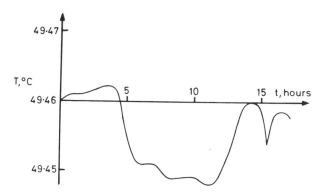

Figure 3.2 Temperature fluctuation of column thermostat of Goedert and Guiochon.[41]

Guiochon noted, such a variation could result from a diaphragm spring-length change of only 1 μ, the results therefore being as good as could be expected. (The pressures were measured with a Texas Instruments Model 141 quartz Bourdon-tube gauge which was found to be accurate to 1.5×10^{-4} times the nominal value plus 0.035 mbar.) The column outlet pressure in this system was controlled above atmospheric pressure with a third R 182 regulator (referenced against a vacuum), although for low flow rates (e.g., 10 ml/min) some difficulty was encountered; a flow rate of roughly 100 ml/min was found to be more suited to this method of outlet (hence pressure drop and flow rate) control. In later refinements[38] the second inlet pressure regulator was referenced against the output pressure of the column outlet regulator which was stable to 0.2–0.3 mbar at a nominal value of 1100 mbar. A Texas Instruments pressure controller was also found to be satisfactory for column inlet pressure control[38] as long as the unit was referenced against the stabilized column outlet pressure. Oberholtzer and Rogers[56] have, alternatively, used a Brooks Model 8743 ELF flow controller for inlet pressure/flow control, the input pressure to which was supplied by an R 182 NC regulator. The column outlet pressure was ambient and all components were thermostated at $35 \pm 0.05°C$. This system was tested by eluting 10 replicate samples of ethane from a column containing SE-30 silicone gum at 80°C; the mean retention time was found to be 26.673 ± 0.005 sec, a standard deviation of 0.02%. Wicarova and co-workers[37] used two precision needle valves in conjunction with a Cartesian manostat[85] (Hoke-Manostat Corp.) which were all enclosed in a polystyrene box. They also used ambient pressure at the outlet and estimated a standard deviation of 0.01% on p_i (as measured with a multiple-arm mercury manometer[86,87]) and 0.06% on F (Table 3.1).

Bowen and co-workers[88] reported that a Veriflow Model PN-41200649 "balanced" regulator was superior to ordinary tank regulators; for example, a drop of 100 psi at the inlet of their unit produced an output fluctuation of less than 0.02 psi. The column inlet pressure stability was further improved by the addition of a Veriflow Model 41300451 single-stage regulator and a Veriflow Model 4230080 flow controller. Rijks and Cramers[61] used two Becker-Delft Model MB-19936 inlet pressure controllers in series to obtain a column pressure drop said to be constant to about 0.002 atm as measured with a Wallace and Tiernan Model FA 145 precision manometer. Jonsson and co-workers[89,90] obtained excellent results with a Brooks Model 8743 flow controller and Rosemount Model 831A-18 differential[91–93] and Model 830A-7 absolute pressure sensors. A Negretti-Zambra Model M2691 precision manometer also used in their system was found to be accurate to ± 0.1 torr.

In light of these data cited for systems with no outlet pressure control it is questionable whether such a feature is entirely necessary. Without question, if an elevated column pressure coupled with a small pressure drop (as in the case of virial coefficient measurements) is to be used, the outlet pressure must be controlled above ambient. If, on the other hand, retention times are to be measured in order, for example, to determine activity coefficients for a few solutes with a given phase, the running time of the experiment will require at most a few hours during which atmospheric pressure will generally be constant to ± 10 torr. The error introduced by such fluctuation amounts to less than 0.2% on the *j*-correction at an inlet pressure of 1000 torr, which is entirely adequate for all but the most demanding of physicochemical measurements.

High (>5 atm) pressure GC has been exploited by several groups including Kobayashi and co-workers,[94-99] Giddings and co-workers,[100-105] Pretorius and Smuts,[106] Karayannis and co-workers,[107, 108] and Wicar and Novak.[109] Many low-molecular-weight compounds exhibit critical points at pressures of 10–30 atm and at temperatures of 50–300°C so that high-pressure GC can also be used for the investigation of supercritical phenomena, as demonstrated by Klesper Corwin, and Turner,[110] Sie, Rijnders, and co-workers,[111-113] and Gouw and Jentoft.[114] Apparatus for such studies must, in general, be designed and fabricated by the investigator; the above references (for pressures up to 30,000 psi) will serve as a useful guide.

Reduced-pressure GC appears so far to have been examined solely with respect to analytical separations. It seems likely,[115] however, that it may also prove useful for a variety of physicochemical studies; for instance, as a method of investigating mass transport (flow) properties and porous media[116] (cf. Purnell,[55] Chapter 6). Various instrumental designs have been presented for reduced-pressure operation[117-127]; Findlay[128] has described a particularly simple manostat that is useful from 10 to 760 (and, with slight modification, higher) torr and is stable to better than 0.1 torr.

3.2.3 Flowmeters

Once the inlet (and/or outlet) pressure and pressure drop have been controlled, the flow rate will, with constant column temperature, also be controlled. The easiest, simplest, and still one of the most accurate[129] (±0.25%) devices for measuring the flow rate is a thermostated 50–100-ml soap-bubble flowmeter: the time required for a soap bubble to rise in a calibrated tube of uniform diameter gives the gas flow rate at the system outlet. A water-jacket thermostat is particularly useful because it magnifies the inner tube slightly which facilitates observation of the soap film. If the outlet pressure is controlled above ambient, a sealed flowmeter[6] can be inserted directly in the line connecting the detector output and outlet pressure regulator. It is also possible to measure the flow at the column inlet, the appropriate conversion to that at the outlet being

$$F_o = F_i \frac{p_i}{p_o} \frac{(RT + p_i B_{22}^c)}{(RT_{fm} + p_o B_{22}^{fm})} \tag{3.18}$$

where B_{22}^{fm} and B_{22}^c are the carrier gas virial coefficients at the flowmeter and column temperatures, respectively. Making the assumption that $B_{22} \cong 0$ (which is accurate to 0.02% for all common carrier gases except CO_2, where the error is 0.16%), eq. 3.18 reduces to the ideal gas law:

$$F_o = F_i \left(\frac{p_i T}{p_o T_{fm}} \right) \tag{3.19}$$

On the other hand, the inlet flow rate F_i can be used to calculate the average column flow rate directly:

$$j' F_c' \equiv \overline{F}_c = F_i \left(\frac{p_i}{\overline{p}} \right) \left(\frac{T}{T_{fm}} \right) \left(\frac{p_i - p_w}{p_w} \right) \tag{3.20}$$

where the gas compressibility correction factor j' is now given by

$$j' = \frac{p_i}{\overline{p}} = \frac{3}{2} \left[\frac{1 - (p_o/p_i)^2}{1 - (p_o/p_i)^3} \right] \tag{3.21}$$

Levy[129] has presented a critical evaluation of soap-bubble flowmeters and found that for very accurate ($<0.1\%$) work corrections must be made for incomplete carrier saturation by water vapor (which, further, is not identical to the saturation vapor pressure because of the presence of soap in the solution) and the decrease of the flowmeter volume due to the soap film on the inside of the tube. Czubryt and Gesser[130] also pointed out that soap bubbles, if stretched thinly (as happens with a tapered tube), may be permeable to the carrier. As a consequence, other techniques for flow measurement have been developed over the years[130–139]; for example, a mercury-drop flowmeter[138] recently described is said to be accurate to $\pm 0.07\%$.

Jonsson and co-workers[89, 90] have proposed an attractively accurate alternative to the usual method of flow rate (thence retention volume) measurement: a Brooks Model 5810 flow sensor was addressed at 0.05-sec intervals by a computer which resulted in a continuous measurement of the

Table 3.10   Comparison[89] of Total-Flow and $t_R F$-Product Methods of Retention Volume Measurement for the Elution of n-Heptane from a Column Containing n-Octadecane at $60°C$

Run No.	V_R (ml)	$t_R F$ (ml)
1	329.25	329.59
2	329.35	329.88
3	329.32	329.89
4	329.23	329.66
5	329.28	329.39
6	329.42	329.22
7	329.38	328.97
8	329.36	328.81
9	329.27	328.80
Mean	329.32	329.36
% standard deviations	0.019	0.131

flow rate. Integration from the moment of injection to the peak maximum gave the total carrier volume required to elute the solute, that is, the retention volume. Table 3.10 illustrates the success of this approach where V_R values are compared with the data $t_R F$ taken from the same runs.

3.2.4   Injectors   The type[142–159] of sampling device used in GC is of little importance if the moment of injection need be accurate to no better than 1 to 2% of the total retention time and the injection profile is not of major concern. Injection at infinite dilution as well as finite concentration can in these cases be performed with syringes of various types,[160–164] vapor dilution systems,[165–174] ampoules,[175–179] gravimetric sample holders,[180] or automated[181] rotary,[182–184] sliding-seal,[185–188] diaphragm-seal,[189–191] or hybrid-fluidic[192–195] valves.[88, 196–199] Vacuum[200, 201] and high-pressure[202–206] injectors have also been described. For high-precision work at infinite dilution sample-valve injection is the most accurate technique available because the moment and volume of injection can be precisely controlled. Oberholtzer and Rogers[56] and Bowen and co-workers[88] have examined several types of injection system; Table 3.11 gives the plug width, dead time, and percent reproducibility for injections of methane with different automated systems. The Carle valve gives a narrower plug than the other types except for the hybrid-fluidic valve which could be adjusted to

Table 3.11   Comparison of Automated Injector Systems[56, 88]

Valve	Plug Width (msec)	Methane Elution Time (msec)	Reproducibility (%)
Seiscor[a]	67	450 ± 1	0.22
Carle[b] (loop A)[c]	43	340 ± 4	1.18
Carle[b] (loop B)[d]	44	370 ± 4	1.08
Kieselbach[e] (loop A)	65	500 ± 5	1.00
Kieselbach[e] (loop C)[f]	60	250 ± 5	2.00
Hybrid-fluidic	50	227.6 ± 0.41	0.18
Hybrid-fluidic	100	246.6 ± 0.30	0.12
High-pressure	...	117.7 ± 0.39	0.33
Hamilton[g]	...	180.5 ± 1.35	0.75
Hamilton[h]	...	198.7 ± 1.65	0.83

[a]Seiscor Model VIII.
[b]Carle Model 2014 with Model 2050 pneumatic actuator.
[c]Valve to detector connecting-tube size: 0.102 cm I.D.
[d]Valve to detector connecting-tube size: 0.058 cm I.D.
[e]R. Kieselbach, *Anal. Chem.*, **35**, 1342 (1963).
[f]Valve to detector connecting-tube size: 0.051 cm I.D.
[g]Pneumatically operated Hamilton high-pressure liquid chromatography valve; operating temperature: 30°C.
[h]Operating temperature: 100°C.

produce any desired profile duration; all the units except the Hamilton model (which was designed for LC, not GC) give a precision of ±1–5 msec. These data were obtained by connecting the sampling valves directly to a detector with a short length of capillary tube, which, however, does not correspond to realistic GC operating conditions; for example, a 5-ft by 1/4-in. packed column with a flow rate of 20 ml/min helium at 50°C will give a methane peak between 10 and 20 sec so that if the precision of injection is ±5 msec, *irrespective of the sample type*, the *maximum* possible error on any other retention time due to the injection process (assuming nothing elutes before methane) will be ± 2.5–$5 \times 10^{-2}\%$. For a retention time of 1000 sec (16.7 min, which is not uncommon) the percent error is $5 \times 10^{-4}\%$. Therefore *all* the valves in Table 3.11 are adequate for most physicochemical studies, provided that the retention time of the solute is sufficiently long.

For measurements that require very high precision, for example, statistical moments analysis,[207] the shape of the sample input profile[100, 208–212] becomes relevant. Reilley and co-workers[213, 214] have presented a detailed treatment of the peak shape at the column outlet as a function of the solute injection profile and the sample size: briefly, if the solute is placed onto the column as a plug and the corresponding peak shape at the column outlet is described by a Gaussian curve, that is,

$$S(t) = \frac{1}{(2\pi)^{\frac{1}{2}}\sigma} \exp\left[-\frac{(t-t_R)^2}{2\sigma^2}\right] \tag{3.22}$$

the response to a plug input of baseline width s and unit height will be

$$R(t) = \frac{1}{(2\pi)^{\frac{1}{2}}} \int_{(t-t_R-s)/\sigma}^{(t-t_R)/\sigma} e^{-y^2/2} dy \tag{3.23}$$

where $R(t)$ is the ordinate value of the peak height at time t from the moment of injection, σ is the peak standard deviation, and $y = (t-t_R)/\sigma$. For inputs of other shapes Reilley and co-workers[213] found that $R(t)$ was related to the input function $\underline{I}(t)$, the "action" function $A(t)$, and the respective Laplace transforms $\bar{R}(p)$, $\bar{I}(p)$, and $\bar{A}(p)$ by

$$\bar{R}(p) = \bar{I}(p)\bar{A}(p) \tag{3.24}$$

In most cases in GC the "action" function is simply the Gaussian distribution function (eq. 3.22), the Laplace transform of which is

$$\bar{A}(p) = \exp\left(\frac{p^2\sigma^2}{2} - t_R p\right) \tag{3.25}$$

where p is related to the number of theoretical plates N by

$$p = N\frac{t_R'}{t_R} \tag{3.26}$$

and is virtually identical to N for values of $k' > 10$. For a single solute eq. 3.24 therefore becomes

$$\bar{R}(p) = \bar{I}(p)\exp\left[\frac{p^2\sigma^2}{2} - t_R p\right] \tag{3.27}$$

as found earlier by Boeke.[215] For j solutes

$$\bar{R}(p) = \bar{R}_0(p) + \sum_j \bar{I}(p,j)\bar{A}(p,j) \tag{3.28}$$

where $\bar{R}_0(p)$ is zero for all input functions except those involving elution on a plateau.

When the column outlet peak shape is other than Gaussian (or the integral of the Gaussian function as in frontal analysis), eq. 3.28 will still be valid, provided that the appropriate action function is employed.[213]

These relations have been investigated by Bowen and co-workers[88] and Wade and Cram,[198] who used a hybrid-fluidic valve for absolute control of the sample size and injection profile [e.g., finding a relative standard deviation of 0.05–0.08% for the elution of methane from a 46.5-cm length of 1-mm I.D. capillary tube ($t_R \cong 0.2$ sec)]. Reproducibility of the input peak shape and duration is illustrated in Table 3.12 for four configurations of their system,[198] data that represent the highest precision yet achieved in gas chromatography.

Table 3.12   Sample Input Profiles for Four Configurations of a Hybrid-Fluidic Injection System[198]

Configuration	Peak Height (cm) Mean	% Standard Deviations	Baseline Width (msec) Mean	% Standard Deviations	Half-Height Width (msec) Mean	% Standard Deviations
1	10.2	0.53	20.2	4.3	9.6	0
2	6.61	0.56	26.6	2.4	69.0	9.9
3	7.60	3.0	69.0	9.9	63.9	11.0
4	2.62	1.0	131.7	2.9	39.9	2.9

3.2.5   Detectors   Numerous reviews[216–228] and three texts[229–231] deal specifically with GC detectors. In addition to these and the systems cited in Sections 3.2.1 and 3.2.2, detectors capable of operation at vacuum to high pressure[232–234] and temperature,[235,236] for supercritical fluid GC,[237] and with radioactive[238–240] and corrosive or labile[241] materials have also been described. Design considerations such as low dead volume and high sensitivity have long been recognized (e.g., Ref. 230) and need little comment here except to note that the technique of postcolumn sample splitting will enable most types of detectors to be employed in finite-concentration GC; two particularly simple devices are those reported by Gregory[242] and Coduti.[243]

For greatest accuracy the extra-column dead volume and detector response time must be examined, both of which can most easily be determined by the method first reported by Kieselbach[244] (alternative techniques have been described elsewhere[245–249]): a short length of capillary

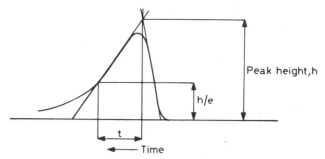

Figure 3.3 Asymmetric peak shape of air used to measure system dead volume and detector time constant.[244] Tangents are drawn to the peak inflection points and a perpendicular dropped from their intersection to the baseline. The asymmetric portion t of the peak dead time is measured from this perpendicular to the trailing edge of the peak at h/e (=0.37 h). The asymmetric portion V of the dead volume is then found from the product of t and the flow rate.

tube is connected between the injector and detector and the flow rate is varied for succeeding injections of a nonretained solute (such as air). The volume asymmetry of the solute peak (illustrated in Fig. 3.3) is plotted versus flow rate; the extrapolated (zero flow) value is the extracolumn dead volume. The detector response time is measured by plotting the time asymmetry versus the inverse of the flow rate and extrapolating to $1/F=0$ ($F=\infty$). Kieselbach found that the dead volume for his system was 0.05 ml, the volume of the injector loop, so that other contributions (such as the detector and fittings) were negligible. The (thermistor) detector response time was 0.07 sec which with an operational amplifier compensating circuit was reduced to 0.01 sec. Symmetrical peaks were then obtained at flow rates as high as several hundred milliliters per minute. In a more recent example Glenn and Cram[197] measured a (flame) detector response time of 8 msec and an extracolumn dead volume of 9.5 μl. Because the sample (methane) was known to occupy a volume of 8.1 μl initially, the true dead volume was 1.4 μl; this value compared favorably with their estimate of 1 μl for the detector jet.

3.2.6 Data Acquisition Systems By far the most common data acquisition device used in GC is a strip-chart recorder, most modern units being capable of at least 0.1% chart-speed reproducibility and 0.2-sec full-scale response to an input signal step-change. However, recorders have on occasion[250,251] been found to contribute to peak asymmetry due mainly to full-scale nonlinear response and finite response time. These effects can be examined with an electronic peak simulator, two such instruments having been described elsewhere.[253,254] Various types of related recording and readout devices have also been reviewed[255,256] and lists of commercially available units now appear regularly in the literature. Given current state-of-the-art equipment, retention distances measured from a strip-chart recorder will be accurate to about ± 0.2 mm, which is trivial, for retention distances can be made any length desired and, in any event, in physicochemical studies the relevant retention distance is l'_R ($= l_R - l_A$) which is measured from the "air" peak to the solute peak maximum. Therefore, because absolute measurement of the retention time is not required, a modern strip-chart recorder will suffice to the extent indicated; for example, Wicarova and co-workers[37] found that their recorder and manual data acquisition contributed a relative standard deviation of only 0.06% to the V_g^0 value of n-hexane (Table 3.1), where l'_R was 437.9 mm and the standard deviations of the difference $l_R - l_A$ and the chart speed were, respectively, 0.28 mm and 2×10^{-4} mm/sec.

Digital data acquisition in GC has been reviewed comprehensively by Leathard[256] and Cram and Leitner[257]; additional investigations[258–294] including moments analysis[295–309] have also been reported to which the reader is referred for details. It is a difficult task to assess the usefulness of such systems for a given physicochemical study other than in terms of enhanced accuracy in retention measurements. Clearly, if, for example, cost is no object and peak-broadening experiments similar to those of Glenn and Cram[197] are to be conducted, a computerized data acquisition system appears to be desirable. If, on the other hand, the project of interest commands a limited budget and involves the determination of infinite-dilution activity coefficients to an accuracy of around 1%, a strip-chart recorder will be sufficient. To this extent the matter of analog versus digital data acquisition remains largely subjective.

3.2.7 Examples of Precision Apparatus A wide range of designs and components has been employed in precision gas chromatographic systems; a few examples which have proved to be useful for a variety of physicochemical studies are presented here as guidelines.

Figure 3.4 shows a schematic diagram of the system used by Cruickshank, Windsor, and Young[6] for the measurement of infinite-dilution activity and virial coefficients at pressures up to 30 atm. The valves were conventional bronze high-pressure on-off controls for leak testing, evacuation, and system flushing. Inlet pressure control was achieved with a single-stage high-precision regulator and needle valve. The pressure gauge was a calibrated Bourdon-tube unit and a sealed soap-bubble flowmeter was used for flow measurement at the column inlet. The outlet pressure was maintained above ambient with a thermostated needle valve and the column effluent split with a Y-junction, one arm of which was a length of

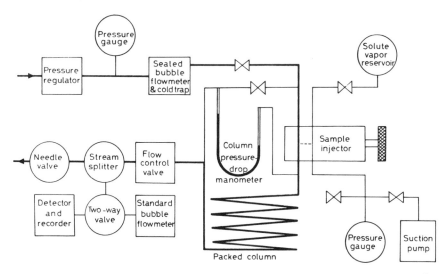

Figure 3.4 High (30 atm) pressure precision GC of Cruickshank, Windsor, and Young.[6]

constricted capillary tube connected to a flame detector. The pressure drop was monitored with a differential mercury manometer. The column thermostat was a water bath. Sample injection was performed with a laboratory-constructed valve system that was essentially an adjustable-volume gas storage cylinder fitted with a displacer piston.

The precision GC of Wicarova, Novak, and Janak[37] used to measure specific retention volumes is illustrated in Fig. 3.5. The column outlet was

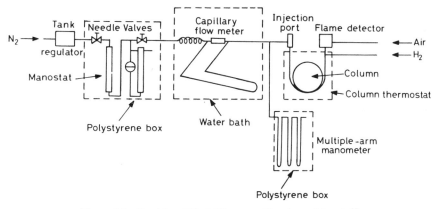

Figure 3.5 Precision GC of Wicarova, Novak, and Janak.[37]

at ambient pressure and the inlet pressure was controlled with two thermo-stated precision needle valves and a manostat. The flow rate at the inlet was measured with a capillary flowmeter and the inlet pressure, with a multiple U-tube mercury manometer. A flame detector and strip-chart recorder constituted the detection system. Injection was performed with a syringe into a low dead-volume port and the column thermostated with a glycerol bath.

Figure 3.6 shows the system of Goedert and Guiochon[41] which employed digital data acquisition and automated sample injection. Two precision pressure regulators contained in an evacuated thermostat controlled the inlet pressure and a third regulator and needle valve maintained the column outlet above ambient. Pressures were measured with a precision quartz Bourdon-tube gauge. The column thermostat was an ethylene glycol bath and the detector was a twin-cell katharometer. Replicate (approximately 100 per night) injection and data (retention time, peak height, peak area) acquisition were performed automatically and included a print-out of the inlet, outlet, sample, and atmospheric pressures, liquid

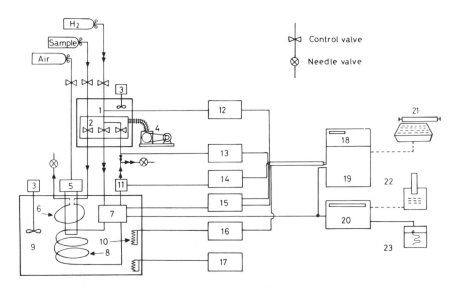

Figure 3.6 Precision GC of Goedert and Guiochon[41]: 1. Control valves in thermostated liquid bath. 2. Vacuum container. 3. Pump and heater systems. 4. Vacuum pump. 5. Sampling valve. 6. Gas sample loop. 7. Katharometer. 8. Column. 9. Column thermostat. 10. Temperature sensor. 11. Flowmeter sensor. 12. Pressure gauge. 13. Precision pressure gauge. 14. Flowmeter. 15. Dc power supply. 16. Temperature bridge. 17. Proportional temperature controller. 18. Digital voltmeter. 19. Data logging system. 20. Integrator. 21. Electric typewriter. 22. Printer. 23. Recorder.

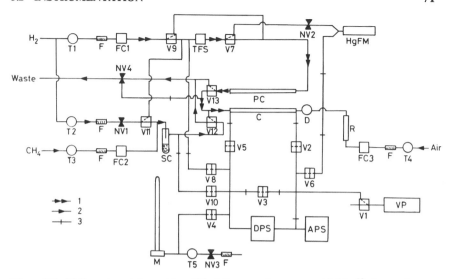

Figure 3.7 Fully automated precision GC of Jonsson, Jonsson, and Malm[90]: 1 = carrier gas flow; 2 = sample gas flow; 3 = pressure-sensing lines. V1-V13, solenoid valves. T, toggle valves. FC1-FC3, flow controllers. NV1-NV4, needle valves. TFS, thermal-flow sensor. DPS, differential pressure sensor. APS, absolute pressure sensor. F, filters. PC, precolumn. C, column. M, manometer. VP, vacuum pump. SC, sample chamber. D, detector. HgFM, mercury-drop flowmeter.

bath, detector block, and ambient temperatures, the inlet flow rate, and the detector bridge current.

The pneumatic circuit of what is perhaps the most automated precision GC system yet devised, developed by Jonsson, Jonsson, and Malm,[90] is shown in Fig. 3.7. A flow regulator controlled the carrier at the column inlet, at which point the flow rate was determined with a thermal-flow sensor. Carrier was also fed to a sample chamber (to which methane could be added) and then through either one of two injection valves to waste. Differential and absolute pressure sensors were used to measure the column pressure drop and inlet pressure, respectively. A flame detector, the outlet of which was at ambient pressure, was employed. The pressure and flow sensors as well as a mercury-drop flowmeter were read and calibrated automatically. Injection, signal acquisition, pressure, flow, and temperature measurement and control were also computerized. Retention volumes were calculated via the total-flow method already described.

Finally, Fig. 3.8 presents a diagram of the finite-concentration GC designed by Conder and Purnell[310] which, with an injection port, could also be used for infinite-dilution studies. The principal feature of this instrument was a specially designed total-stream switching valve.[311]

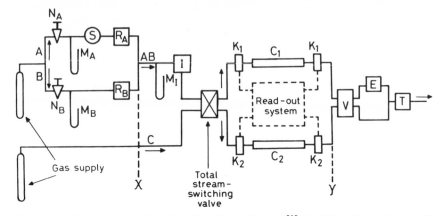

Figure 3.8 Finite-concentration GC of Conder and Purnell[310]: A, AB, B, C, gas streams. C_1, C_2, columns. E, flowmeter. I, injection port. K_1, K_2, twin-channel katharometers. M_A, M_B, M_I, manometers. N_A, N_B, needle valves. R_A, R_B, capillary restrictors. S, saturator. T, traps. V, tap assembly directing gas streams to flowmeter or bypass. Tubing between x and y is 1-mm I.D. stainless steel.

3.3 REFERENCES

1. K. Olah and G. Schay, *Acta Chim.*, **14**, 453 (1958).

2. P. Fejes and G. Schay, *Acta. Chim.*, **17**, 377 (1958).

3. G. Schay, A. Petho, and P. Fejes, *Acta Chim.*, **22**, 285 (1960).

4. A. J. B. Cruickshank, D. H. Everett, and M. T. Westaway, *Trans. Faraday Soc.*, **61**, 235 (1965).

5. M. S. Vigdergauz, *Neftekhim.*, **5**, 425 (1965).

6. A. J. B. Cruickshank, M. L. Windsor, and C. L. Young, *Proc. Roy. Soc. Ser. A*, **295**, 259, 271 (1966).

7. D. A. Tourres, *J. Gas Chromatogr.*, **5**, 35 (1967); *J. Chromatogr.*, **30**, 357 (1967).

8. L. Mikkelsen, *J. Gas Chromatogr.*, **5**, 601 (1967).

9. M. B. Evans and J. F. Smith, *J. Chromatogr.*, **28**, 277 (1967); **30**, 325 (1967).

10. L. S. Ettre and K. Billeb, *J. Chromatogr.*, **30**, 1 (1967).

11. A. J. B. Cruickshank, B. W. Gainey, and C. L. Young, *Trans. Faraday Soc.*, **64**, 337 (1968); in *Gas Chromatography 1968*, C. L. A. Harbourn, Ed., Institute of Petroleum, London, 1969, p. 76.

12. B. W. Gainey and C. L. Young, *Trans. Faraday Soc.*, **64**, 349 (1968).

13. C. P. Hicks and C. L. Young, *Trans. Faraday Soc.*, **64**, 2675 (1968).

14. G. Schomburg and D. Henneberg, *Chromatographia*, **1**, 23 (1968).

15. J. J. Walraven, A. W. Ladon, and A. I. M. Keulemans, *Chromatographia*, **1**, 195, 433 (1968).

16. J. H. Groenendijk and A. H. C. van Kemenade, *Chromatographia*, **1**, 472 (1968).

17. J. Bonastre, P. Grenier, and P. Casenave, *Bull. Soc. Chim. Fr.*, 3885 (1968).

18. R. A. Hively and R. E. Hinton, *J. Gas Chromatogr.*, **6**, 203 (1968).

19. H. Veening and J. F. K. Huber, *J. Gas Chromatogr.*, **6**, 326 (1968).

20. A. H. C. van Kemenade and J. H. Groenendijk, *Chromatographia*, **2**, 148, 316 (1969).

21. K. Tesarik and M. Novotny, *Chromatographia*, **2**, 384 (1969).

22. A. J. B. Cruickshank, B. W. Gainey, C. P. Hicks, T. M. Letcher, R. W. Moody, and C. L. Young, *Trans. Faraday Soc.*, **65**, 1014 (1969).

23. R. Kaiser, *Chromatographia*, **3**, 127, 383 (1970).

24. W. Ebing and A. Kossman, *Chromatographia*, **3**, 418 (1970).

25. J. Takacs and D. Kralik, *J. Chromatogr.*, **50**, 379 (1970).

26. C. A. Cramers, J. A. Rijks, V. Pacakova, and R. de Andrade, *J. Chromatogr.*, **51**, 13 (1970).

27. J. Uhdeova, *J. Chromatogr.*, **51**, 23 (1970).

28. P. G. Robinson and A. L. Odell, *J. Chromatogr.*, **57**, 11 (1971).

29. V. G. Berezkin and V. M. Fateeva, *J. Chromatogr.*, **58**, 73 (1971).

30. J. F. K. Huber and R. G. Gerritse, *J. Chromatogr.*, **58**, 137 (1971); **80**, 25 (1973).

31. A. N. Korol, *J. Chromatogr.*, **67**, 213 (1972).

32. J. Serpinet, *J. Chromatogr.*, **68**, 9 (1972); *Chromatographia*, **8**, 18 (1975).

33. C. Eon, A. K. Chatterjee, and B. L. Karger, *Chromatographia*, **5**, 28 (1972).

34. J. A. Rijks and C. A. Cramers, *Chromatographia*, **7**, 99 (1974).

35. J. C. Giddings, *Anal. Chem.*, **34**, 458 (1962).

36. G. Guiochon and M. Goedert, *Chim. Anal.*, **53**, 214 (1971).

37. O. Wicarova, J. Novak, and J. Janak, *J. Chromatogr.*, **51**, 3 (1970).

38. M. Goedert and G. Guiochon, *Anal. Chem.*, **42**, 962 (1970); **45**, 1180, 1188 (1973).

39. H. D. Young, *Statistical Treatment of Experimental Data*, McGraw-Hill, New York, 1962, Chapter IV.

40. G. Guiochon, *Chromatogr. Rev.*, **8**, 1 (1967).

41. M. Goedert and G. Guiochon, *J. Chromatogr. Sci.*, **7**, 323 (1969).

42. S. A. Green, *J. Chem. Educ.*, **34**, 194 (1957).

43. J. A. Perry, *Dev. Appl. Spectrosc.*, **4**, 347 (1965).

44. J. M. Miller, *J. Chem. Educ.*, **47**, 306 (1970).

45. J. C. Hawkes, in *Vapour Phase Chromatography*, D. H. Desty, Ed., Butterworths, London, 1957, p. 266.

46. J. S. Lewis and H. W. Patton, in *Gas Chromatography*, V. J. Coates, H. J. Noebels, and I. S. Fagerson, Eds., Academic, New York, 1958, p. 145.

47. R. S. Gohlke, *Anal. Chem.*, **29**, 1723 (1957).

48. H. R. Felton, in *Gas Chromatography*, V. J. Coates, H. J. Noebels, and I. S. Fagerson, Eds., Academic, New York, 1958, p. 131.

49. J. N. Roper, Jr., *Anal. Chem.*, **32**, 447 (1960).

50. G. B. Ceresia and C. A. Brusch, *Am. J. Pharm. Educ.*, **28**, 194 (1964).

51. F. Sicilio, H. Bull, R. C. Palmer, and J. A. Knight, *J. Chem. Educ.*, **38**, 506 (1961).

52. K. P. Hupe and E. Bayer, in *Gas Chromatography 1964*, A. Goldup, Ed., Institute of Petroleum, London, 1965, p. 62.

53. R. J. Laub and R. L. Pecsok, *Anal. Chem.*, **46**, 1214 (1974).

54. G. K. Ashbury, A. J. Davies, and J. W. Drinkwater, *Anal. Chem.*, **29**, 918 (1957).

55. J. H. Purnell, *Gas Chromatography*, Wiley, New York, 1962, pp. 255 to 260.

56. J. E. Oberholtzer and L. B. Rogers, *Anal. Chem.*, **41**, 1234 (1969).

57. R. S. Swingle and L. B. Rogers, *Anal. Chem.*, **43**, 810 (1971).

58. R. A. Culp, C. H. Lockmüller, A. K. Moreland, R. S. Swingle, and L. B. Rogers, *J. Chromatogr. Sci.*, **9**, 6 (1971).

59. J. F. Parcher, P. H. Weiner, C. L. Hussey, and T. N. Westlake, *J. Chem. Eng. Data*, **20**, 145 (1975).

60. L. Premru and M. Kremser, *J. Chromatogr.*, **51**, 65 (1970).

61. J. A. Rijks and C. A. Cramers, *Chromatographia*, **7**, 99 (1974).

62. R. J. Laub, D. E. Martire, and J. H. Purnell, *J. Chem. Soc. Faraday Trans. II*, **74**, 213 (1978).

63. F. E. De Boer, *Nature*, **185**, 915 (1960).

64. J. W. Barrett, U.S. Patent 3,237,380 (1966).

65. E. Cremer and F. Deutscher, *Aluminium*, **46**(2), 174 (1970).

66. D. N. Sokolov, M. A. Baydarovtseva, and N. A. Vakin, *Izv. Akad. Nauk SSSR, Ser. Khim.*, 1396 (1968).

67. D. N. Sokolov, N. A. Vakin, and Y. B. Kalmykov, *Zavod. Lab.*, **35**, 150 (1969).

68. D. N. Sokolov, *J. Chromatogr.*, **47**, 320 (1970).

69. D. N. Sokolov and N. A. Vakin, *J. Chromatogr. Sci.*, **10**, 417 (1972); *Zavod. Lab.*, **39**, 274 (1973).

70. C. Merritt, Jr., T. Walsh, D. A. Forss, P. Angelini, and S. M. Swift, *Anal. Chem.*, **36**, 1502 (1964).

71. F. Bruner, G. P. Cartoni, and A. Liberti, *Anal. Chem.*, **38**, 298 (1966).

72. M. L. Conti and M. Lesimple, *J. Chromatogr.*, **29**, 32 (1967).

73. W. J. Haubach, C. M. Knobler, A. Katorski, and D. White, *J. Phys. Chem.*, **71**, 1398 (1967).

74. J. A. Giannovario, R. J. Gondek, and R. L. Grob, *J. Chromatogr.*, **89**, 1 (1974).

75. D. R. Deans, in *Gas Chromatography 1968*, C. L. A. Harbourn, Ed., Institute of Petroleum, London, 1969, p. 447.

76. A. T. James and A. J. P. Martin, *Biochem. J.*, **50**, 679 (1952).

77. D. H. James and C. S. G. Phillips, *J. Sci. Instr.*, **29**, 362 (1952).

78. L. Guild, S. Bingham, and F. Aul, in *Gas Chromatography 1958*, D. H. Desty, Ed., Butterworths, London, 1958, p. 226.

79. F. van de Craats, in *Gas Chromatography 1958*, D. H. Desty, Ed., Butterworths, London, 1958, p. 248.

80. E. P. Atkinson and G. A. P. Tuey, in *Gas Chromatography 1958*, D. H. Desty, Ed., Butterworths, London, 1958, p. 270.

81. J. H. Knox, *Chem. Ind.*, 1085 (1959).

82. H. W. Habgood and W. E. Harris, *Anal. Chem.*, **32**, 450 (1960).

83. J. Novak, *Collect. Czech. Chem. Commun.*, **27**, 411 (1962).

84. O. Knapp and M. Lochner, *Chromatographia*, **3**, 378 (1970).

85. C. E. Fawsitt, *Proc. Roy. Soc. Ser. A*, **80**, 290 (1908).

86. F. M. Tiller, *Anal. Chem.*, **26**, 1252 (1954).

87. J. R. Roebuck and H. W. Ibser, *Rev. Sci. Instrum.*, **25**, 46 (1954).

88. B. E. Bowen, S. P. Cram, J. E. Leitner, and R. L. Wade, *Anal. Chem.*, **45**, 2185 (1973).

89. J. A. Jonsson and R. Jonsson, *J. Chromatogr.*, **111**, 265 (1975).

90. J. A. Jonsson, R. Jonsson, and K. Malm, *J. Chromatogr.*, **115**, 57 (1975).

91. J. C. Lilly, V. Legallais, and R. Cherry, *J. Appl. Phys.*, **18**, 613 (1947).

92. J. J. Opstelten and N. Warmoltz, *Appl. Sci. Res., Sec. B*, **4**, 329 (1955).

93. J. O. Cope, *Rev. Sci. Instrum.*, **33**, 980 (1962).

94. F. I. Stalkup and R. Kobayashi, *AIChE J.*, **9**, 121 (1963); *J. Chem. Eng. Data*, **8**, 564 (1963).

95. K. T. Koonce and R. Kobayashi, *J. Chem. Eng. Data*, **9**, 494 (1964).

96. K. T. Koonce, H. A. Deans, and R. Kobayashi, *AIChE J.*, **11**, 259 (1965).

97. L. D. Van Horn and R. Kobayashi, *J. Chem. Eng. Data*, **12**, 294 (1967).

98. K. Asano, T. Nakahara, and R. Kobayashi, *J. Chem. Eng. Data*, **16**, 16 (1971).

99. W. Ruska, J. T. Kao, S.-Y. Chuang, and R. Kobayashi, *Rev. Sci. Instrum.*, **39**, 1889 (1968).

100. M. N. Myers and J. C. Giddings, *Anal. Chem.*, **37**, 1453 (1965); *Sep. Sci.*, **1**, 761 (1966); in *Progress in Separation and Purification*, Vol. III, E. S. Perry and C. J. Van Oss, Eds., Wiley-Interscience, New York, 1970, p. 133.

101. P. D. Schettler, M. Bikelberger, and J. C. Giddings, *Anal. Chem.*, **39**, 146 (1967).

102. L. McLaren, M. N. Myers, and J. C. Giddings, *Science*, **159**, 197 (1968).

103. J. C. Giddings, M. N. Myers, L. McLaren, and R. A. Keller, *Science*, **162**, 67 (1968).

104. J. C. Giddings, M. N. Myers, and J. W. King, *J. Chromatogr. Sci.*, **7**, 276 (1969).

105. J. J. Czubryt, M. N. Myers, and J. C. Giddings, *J. Phys. Chem.*, **74**, 4260 (1970).

106. V. Pretorius and T. W. Smuts, *Anal. Chem.*, **38**, 274 (1966).

107. N. M. Karayannis and A. H. Corwin, *Anal. Biochem.*, **26**, 34 (1968).

108. N. M. Karayannis, A. H. Corwin, J. A. Walter, E. W. Baker, and E. Klesper, *Anal. Chem.*, **40**, 1736 (1968).

109. S. Wicar and J. Novak, *J. Chromatogr.*, **95**, 13 (1974).

110. E. Klesper, A. H. Corwin, and D. A. Turner, *J. Org. Chem.*, **27**, 700 (1962).

111. S. T. Sie, W. van Beersum, and G. W. A. Rijnders, *Sep. Sci.*, **1**, 459 (1966).

112. S. T. Sie, J. P. A. Bleumer, and G. W. A. Rijnders, in *Gas Chromatography 1968*, C. L. A. Harbourn, Ed., Institute of Petroleum, London, 1969, p. 235; *Sep. Sci.*, **3**, 165 (1968).

113. S. T. Sie and G. W. A. Rijnders, *Anal. Chim. Acta*, **38**, 3, 31 (1967); *Sep. Sci.*, **2**, 699, 729, 755 (1967).

114. T. H. Gouw and R. E. Jentoft, *J. Chromatogr.*, **68**, 303 (1972); *Adv. Chromatogr.*, **13**, 1 (1975).

115. C. S. Allen, Ph. D. Dissertation, University College of Swansea, Swansea, Wales, 1975; C. S. Allen and J. H. Purnell, *unpublished work*.

116. P. C. Carmen, *Trans. Inst. Chem. Eng.*, **15**, 150 (1937).

117. E. R. Adlard and B. T. Whitham, in *Gas Chromatography 1958*, D. H. Desty, Ed., Butterworths, London, 1958, p. 351.

118. D. A. Vyakhirev and P. F. Komissarov, *Proc. Acad. Sci. USSR, Phys. Chem. Sec. (Engl. transl.)*, **129**, 901 (1959).

119. T. Haruki, *Jap. Anal.*, **9**, 865 (1960).

120. M. Rogozinksi, M. Shorr, and K. Juszkiewicz, *Bull. Res. Counc. Isr., Sec. A*, **10** (3), 15 (1961).

121. P. F. Varadi and K. Ettre, *Anal. Chem.*, **35**, 410 (1963).

122. D. C. Locke and W. W. Brandt, in *Gas Chromatography*, L. Fowler, Ed., Academic, New York, 1963, p. 55.

123. R. Teranishi, R. G. Buttery, W. H. McFadden, T. R. Mon, and J. Wasserman, *Anal. Chem.*, **36**, 1509 (1964).

124. L. V. Kondakova and D. A. Vyakhirev, *Neftekhim.*, **8**, 297 (1968).

125. E. Findl and K. Lui, *J. Gas Chromatogr.*, **6**, 165 (1968).

126. S. R. Palamand and D. O. Thurow, *J. Chromatogr.*, **40**, 152 (1969).

127. A. F. Shlyakhov, *Sb. Nauchn. Tr. Gazov. Khromatogr., N. I. Fiz. Khim. Inst.*, **19**, 4 (1973).

128. A. Findlay, *Practical Physical Chemistry*, eighth ed., Longmans, London, 1965, p. 81.

129. A. Levy, *J. Sci. Instrum.*, **41**, 449 (1964).

130. J. J. Czubryt and H. D. Gesser, *J. Gas Chromatogr.*, **6**, 528 (1968).

131. A. Zlatkis, L. O'Brien, and P. R. Schooly, *Nature*, **181**, 1794 (1958).

132. D. M. Smith and R. G. Campbell, *Chemist-Analyst*, **50**, 80 (1961).

133. M. van Swaay, *J. Chromatogr.*, **12**, 99 (1963).

134. C. J. Frisone, *Chemist-Analyst*, **54**, 56 (1965).

135. G. W. Munns, Jr., and J. V. Frilette, *J. Gas Chromatogr.*, **3**, 145 (1965).

136. R. G. Thurman and M. F. Burke, *J. Chromatogr. Sci.*, **9**, 181 (1971).

137. Anon, *Instrum. Control Syst.*, **42**(3), 115 (1969); **42**(7), 100 (1969).

138. J. A. Jonsson, *J. Chromatogr.*, **111**, 271 (1975).

139. Y. Hori and R. Kobayashi, *Bull. Chem. Soc. Jap.*, **47**, 1791 (1974).

140. C. E. Hawk and W. C. Baker, *J. Vac. Sci. Technol.*, **6**, 255 (1969).

141. J. M. Benson, W. C. Baker, and E. Easter, *Instrum. Control Syst.*, **43**(2), 85 (1970).

142. D. W. Carle, in *Gas Chromatography*, V. J. Coates, H. J. Noebels, and I. S. Fagerson, Eds., Academic, New York, 1958, p. 67.

143. B. O. Ayers, in *Gas Chromatography*, V. J. Coates, H. J. Noebels, and I. S. Fagerson, Eds., Academic, New York, 1958, p. 93.

144. J. C. Lamkin, U.S. Patent 2,972,888 (1961).

145. A. B. Broerman, U.S. Patent 3,111,849 (1963).

146. R. Blomstrand and J. Gurtler, *Acta Chem. Scand.*, **18**, 276 (1964).

147. D. A. Forss, M. L. Bazinet, and S. M. Swift, *J. Gas Chromatogr.*, **2**, 134 (1964).

148. E. R. Fenoke and J. H. McLaughlin, U.S. Patent 3,266,321 (1966).

149. H. G. Boettger, U.S. Patent 3,267,736 (1966).

150. C. R. Ferrin, U.S. Patent 3,306,111 (1967).

151. J. R. Rendina, U.S. Patent 3,318,154 (1967).

152. W. H. Topham, U.S. Patent 3,321,977 (1967).

153. D. Jentzsch and W. Schumann, U.S. Patent 3,365,951 (1968).

154. G. R. Harvey, Jr., U.S. Patent 3,368,385 (1968).

155. E. L. Szonntagh, U.S. Patent 3,386,472 (1968).

156. C. H. Hamilton, in *Instrumentation in Gas Chromatography*, J. Krugers, Ed., Centrex, Eindhoven, 1968, p. 33.

157. R. D. Condon and L. S. Ettre, in *Instrumentation in Gas Chromatography*, J. Krugers, Ed., Centrex, Eindhoven, 1968, p. 87.

158. J. W. Todd and C. G. Courneya, U.S. Patent 3,393,557 (1968).

159. W. F. Gerdes, U.S. Patent 3,435,661 (1969).

160. H. Abegg, *J. Chromatogr.*, **9**, 519 (1962).

161. M. Taramasso and A. Guerra, U.S. Patent 3,094,155 (1963).

162. R. J. Harris, Jr., U.S. Patent 3,355,950 (1967).

163. M. Ellison, *Analyst*, **93**, 264 (1968).

164. P. Pitt, *Chromatographia*, **1**, 252 (1968).

165. J. E. Lovelock, in *Gas Chromatography 1960*, R. P. W. Scott, Ed., Butterworths, London, 1960, p. 26.

166. D. H. Desty, C. J. Geach, and A. Goldup, in *Gas Chromatography 1960*, R. P. W. Scott, Ed., Butterworths, London, 1960, p. 46.

167. I. A. Fowlis and R. P. W. Scott, *J. Chromatogr.*, **11**, 1 (1963).

168. M. van Swaay and J. R. Bacon, *J. Chromatogr.*, **19**, 604 (1965).

169. C. H. Hartmann and K. P. Dimick, *J. Gas Chromatogr.*, **4**, 163 (1966).

170. H. P. Williams and J. D. Winefordner, *J. Gas Chromatogr.*, **4**, 271 (1966).

171. L. J. Lorenz, R. A. Culp, and R. T. Dixon, *Anal. Chem.*, **42**, 1119 (1970).

172. H. D. Axelrod, R. J. Teck, J. P. Lodge, and R. H. Allen, *Anal. Chem.*, **43**, 496 (1971).

173. F. Bruner, A. Liberti, M. Possanzini, and I. Allegrini, *Anal. Chem.*, **44**, 2070 (1972).

174. F. Bruner, C. Canulli, and M. Possanzini, *Anal. Chem.*, **45**, 1790 (1973).

175. S. F. Michelleti, U.S. Patent 3,002,387 (1961).

176. A. G. Nerheim, U.S. Patent 3,063,286 (1962); U.S. Patent 3,119,252 (1964).

177. E. S. Khuhovskii and A. G. Sharonov, Br. Patent 1,042,652 (1964).

178. D. S. Berry, *Anal. Chem.*, **39**, 692 (1967).

179. R. A. Back, N. J. Friswell, J. C. Boden, and J. M. Parsons, *J. Chromatogr. Sci.*, **7**, 708 (1969).

180. R. S. Porter, A. S. Hoffman, and J. F. Johnson, *Anal. Chem.*, **34**, 1179 (1962).

181. J. E. Oberholtzer, *Anal. Chem.*, **39**, 959 (1967).

182. G. L. Pratt and J. H. Purnell, *Anal. Chem.*, **32**, 1213 (1960).

183. C. J. Penther, *Control Eng.*, **10**, 78 (1963).

184. E. L. Szonntagh, in *Gas Chromatography*, L. Fowler, Ed., Academic, New York, 1963, p. 233.

185. M. L. Marks and F. C. Calcaprina, U.S. Patent 3,000,218 (1961).

186. D. L. Peterson and G. W. Lundberg, *Anal. Chem.*, **33**, 652 (1961).

187. C. J. Penther and J. W. Hickling, *Oil Gas J.*, **59**, 130 (1961).

188. G. S. Turner and R. Villalobos, in *Gas Chromatography*, N. Brenner, J. E. Callen, and M. D. Weiss, Eds., Academic, New York, 1962, p. 363.

189. J. Hooimeijer, S. Kwantes, and F. van de Craats, in *Gas Chromatography 1958*, D. H. Desty, Ed., Butterworths, London, 1958, p. 288.

190. W. H. Topham, Br. Patent 1,111,443 (1968).

191. A. P. Broerman, U.S. Patent 3,387,496 (1968).

192. H. Coanda, Fr. Patent 788,140 (1934).

193. J. Young, *Design Eng.*, **10**, 26 (1964).

194. E. J. Kompass, *Control Eng.*, **11**, 73 (1964).

195. R. E. Olson, *Fluid. Qt.*, **1**, 85 (1967).

196. R. C. Palmer, *Control Eng.*, **8**, 121 (1961).

197. T. H. Glenn and S. P. Cram, *J. Chromatogr. Sci.*, **8**, 46 (1970).

198. R. L. Wade and S. P. Cram, *Anal. Chem.*, **44**, 131 (1972).

199. R. Jonsson and J. A. Jonsson, *J. Chromatogr.*, **120**, 197 (1976).

200. J. R. Anderson and G. H. McConkey, *J. Chromatogr.*, **27**, 480 (1967).

201. R. V. Hems and J. B. Adams, *Lab. Pract.*, **21**, 430 (1972).

202. R. P. W. Scott, in *Gas Chromatography 1958*, D. H. Desty, Ed., Butterworths, London, 1958, p. 189.

203. W. M. Langdon, V. R. Ivanuski, R. E. Putscher, and H. J. O'Neill, *J. Gas Chromatogr.*, **4**, 269 (1966).

204. R. L. Pecsok and M. L. Windsor, *Anal. Chem.*, **40**, 1238 (1968).

205. B. Pearce and W. Ll. Thomas, *Anal. Chem.*, **44**, 1107 (1972).

206. F. Khoury and D. B. Robinson, *J. Chromatogr. Sci.*, **10**, 683 (1972).

207. B. E. Bowen and S. P. Cram, *J. Chromatogr. Sci.*, **12**, 579 (1974).

208. P. C. Haarhoff, P. C. van Berge, and V. Pretorius, *J. S. Afr. Chem. Inst.*, **14**, 82 (1961).

209. G. Guiochon, *Anal. Chem.*, **35**, 399 (1963).

210. J. H. Purnell and D. T. Sawyer, *Anal. Chem.*, **36**, 668 (1964).

211. E. Wicke, *Ber. Bunsenges. Phys. Chem.*, **69**, 761 (1965).

212. K. DeClerk, T. S. Buys, and V. Pretorius, *Sep. Sci.*, **6**, 733 (1971).

213. C. N. Reilley, G. P. Hildebrand, and J. W. Ashley, *Anal. Chem.*, **34**, 1198 (1962).

214. J. W. Ashley, G. P. Hildebrand, and C. N. Reilley, *Anal. Chem.*, **36**, 1369 (1964).

215. J. Boeke, in *Gas Chromatography 1960*, R. P. W. Scott, Ed., Butterworths, London, 1960, p. 88.

216. R. B. Seligman and F. L. Gazer, Jr., *Adv. Anal. Chem. Instrum.*, **1**, 119 (1960).

217. A. Karmen, *Adv. Chromatogr.*, **2**, 293 (1966).

218. A. E. Lamson, Jr., and J. M. Miller, *J. Gas Chromatogr.*, **4**, 273 (1966).

219. J. D. Winefordner and T. H. Glenn, *Adv. Chromatogr.*, **5**, 263 (1968).

220. R. P. Tye, Ed., *Thermal Conductivity*, Vol. 2, Academic, New York, 1969.

221. T. A. Gough and E. A. Walker, *Analyst*, **95**, 1 (1970).

222. A. B. Littlewood, in *Aspects in Gas Chromatography*, H. G. Struppe, Ed., Akadamie Verlag, Berlin, 1971, p. 81.

223. A. B. Richmond, *J. Chromatogr. Sci.*, **9**, 92 (1971).

224. C. H. Hartmann, *Anal. Chem.*, **43**(2), 113A (1971).

225. R. A. Keller, *J. Chromatogr. Sci.*, **11**, 223 (1973).

226. T. Johns and C. Stapp, *J. Chromatogr. Sci.*, **11**, 234 (1973).

227. W. A. Aue, *Res./Dev.*, **25**(1), 25 (1974).

228. E. R. Adlard, *Crit. Rev. Anal. Chem.*, **5**, 1, 13 (1975).

229. V. V. Brazhnikov, *Differential Detectors for Gas Chromatography*, Nauka, Moscow, 1974.

230. D. J. David, *Gas Chromatographic Detectors*, Wiley-Interscience, New York, 1974.

231. J. Sevcik, *Detectors in Gas Chromatography*, Elsevier, Amsterdam, 1976.

232. S. J. Hawkes and M. D. Wheaton, *J. Gas Chromatogr.*, **5**, 380 (1967).

233. P. Bocek, J. Novak, and J. Janak, *J. Chromatogr.*, **48**, 412 (1970).

234. G. Blu, F. Lazarre, and G. Guiochon, *Anal. Chem.*, **45**, 1375 (1973).

235. A. Karmen, *J. Gas Chromatogr.*, **3**, 180 (1965).

236. D. N. Sokolov and N. A. Vakin, *Zavod. Lab.*, **36**, 1314 (1970).

237. J. L. Cashaw, R. Segura, and A. Zlatkis, *J. Chromatogr. Sci.*, **8**, 363 (1970).

238. D. C. Nelson, R. A. Hawes, and P. C. Ressler, Jr., *Dev. Appl. Spectrosc.*, **4**, 323 (1965).

239. A. Karmen, *J. Am. Oil Chem. Soc.*, **44**, 18 (1967).

240. P. Volpe and M. Castiglioni, *J. Chromatogr.*, **114**, 23 (1975).

241. H. Rotzsche, in *Column Chromatography*, E. sz. Kovats, Ed., Swiss Chemists Association, 1970, p. 206.

242. N. L. Gregory, *J. Chromatogr.*, **13**, 33 (1964).

243. P. L. Coduti, *J. Chromatogr. Sci.*, **14**, 423 (1976).

244. R. Kieselbach, *Anal. Chem.*, **35**, 1342 (1963).

245. L. J. Schmauch, *Anal. Chem.*, **31**, 225 (1959).

246. I. G. McWilliam and H. C. Bolton, *Anal. Chem.*, **41**, 1755 (1969).

247. V. N. Lipavskii and V. G. Berezkin, *Zavod. Lab.*, **38**, 657 (1972).

248. V. N. Lipavskii, *Zavod. Lab.*, **39**, 788 (1973).

249. N. G. Farzane, L. V. Illyasov, and A. Y. Azim-Zade, *Zavod. Lab.*, **40**, 782 (1974); *J. Chromatogr.*, **104**, 231 (1975).

250. I. G. McWilliam, *J. Appl. Chem.*, **9**, 397 (1959).

251. I. G. McWilliam and H. C. Bolton, *Anal. Chem.*, **41**, 1762 (1969).

252. A. E. Thompson, *J. Chromatogr.*, **6**, 539 (1961).

253. R. D. Johnson, *J. Gas Chromatogr.*, **6**, 43 (1968).

254. S. Z. Lewin, *J. Chem. Educ.*, **39**, A161 (1962).

255. D. H. Carter, *J. Gas Chromatogr.*, **5**, 612 (1967).

256. D. A. Leathard, *Adv. Anal. Chem. Instrum.*, **11**, 29 (1973).

257. S. P. Cram and J. E. Leitner, in *Computers in Chemical and Biochemical Research*, C. E. Klopfenstein and C. L. Wilkins, Eds., Academic, New York, 1974, p. 235.

258. J. T. Shank and J. E. Persinger, *J. Gas Chromatogr.*, **5**, 631 (1967).

259. J. Munnik and C. C. M. Fabrie, *Z. Anal. Chem.*, **236**, 51 (1968).

260. E. K. Clardy, U.S. Patent 3,468,156 (1969).

261. J. F. K. Huber and H. C. Smit, *Z. Anal. Chem.*, **245**, 84 (1969).

262. H. A. Hancock and I. Lichenstein, *J. Chromatogr. Sci.*, **7**, 290 (1969).

263. O. Knapp and M. Keller, *Chromatographia*, **2**, 500 (1969).

264. D. E. Smith, *J. Assoc. Off. Anal. Chem.*, **52**, 206 (1969).

265. H. D. Metzger, *Chromatographia*, **3**, 64 (1970).

266. H. Pauschmann, *Chromatographia*, **3**, 84 (1970).

267. W. Schneeweiss, *Chromatographia*, **3**, 341 (1970).

268. E. Brawell, *Anal. Biochem.*, **44**, 58 (1971).

269. P. C. Kelly and W. E. Harris, *Anal. Chem.*, **43**, 1170, 1184 (1971).

270. D. F. Nenarokov, S. S. Karavaev, V. V. Brazhnikov, and K. I. Sakodinski, *Chromatographia*, **3**, 106 (1971).

271. M. Dressler and M. Deml, *J. Chromatogr.*, **56**, 23 (1971).

272. D. Ford and K. Weihman, in *Recent Advances in Gas Chromatography*, I. I. Domsky and J. A. Perry, Eds., Marcel Dekker, New York, 1971, p. 377.

273. R. S. Swingle and L. B. Rogers, *Anal. Chem.*, **44**, 1415 (1972).

274. D. Macnaughton, L. B. Rogers, and G. Wernimont, *Anal. Chem.*, **44**, 1421 (1972).

275. G. C. Carle, *Anal. Chem.*, **44**, 1905 (1972).

276. J. M. Gill, *J. Chromatogr. Sci.*, **10**, 1 (1972).

277. H. M. McNair and W. M. Cooke, *J. Chromatogr. Sci.*, **10**, 27 (1972); *Am. Lab.*, **5**(2), 12 (1973).

278. H. Clough, T. C. Gibb, and A. B. Littlewood, *Chromatographia*, **5**, 351 (1972).

279. B. A. Rudenko, *J. Chromatogr. Sci.*, **10**, 230 (1972).

280. G. A. Schleterh and K. Greiner, *Chromatographia*, **5**, 70 (1972).

281. J. M. Gill and J. Henselman, *Chromatographia*, **5**, 108 (1972).

282. G. Charrier, M. C. Dupuis, J. C. Merlivat, J. Pons, and R. Sigelle, *Chromatographia*, **5**, 119 (1972).

283. A. Fozard, J. J. Framses, and A. J. Wyatt, *Chromatographia*, **5**, 130, 377 (1972).

284. P. Sutre and J. P. Malenge, *Chromatographia*, **5**, 141 (1972).

285. G. Bischet and G. Knechtel, *Chromatographia*, **5**, 166 (1972).

286. E. Malan and B. Brink, *Chromatographia*, **5**, 182 (1972).

287. K. Derge, *Chromatographia*, **5**, 284 (1972).

288. A. B. Rozenblit, Y. Y. Silis, A. A. Avots, and V. D. Shats, *Zh. Anal. Khim.*, **28**, 22 (1973).

289. M. Goedert and G. Guiochon, *Chromatographia*, **6**, 76 (1973).

290. R. W. Dwyer, *Anal. Chem.*, **45**, 1380 (1973).

291. G. Brouwer and J. A. J. Jansen, *Anal. Chem.*, **45**, 2239 (1973).

292. E. Cuso, X. Guardino, J. M. Riera, and M. Gassiot, *J. Chromatogr.*, **95**, 147 (1974).

293. A. Sabatier, M. Goedert, and G. Guiochon, *Chromatographia*, **7**, 560 (1974).

294. C. Way-Jones and L. Glasser, *Anal. Chem.*, **48**, 1426 (1976).

295. V. A. Kaminskii, S. F. Timashev, and N. N. Tunitskii, *Zh. Fiz. Khim.*, **39**, 2540 (1965).

296. J. W. Ashley and C. N. Reilley, *Anal. Chem.*, **37**, 626 (1965).

297. J. C. Sternberg, *Adv. Chromatogr.*, **2**, 205 (1966).

298. O. Grubner, *Adv. Chromatogr.*, **6**, 173 (1968).

299. E. Grushka, M. N. Myers, and J. C. Giddings, *Anal. Chem.*, **42**, 21 (1970).

300. D. W. Underhill, *Sep. Sci.*, **5**, 219 (1970).

301. O. Grubner and D. W. Underhill, *Sep. Sci.*, **5**, 555 (1970); *J. Chromatogr.*, **73**, 1 (1972).

302. T. S. Buys and K. DeClerk, *Sep. Sci.*, **5**, 543 (1970); **7**, 441 (1972); *J. Chromatogr.*, **69**, 87 (1972); *J. Chromatogr. Sci.*, **10**, 722 (1972); *Anal. Chem.*, **44**, 1273 (1972).

303. K. DeClerk and T. S. Buys, *J. Chromatogr.*, **63**, 193 (1971).

304. J. E. Funk and P. R. Rony, *Sep. Sci.*, **6**, 365 (1971).

305. P. R. Rony and J. E. Funk, *Sep. Sci.*, **6**, 383 (1971); *J. Chromatogr. Sci.*, **9**, 215 (1971).

306. E. Grushka, *Anal. Chem.*, **44**, 1733 (1972).

307. Y. Yamaoka and T. Nakagawa, *J. Chromatogr.*, **92**, 213 (1974); **93**, 1 (1974); **103**, 221 (1975); **105**, 225 (1975).

308. T. Peticlerc and G. Guiochon, *Chromatographia*, **7**, 10 (1974); *J. Chromatogr. Sci.*, **14**, 531 (1976).

309. S. P. Cram and T. H. Glenn, *J. Chromatogr.*, **112**, 329 (1975).

310. J. R. Conder and J. H. Purnell, *Trans. Faraday Soc.*, **65**, 839 (1969).

311. J. R. Conder, J. H. Purnell, and R. Walsh, *Talanta*, **15**, 145 (1968).

PART TWO
Thermodynamics

CHAPTER 4

Virial Coefficients

4.1 REAL GASES AT MODERATE PRESSURES

Derivation of the partition and activity coefficient equations in Chapter 2 required correction for virial effects for permanent carriers at low ($p_i < 5$ atm at $p_i - p_o < 1.2$ atm) column pressures. The relations are now extended to include medium- and high-pressure GC and the possibility of mobile-phase solubility in the stationary phase. Development of the theory of virial and fugacity corrections (the magnitudes of which depend on the solutes, carriers, and system pressure) to partition coefficient data has, in addition, led to the use of GC for the direct measurement of the former and in at least one case indicated that the latter may be available from purely chromatographic data.

4.1.1 Fugacity Effects At constant temperature the fugacity f_i and absolute activity a_i of the components of a binary gaseous mixture are related by[1]

$$\frac{f_i}{a_i} = \text{const}$$

$$\frac{f_i}{x_i p_i} \rightarrow 1 \quad \text{as} \quad p_i \rightarrow 0$$

In terms of a mixture of solute, 1, and carrier, 2,

$$f_1 = x_1 P_{12} \exp\left[\left(B_{11} + 2x_2^2 \delta_{12} \right) \frac{P_{12}}{RT} \right] \tag{4.1}$$

where

$$\delta_{12} = B_{12} - \tfrac{1}{2}(B_{11} + B_{22}) \tag{4.2}$$

83

B_{11} and B_{22} are the solute-solute and carrier-carrier fugacity coefficients and B_{12} is termed the solute-carrier mixed second-interaction virial[2] coefficient.

As $x_2 \to 0$, eq. 4.1 becomes

$$f_1^0 = p_1^0 \exp\left(\frac{B_{11} p_1^0}{RT}\right) \tag{4.3}$$

The pressure-dependent infinite-dilution activity coefficient γ_p^∞ was shown earlier (eq. 2.35) to be given by

$$\gamma_p^\infty = \frac{RT}{V_g^T p_1^0 MW_L} \tag{4.4}$$

In terms of fugacity

$$\gamma_1^\infty = \frac{RT}{V_g^T f_1^0 MW_L} \tag{4.5}$$

Equations 4.4 and 4.5 are thus related through eq. 4.3:

$$\ln \gamma_1^\infty = \ln \gamma_p^\infty - \frac{B_{11} p_1^0}{RT} \tag{4.6}$$

which is identical to the approximation (eq. 2.65) mentioned in Section 2.2.3. However, the solute saturation fugacity f_1^0 was identified there implicitly with that at infinite dilution f_1^∞, which, as Everett and Stoddart[3] noted, is not strictly correct. The effect of the change of the solute chemical potential on being diluted with carrier must, as shown by Guggenheim,[1] also be included:

$$\ln \gamma_1^\infty = \ln \gamma_p^\infty - \frac{p_1^0 (B_{11} - v_1^0)}{RT} \tag{4.7}$$

where v_1^0 is the solute bulk molar volume. Equations 4.6 and 4.7 differ by an amount, $p_1^0 v_1^0 / RT$, that is often trivial (e.g., being 0.0013 for n-hexane at 30°C) and accounts for the success of the approximation (eq. 4.6). B_{11} data for common hydrocarbon solutes are calculable from the McGlashan-Potter[4] modification of the Beattie-Bridgeman[5] equation:

$$\frac{B_{11}}{V_{11}^c} = 0.430 - 0.886 \left(\frac{T_{11}^c}{T}\right) - 0.694 \left(\frac{T_{11}^c}{T}\right)^2 - 0.0375(n-1) \left(\frac{T_{11}^c}{T}\right)^{4.5} \tag{4.8}$$

where a superscript, c, indicates a critical value and n is the number of carbon atoms per molecule (with the exception of alicyclic and aromatic hydrocarbons for which $n = 4.5$ for a six-membered ring[6,7]). Table 4.1 lists calculated fugacity coefficients for a few common solutes.

A relatively simple apparatus with which B_{11} values can be measured directly has been described by Feng and Melzer[8] and a device capable of high precision, reported by McGlashan and Potter.[4] Experimental details of earlier designs have also been reviewed.[9] Spertell and Chang[10] have further indicated that the GLC technique may be used to measure B_{11} values but the solute must be an isotope of the carrier which itself must be soluble in the stationary phase. The method has not achieved widespread use because of the simplicity and ease of operation of static apparatus.

Broadly speaking, eq. 4.7 accounts only for gas-phase imperfections due to the solute fugacity and ignores nonideal behavior caused by termolecular and higher order solute interactions. That this, nonetheless, is a reasonable and, indeed, quantitatively successful approach is a consequence of the advantage of the GC technique in which the solute is maintained at infinite dilution, irrespective of the pressure of the system.

4.1.2 Mixed Virial Coefficients It was early[3,11-14] recognized in gas chromatography that the use of different carrier gases altered elution times; in some cases the effect can produce reversals in relative retention behavior.[14] A variety of equations has been proposed which account for gas-phase solute-carrier (mixed) second virial interactions[15,16] in addition to those that describe the fugacity of the carrier[17] and the fugacity correction to the compressibility factor.[18] The first comprehensive study of the effects of the carrier gas and the column pressure was made by Desty and co-workers[19] in 1962. Beginning with eqs. 4.1 and 4.3, they derived a relation that included fugacity as well as solute-carrier mixed virial effects:

$$\ln K_R = \ln \frac{RT}{\overline{V}_L \gamma_1^\infty p_1^0} - \frac{p_1^0}{RT}\left(B_{11} - v_1^0\right) + \frac{\overline{p}}{RT}\left(2B_{12} - v_1^0\right) \qquad (4.9)$$

where $\overline{p}$ is the mean column pressure and B_{12} is the mixed virial coefficient. Because $k' = K_R V_L / V_M$, eq. 4.9 can be cast in the form

$$\ln k' = A + \frac{\overline{p}}{RT}\left(2B_{12} - v_1^0\right) \qquad (4.10)$$

where A is a constant and $\overline{p}$ was said to be given by $p_o J_2^3$. Thus a plot of $\ln k'$ versus $p_o J_2^3$ should produce a straight line of slope $(2B_{12} - v_1^0)/RT$, from which the solute-carrier B_{12} value can be calculated. Further,

Table 4.1 Fugacity Coefficients for Common Hydrocarbon Solutes[a]

Solute	V^c (ml/mole)	T^c (°K)	25°C			50°C		
			v_1^0 (ml/mole)	p_1^0 (torr)	$-B_{11}$ (ml/mole)	v_1^0 (ml/mole)	p_1^0 (torr)	$-B_{11}$ (ml/mole)
n-Butane	254.97	425.16	101.43	1823.	713.95	106.94	3703.	592.46
2-Methylpropane	262.99	408.13	105.48	2611.	669.42	111.56	5058.	556.93
n-Pentane	311.02	469.77	116.10	512.5	1197.2	120.87	1194.	974.22
2-Methylbutane	307.99	460.95	117.38	688.1	1128.5	122.47	1537.	920.14
2,2-Dimethylpropane	303.01	433.75	123.31	1285.	950.91	129.85	2661.	779.85
n-Hexane	368.04	507.85	131.60	151.3	1896.4	136.37	405.3	1512.7
2-Methylpentane	367.01	498.05	132.88	211.8	1788.7	137.83	541.8	1430.4
3-Methylpentane	367.01	504.35	130.61	189.8	1853.9	135.30	490.5	1480.2
2,2-Dimethylbutane	358.99	489.35	133.71	319.1	1664.6	138.82	766.3	1334.2
2,3-Dimethylbutane	357.96	500.25	131.16	234.6	1766.6	135.97	586.5	1412.0
n-Heptane	426.04	540.16	147.46	45.81	2861.2	152.23	141.6	2241.4
2-Methylhexane	428.06	531.05	148.58	65.88	2727.6	153.32	193.9	2142.0
3-Methylhexane	418.03	535.55	146.71	61.59	2733.9	151.35	182.4	2144.3
3-Ethylpentane	416.02	540.75	144.39	58.05	2803.4	148.82	172.7	2195.8
2,2-Dimethylpentane	404.00	520.85	149.65	105.2	2426.2	154.67	288.8	1910.4
2,3-Dimethylpentane	405.00	537.75	145.02	68.87	2682.5	149.55	199.2	2102.7

2,4-Dimethylpentane	420.03	520.25	149.93	98.40	2513.7	150.92	274.0	1979.6
n-Octane	486.02	569.35	163.53	14.04	4187.7	168.40	50.36	3228.6
2-Methylheptane	488.08	561.15	164.61	20.64	4009.5	169.51	70.03	3097.6
3-Methylheptane	478.03	565.15	162.77	19.58	4019.6	167.64	66.88	3102.2
n-Nonane	525.06	595.15	179.67	4.35	5688.2	184.61	18.06	4326.4
1-Butene	241.02	419.15	95.29	2217.	652.73	100.67	4461.	542.16
cis-2-Butene	235.97	428.15	91.17	1604.	671.75	95.84	3342.	557.17
trans-2-Butene	235.97	428.15	93.76	1753.	671.75	98.62	3601.	557.17
2-Methylpropene	240.01	417.88	95.43	2278.	645.40	100.93	4549.	536.18
1-Pentene	309.20	461.15	110.38	648.8	1134.2	115.10	1470.	924.75
1-Hexene	365.41	501.15	125.89	187.2	1812.6	130.49	487.9	1448.4
1-Heptene	416.98	535.15	141.74	56.33	2720.7	146.37	169.5	2134.2
1-Octene	468.24	565.15	157.85	17.38	3937.2	162.51	60.48	3038.7
1,3-Butadiene	220.79	425.15	87.96	2105.	618.21	92.71	4250.	513.01
Benzene	260.33	562.60	89.40	95.18	1561.7	92.21	271.3	1251.4
Toluene	319.98	593.95	106.85	28.44	2508.7	109.86	92.10	1969.1
Ethylbenzene	366.04	619.55	123.06	9.57	3642.6	126.27	35.16	2810.5
Cyclohexane	309.63	554.15	108.74	97.58	1780.2	112.13	271.8	1429.4
Methylcyclohexane	344.52	572.28	128.33	46.33	2412.0	132.08	138.3	1903.3

[a]P. S. Williams, *unpublished work*

extrapolation to the intercept provides a means of calculating k' (hence K_R^0) at what amounts to zero column pressure drop:

$$\ln K_R \rightarrow \ln K_R^0 = \ln \frac{RT}{\bar{V}_L \gamma_1^\infty p_1^0} - \frac{p_1^0}{RT}(B_{11} - v_1^0) \quad \text{as} \quad \bar{p} \rightarrow 0 \qquad (4.11)$$

Table 4.2 gives the B_{12} values determined via GLC (eq. 4.10) by Desty and co-workers.[19]

Table 4.2 Second-Interaction Virial Coefficients Determined at 25°C via GLC[19]

Solute	$-B_{12}$ (ml/mole)			
	H_2	N_2	O_2	CO_2
iso-Pentane	−26	105	125	163
n-Pentane	−7	105	152	173
2,2-Dimethylbutane	−26	105	145	168
Cyclopentane	−5	91	152	197
2,3-Dimethylbutane	−9	112	153	203
2-Methylpentane	7	127	155	206
3-Methylpentane	−3	117	163	211
n-Hexane	0	128	163	233
Methylcyclopentane	12	128	...	234
2,2-Dimethylpentane	14	133	...	245
2,4-Dimethylpentane	5	−30	165	249
Benzene	13	117	157	251
2,2,3-Trimethylbutane	6	123	160	222
3,3-Dimethylpentane	12	131	166	242
Cyclohexane	18	120	163	236
2-Methylhexane	35	135	175	272
2,3-Dimethylpentane	6	133	167	264
3-Methylhexane	32	136	180	276
3-Ethylpentane	36	144	176	292
n-Heptane	36	154	183	294

Everett[20] later derived an equation that appeared to differ slightly from eq. 4.9; in terms of $\ln \gamma_1^\infty$ eq. 4.9 is

$$\ln \gamma_1^\infty = \ln \gamma_p^\infty - \frac{p_1^0}{RT}(B_{11} - v_1^0) + \frac{\bar{p}}{RT}(2B_{12} - v_1^0) \qquad (4.12)$$

whereas Everett's relation was

$$\ln \gamma_1^\infty = \ln \gamma_p^\infty - \frac{p_1^0}{RT}(B_{11} - v_1^0) + \frac{\bar{p}}{RT}(2B_{12} - B_{22} - v_1^\infty) \qquad (4.13)$$

where γ_p^∞ is now identified as

$$\gamma_p^\infty = \frac{RT}{K_R \bar{V}_L p_1^0}\left(1 + \frac{B_{22}\bar{p}}{RT}\right) \qquad (4.14)$$

and v_1^∞ is the solute molar volume at infinite dilution in the solvent. Equations 4.12 and 4.13 are formally identical, however, if the approximations

$$v_1^\infty \simeq v_1^0 \qquad (4.15a)$$

$$\ln\left(1 + \frac{B_{22}\bar{p}}{RT}\right) \simeq \frac{B_{22}\bar{p}}{RT} \qquad (4.15b)$$

are made. A more important difference is the method of calculation of $\bar{p}$ which Everett claimed was given by $p_o J_3^4$, whereas Desty and coworkers used $p_o J_2^3$. In addition, neither of these workers anticipated that fixed-gas carrier solubility in the stationary phase could be significant at low column pressures, although Desty noted that condensable gases with stationary phases other than that employed (squalane) might well abrogate this view.

Cruickshank, Windsor, and Young[21] reconsidered these relations from the point of view of solute partitioning; their treatment was later modified by Everett, Gainey, and Young[22] and in greater detail by Cruickshank, Gainey, and Young[23] and Cruickshank and co-workers.[24] First, the rate of travel dl/dt of solute molecules through an infinitely thin cross section of a column was expressed (eq. 2.7) by these workers as

$$\frac{dl}{dt} = uv_M\left(\frac{1}{v_M + K_R v_L}\right) \qquad (4.16)$$

where v_M and v_L are the mobile- and stationary-phase areas within the segment and u is the linear carrier velocity. Because $uv_M = f$, the cross-sectional volume flow rate,

$$f dt = (v_M + K_R v_L) dl \qquad (4.17)$$

The mobile-phase flow rate varies along the column as the inverse of the pressure and is conveniently described by

$$f = \frac{RT + B_{22}p}{p} \tag{4.18}$$

At the column outlet

$$f_o = \frac{RT + B_{22}p_o}{p_o} \tag{4.19}$$

Combining eqs. 4.18 and 4.19, substituting the result into eq. 4.17, and letting $B_{22}/RT = b$ provides

$$f_o\,dt = \frac{p(K_R v_L + v_M)(1 + bp_o)}{p_o(1 + bp)}\,dl \tag{4.20}$$

Thus the volume of carrier is $f_o\,dt$ which passes out of the column, whereas the solute peak maximum passes through dl.

The term

$$v_M\left[\frac{p(1 + bp_o)}{p_o(1 + bp)}\right]dl$$

is just the dead volume in the segment corrected to conditions at the outlet so that

$$f_o\,dt - v_M\left[\frac{p(1 + bp_o)}{p_o(1 + bp)}\right]dl = dV_N = K_R v_L\left[\frac{p(1 + bp_o)}{p_o(1 + bp)}\right]dl \tag{4.21}$$

Recalling Darcy's law (eq. 2.18),

$$\frac{dp}{dl} = -\frac{u\varepsilon\eta}{K} \tag{4.22}$$

and noting that the pressure-dependence of the carrier viscosity is described adequately below $p = 50$ atm at $p_i - p_o < 5$ atm by[21,25]

$$\eta = \eta^0(1 + ap) \tag{4.23}$$

where

$$a = \frac{0.175\,B_{22}}{RT} \tag{4.24}$$

gives

$$dl = -\left[\frac{\varepsilon/K}{\eta^0(1+ap)}\right]\left(\frac{1+bp_o}{p_o u_o}\right)\left(\frac{p}{1+bp}\right)dp \qquad (4.25)$$

Dividing dl by the column length L yields

$$\frac{dl}{L} = \frac{\dfrac{p\,dp}{(1+ap)(1+bp)}}{\displaystyle\int_{p_o}^{p_i}\frac{p\,dp}{(1+ap)(1+bp)}} \qquad (4.26)$$

The term dl given earlier by eq. 4.21 could at this stage be substituted into eq. 4.26. However, no account has yet been made of solute fugacity or solute-carrier virial effects. Cruickshank and co-workers[21] combined Everett's earlier treatment[20] with the more general relations of Buckingham[26] to relate the solute partition coefficient to that at zero pressure drop; namely,

$$\ln K_R = \ln K_R^0 + \beta'\bar{p} + \zeta'\bar{p}^2 + \cdots \qquad (4.27)$$

where $\ln K_R^0$ is now given by

$$\ln K_R^0 = \ln\frac{RT}{\overline{V}_L p_1^0 \gamma_1^\infty} - \frac{p_1^0(B_{11}-v_1^0)}{RT} - \left(\frac{p_1^0}{RT}\right)^2(B_{11}^2 - C_{111}) + \cdots \qquad (4.28)$$

and where

$$\beta' = \frac{2B_{12}-v_1^\infty}{RT} + \lambda\left[1-\left(\frac{\partial\ln\gamma_1^\infty}{\partial x_2}\right)_o\right] \qquad (4.29)$$

$$\zeta' = \frac{3C_{122}-4B_{12}B_{22}}{2(RT)^2} + \phi\left[1-\left(\frac{\partial\ln\gamma_1^\infty}{\partial x_2}\right)_o\right] + \frac{\lambda^2}{2}\left[1-\left(\frac{\partial^2\ln\gamma_1^\infty}{\partial x_2^2}\right)_o\right] + \frac{K'v_1^\infty}{RT}$$

$$(4.30)$$

C_{111} and C_{122} are third virial coefficients that take into account trimolecular interactions, λ and ϕ represent the carrier molal solubility in the stationary phase, and the final term concerns the effects of pressure and composition on v_1^∞ (which will in most cases be negligible).

Combining eqs. 4.21 and 4.26–4.30 now yields

$$V_N = \frac{K_R^0 V_L \left(\dfrac{1 + bp_o + cp_o^2 + \cdots}{p_o} \right) \displaystyle\int_{p_o}^{p_i} \dfrac{p^2 \exp(\beta'p + \zeta'p^2 + \cdots)}{(1+ap)(1+bp+cp^2+\cdots)^2}\, dp}{\displaystyle\int_{p_o}^{p_i} \dfrac{p}{(1+ap)(1+bp+cp^2+\cdots)}\, dp}$$

(4.31)

Equation 4.31, although appearing to be formidable, does nothing more than express the pressure dependence of V_N (hence K_R) as a function of K_R^0, b, β', and ζ'. Several approximations further aid in rendering it less cumbersome and these are now considered in turn.

If the carrier behaves ideally and is insoluble in the stationary phase, $B_{22} = \lambda = \phi = 0$, $\beta' = \beta$, $\zeta' = \zeta$, and, in terms of the net retention volume,

$$\ln V_N = \ln V_N^0 + \beta p_o J_3^4 + \zeta p_o^2 J_3^5 \qquad (4.32)$$

which predicts that $\ln V_N$ will vary quadratically as $p_o J_3^4$. However, Cruickshank and co-workers noted that curvature in $\ln V_N$ versus $p_o J_3^4$ plots will in all likelihood amount to less than the experimental error of the GLC technique at inlet pressures of less than 20 atm for permanent-gas carriers; β (hence B_{12}) values can, in any event, be found from the limiting slope of such plots; an example is presented in Fig. 4.1 and a variety of GLC-determined B_{12} data is given in Table 4.3. It was also pointed out that even when $B_{12} = 0$ the term $-v_1^\infty / RT$ will not be zero under any conditions conceivable in gas chromatography; thus $\ln V_N$ versus $p_o J_3^4$ will have a nonnegligible gradient except when ζ and $(B_{12} - v_1^\infty)/RT \simeq 0$ which, as it happens, is approximately obeyed for helium with hydrocarbon solutes.

Expressions for nonideal carriers that are insoluble in the stationary phase require that the quotient of the integrals of eq. 4.31 be evaluated

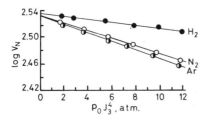

Figure 4.1 Plot of $\log V_N$ versus $p_o J_3^4$ for 2,2-dimethylbutane with di-nonyl phthalate stationary phase and indicated carriers at 25°C.[21]

analytically; if powers of J higher than J_2^3 outside the integrals are neglected,

$$\ln V_N + \ln\left[\frac{1 + bp_o J_2^3}{1 + bp_o}\right] = \ln V_N^0 + \beta p_o J_3^4 + \zeta p_o^2 J_3^5 \qquad (4.33)$$

which will be obeyed quantitatively provided that $|B_{22}| < 20$ ml/mole. If this is not the case, an expanded form of eq. 4.33 must be employed:

$$\ln V_N$$

$$+ \ln\left[\frac{1 + bp_o\left(2J_3^4 - J_2^3\right) + ap_o\left(J_3^4 - J_2^3\right) - b^2 p_o^2\left(J_4^6 - J_3^5\right) + cp_o^2\left(2J_4^6 - J_3^5\right)}{1 + bp_o + cp_o^2}\right]$$

$$= \ln K_R^0 V_L + \beta p_o J_3^4 + \zeta p_o^2 J_3^5 \qquad (4.34)$$

The treatment of nonideal *sorbed* carriers requires that β and ζ in eq. 4.34 be replaced by β' and ζ' and is discussed separately.

In contrast to the foregoing, Vigdergauz and Semkin[30] have used homologous series behavior to determine B_{12} values. First, they demonstrated that plots of $-B_{12}$ versus solute carbon number Z are linear, as shown in Fig. 4.2. The method may, in addition to n-alkanes, be applicable to other chemical classes; for example, data for 1-alkenes with H_2, N_2, and Ar carriers were also determined. The second method of Vigdergauz and Semkin[30] relies on the change of Kovats retention indices[31] I_i, defined for solute i:

$$I_i = \frac{100 \log V_g^i - \log V_g^z}{\log V_g^{z+1} - \log V_g^z} + 100z \qquad (4.35)$$

where z and $z+1$ are n-alkane standards (which have z and $z+1$ number of carbon atoms) whose specific retention volumes bracket the solute of interest. They showed that

$$\frac{\Delta I_i}{\Delta \bar{p}} = \frac{(I_i - 100z)(\beta_{z+1} - \beta_z) - 100(\beta_i - \beta_z)}{-b_2} \qquad (4.36)$$

where $\Delta I_i = I_2 - I_1$ (the ith solute retention indices at the mean column

Table 4.3 Second-Interaction Virial Coefficients for Listed Systems Determined by GLC

Solute	

A. Nitrogen Carrier[27] (40°C)

	$-B_{12}$ (ml/mole)
n-Pentane	86
n-Hexane	110
n-Heptane	110
n-Octane	134
2-Methylbutane	92
2-Methylpentane	106
3-Methylpentane	96
2,3-Dimethylbutane	99
2,2-Dimethylbutane	83
2,2-Dimethylpentane	117
2,2-Dimethylhexane	150
2-Methylhexane	115
3-Methylhexane	105
2,3-Dimethylpentane	108
2,4-Dimethylpentane	115
3,3-Dimethylpentane	117
2,2,3-Trimethylbutane	100
2,2,3,3-Tetramethylbutane	128
Cyclohexane	116
Benzene	100
Hexafluorobenzene	126

B. Nitrogen Carrier[23]

	$-B_{12}$ (ml/mole)		
	32°C	40°C	50°C
Benzene	103	96	90
Fluorobenzene	117	93	96
1,2-Difluorobenzene	...	...	95
1,3-Difluorobenzene	...	...	97
1,4-Difluorobenzene	...	...	94
1,3,5-Trifluorobenzene	110	111	106
1,2,3,4-Tetrafluorobenzene	...	...	110
1,2,4,5-Tetrafluorobenzene	...	...	105
Pentafluorobenzene	...	...	110
Hexafluorobenzene	150	130	111

Table 4.3 (*Continued*)

Solute

C. Nitrogen Carrier[28] (? 200–280°C)

	$-B_{12}$ (ml/mole)
BF_3	174
$AsCl_3$	628
$SbCl_3$	1745
CCl_4	920
$SiCl_4$	2675
$GeCl_4$	1433
$SnCl_4$	1192
$TiCl_4$	3193
$ZrCl_4$	575
$HfCl_4$	824
VCl_3	1493
$HgCl_2$	701

D. Carbon Dioxide Carrier[29] (40°C)

Propane	99
n-Butane	153
n-Pentane	198
1,3-Butadiene	155
Benzene	288
Methanol	236
Ethanol	303
1,2-Dichloroethane	215
Acetone	343

E. Carbon Dioxide (80°C), Nitrogen (80°C), and Argon (50°C) Carriers[30]

	CO_2	N_2	Ar
n-Pentane	76	60	82
n-Hexane	147	69	106
n-Heptane	177	81	136
n-Octane	227	98	160
n-Nonane	249	117	181
n-Decane	...	130	...
Cyclohexane	163	63	102
2,2,4-Trimethylpentane	147	47	128

Table 4.3 (*Continued*)

Solute

	CO_2	N_2	Ar
Benzene	216	74	117
Toluene	235	94	135
Styrene	300	117	169
Pyridine	207	42	115
Nitromethane	188	41	91
Nitroethane	247	71	125
Chloroform	169	12	60
Carbon tetrachloride	154	24	115
Ethanol	95	10	118
iso-Propyl alcohol	127	...	80
Acetone	129	...	...
Methyl ethyl ketone	184	36	86
Dioxane	136	67	124
Butyl acetate	260	79	138

F. Argon Carrier[30] (25°C)	$-B_{12}$ (ml/mole)
2,2-Dimethylbutane	108
Cyclopentane	108
2,3-Dimethylbutane	112
2-Methylpentane	115
3-Methylpentane	116
Methylcyclopentane	127
2,2-Dimethylpentane	128
2,4-Dimethylpentane	130
2,2,3-Trimethylbutane	127
Benzene	112
3,3-Dimethylpentane	132
Cyclohexane	130
1,1-Dimethylcyclopentane	134
2-Methylhexane	137
2,3-Dimethylpentane	135
3-Methylhexane	138
cis-1,3-Dimethylcyclopentane	136
trans-1,3-Dimethylcyclopentane	136
trans-1,2-Dimethylcyclopentane	136

Table 4.3 (*Continued*)

Solute		

G. Argon and Carbon Dioxide Carriers[30] (80°C)

	$-B_{12}$ (ml/mole)	
	Ar	CO_2
Toluene	123	248
Ethylbenzene	138	271
p-Xylene	143	284
m-Xylene	141	282
iso-Propylbenzene	148	286
o-Xylene	146	289
n-Propylbenzene	152	292
1-Methyl-3-ethylbenzene	154	309
iso-Butylbenzene	159	...
1,3,5-Trimethylbenzene	160	325
sec-Butylbenzene	160	312
1-Methyl-2-ethylbenzene	158	310
1,2,4-Trimethylbenzene	156	340
1-Methyl-3-propylbenzene	167	328
1-Methyl-2-iso-propylbenzene	168	329
1-Methyl-4-propylbenzene	169	318
1,4-Diethylbenzene	168	342
n-Butylbenzene	168	345
1,3-Dimethyl-5-ethylbenzene	173	348
1-Methyl-2-propylbenzene	171	327
1,2,3-Trimethylbenzene	165	335
1,4-Dimethyl-2-ethylbenzene	174	348
1,3-Dimethyl-4-ethylbenzene	174	350
1,2-Dimethyl-4-ethylbenzene	177	354
1,3-Dimethyl-2-ethylbenzene	167	352
1,2-Dimethyl-3-ethylbenzene	168	349
1,2,4,5-Tetramethylbenzene	183	376
1,2,3,5-Tetramethylbenzene	183	380
1,2,3,4-Tetramethylbenzene	178	386

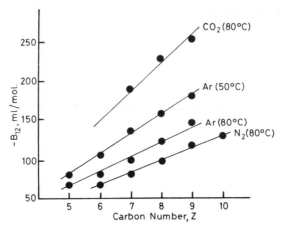

Figure 4.2 Plot of second-interaction virial coefficient versus carbon number for *n*-alkanes with various carriers.[30]

pressures $\bar{p}_2$ and $\bar{p}_1$), β is defined by eq. 4.29 (λ is assumed to be zero), and

$$\Delta\bar{p} = \left(p_o J_3^4\right)_2 - \left(p_o J_3^4\right)_1 \tag{4.37}$$

$$b_2 = \frac{\log V_g^{z+1}}{\log V_g^z} \tag{4.38}$$

B_{12} values for argon with 19 components of a gasoline fraction were determined via eq. 4.36 by these workers (Table 4.3), who used *n*-pentane and *n*-hexane as the internal standards. This method appears to be an attractive one, for only two well-characterized standards and the retention indices and molar volumes of the other solutes are required. The third method of Vigdergauz and Semkin[30] also makes use of interpolation between two well-characterized (but not necessarily homologous) standards, A and B:

$$\frac{\Delta G_i}{\Delta\bar{p}} = \frac{\beta_i - \beta_A - G_1(\beta_B - \beta_A)}{a_2} \tag{4.39}$$

where $\Delta G_i = G_2 - G_1$ (at $\bar{p}_2$ and $\bar{p}_1$) and

$$G_i = \frac{\log t_R'(i) - \log t_R'(A)}{\log t_R'(B) - \log t_R'(A)} \tag{4.40}$$

$$a_2 = \log t_R'(B) - \log t_R'(A) \tag{4.41}$$

Equation 4.39 is based on the assumption that the corrected retention time of a solute relative to those for two standard compounds will vary as its second virial coefficient. Benzene and styrene were chosen as standards for the determination of B_{12} values for 29 aromatic hydrocarbons with Ar and CO_2 carriers (Table 4.3). Although suitable criteria were not indicated, the standards should presumably be similar chemically to the solutes under investigation.

4.2 RELIABILITY OF GLC-DETERMINED B_{12} DATA

4.2.1 Precision of the Method Column-to-column precision of the GC method for the same stationary phase in the same laboratory has been shown by Cruickshank and co-workers[24] to be excellent. Table 4.4 gives the virial coefficients of benzene with N_2 and CO_2 carriers at 50°C obtained with five columns containing packings of different liquid loadings of glycerol on Celite. The results are of particular interest because the retention of benzene was shown to be a function of the stationary phase weight percent; the measured B_{12} data are unaffected, however, by the presence of gas-liquid surface adsorption.

Table 4.4 Second-Interaction Virial Coefficients for Benzene with Nitrogen and Carbon Dioxide at 50°C Obtained from Columns of Various Liquid Loadings of Glycerol on Celite[24]

Col. No.	%L (w/w)	$-B_{12}$ (ml/mole)	
		N_2	CO_2
1	15.7	91.0 ± 6.0	257.5 ± 6.0
2	25.4	104.5 ± 6.0	259.0 ± 6.0
3	33.6	98.5 ± 7.5	256.0 ± 6.0
4	23.3	97.0 ± 3.0	...

Equation 4.32 contains terms that involve the stationary phase only in the case of sorbed carriers; in the absence of these effects GLC B_{12} data are predicted to be independent of the liquid phase, as shown by Everett, Gainey, and Young[22] for several carriers and solvents with benzene. These data are given in Table 4.5. Gainey and Pecsok[27] similarly found no trend in B_{12} values obtained for a variety of solutes with a homologous series of solvents and nitrogen carrier (Table 4.6).

Table 4.5 Comparison[22] of B_{12} Data for Benzene with Various Carriers and Stationary Phases at 50°C

	$-B_{12}$ (ml/mole)		
Carrier	Squalane	n-Octadecane	Di-nonyl Phthalate
Helium	-57 ± 8	-49 ± 8	...
Argon	85 ± 8	79 ± 8	...
Carbon monoxide	113 ± 8	...	122 ± 8

Conder and Langer[32] found that the ratio of $\ln V_N$ values obtained for a given solute with different carriers (A and B) was independent of the stationary phase and fugacity effects in accordance with the relation

$$\frac{\ln V_N(A)}{\ln V_N(B)} = \frac{2p_o J_3^4}{RT}\left[B_{12}(A) - B_{12}(B)\right] \qquad (4.42)$$

Table 4.6 Comparison[27] of B_{12} Data for Named Solutes with Nitrogen Carrier in Various Solvents at 40°C

	$-B_{12}$ (ml/mole)				
Solute	1-Phenyl-decane	1-Phenyl-dodecane	1-Phenyl-tetradecane	1-Phenyl-pentadecane	1-Phenyl-nonadecane
n-Pentane	81	85	95	84	86
n-Hexane	104	110	111	106	109
n-Heptane	110	109	111	114	113
n-Octane	128	133	139	...	132
2-Methylbutane	...	92	...	91	...
2-Methylpentane	...	107	...	105	...
3-Methylpentane	...	96	...	96	...
2,3-Dimethylbutane	...	102	...	96	...
2,2-Dimethylbutane	...	83	...	...	...
2,2-Dimethylpentane	...	...	...	117	...
2,2-Dimethylhexane	...	...	...	150	...
2-Methylhexane	...	116	...	113	...
3-Methylhexane	...	104	...	104	...
2,3-Dimethylpentane	...	108	...	108	...
2,4-Dimethylpentane	...	111	...	108	...
3,3-Dimethylpentane	...	116	...	117	...
2,2,3-Trimethylbutane	...	100	...	103	...
2,2,3,3-Tetramethylbutane	...	...	...	128	...
Cyclohexane	...	118	...	114	...
Benzene	95	98	99	108	101
Hexafluorobenzene	...	124	130	122	...

Table 4.7   Comparison of GLC B_{12} Data for Named
Solutes with Nitrogen Carrier at 40°C

Solute	$-B_{12}$ (ml/mole)	
	Ref. 27	Refs. 33, 34
n-Pentane	86	85
n-Hexane	110	107
n-Heptane	110	111
Hexafluorobenzene	126	128

as long as carrier solubility was negligibly small.

Interlaboratory reproducibility of GLC B_{12} data depends only on the accuracy with which $\ln V_N$ and $p_o J_3^4$ measurements are made and can in general be expected to be on the order of 1–5%. Table 4.7 gives the B_{12} values found by Gainey and Pecsok[27] and those determined independently by Gainey[33] and Young.[34]

4.2.2   Comparison with Nonchromatographic Methods   Three independent investigations have established that GLC-determined second-interaction virial data agree well with those derived from nonchromatographic techniques. In the first Everett and co-workers[22] compared their GLC data for benzene with H_2 and CO_2 at 50°C with those of Connolly[35]; the latter were measured by the solubility of liquids in compressed gases and by gas compressibility studies. The GLC data were -5 and -118 ± 8 ml/mole, respectively, whereas Connolly reported -7.5 and -114 ± 4 ml/mole. Dantzler, Knobler, and Windsor[36] also found the agreement between

Table 4.8   Comparison of Averaged Static[36, 37]
and GLC[21] B_{12} Data for Named Solutes with
Argon at 25°C

Solute	$-B_{12}$ (ml/mole)	
	Static	GLC
n-Pentane	131 ± 20	98 ± 20
n-Hexane	136 ± 62	124 ± 20
2,2-Dimethylbutane	156 ± 26	115 ± 20
2-Methylpentane	147 ± 49	125 ± 20
2-Methylbutane	178 ± 14	94 ± 20

Table 4.9 Comparison[38] of Static and GLC B_{12} Data for Benzene with Named Carriers

Carrier	T (°C)	$-B_{12}$ (ml/mole) Static	GLC (Ref. 22)	GLC (Ref. 40)
Helium	50	-67 ± 4	-57 ± 8	-49 ± 8
Hydrogen	50	-4 ± 3	5 ± 8	...
Nitrogen	35	97 ± 3	...	104 ± 10
Nitrogen	50	85 ± 3	87 ± 8	94 ± 10
Argon	32	122 ± 3	...	135 ± 10
Argon	50	95 ± 3	85 ± 8	90 ± 10
Methane	50	171 ± 3	155 ± 15	...

previously reported GLC and their static[37] data to be within the respective experimental errors of the methods shown in Table 4.8. Finally, Coan and King[38] compared data measured by GLC with those derived from the high-pressure vapor/liquid equilibration method of Prausnitz and Benson[39]; the data are presented in Table 4.9. The GLC procedure is clearly competitive with nonchromatographic methods and, given its comparative simplicity of instrumentation and ease of operation, is an attractive technique for the accurate measurement of second-interaction virial coefficients at moderate pressures.

4.2.3 Comparison with Semiempirical Calculation Methods   Second-interaction virial coefficients can be calculated in an *ab initio* fashion from the (semiempirical) McGlashan-Potter[4,6] relation (eq. 4.8) in which the subscripts 11 are now replaced with 12 and in which [7,41,42]

$$\bar{n} = \frac{n_1 + n_2}{2} \tag{4.43}$$

$$V_{12}^c = \tfrac{1}{8}\left[(V_{11}^c)^{1/3} + (V_{22}^c)^{1/3} \right]^3 \tag{4.44}$$

$$T_{12}^c = 2(T_{11}^c T_{22}^c)^{1/2} \left[\frac{(I_1^d I_2^d)^{1/2}}{I_1^d + I_2^d} \right] \left\{ \frac{64 V_{11}^c V_{22}^c}{\left[(V_{11}^c)^{1/3} + (V_{22}^c)^{1/3} \right]^{1/6}} \right\} \tag{4.45}$$

n_1 and n_2 are the "effective" number of carbon atoms per molecule ($n_2 = 1$ for fixed carriers) and I_i^d, V_i^c, and T_i^c are the ionization potential, critical volume, and critical temperature of the ith species.[43] Equation 4.8 is based on the principle of corresponding states; i.e., specification of two reduced variables (e.g., T/T^c and V/V^c) fixes the third (p/p^c); although one would normally suspect that eq. 4.45 could be replaced by the simpler geometric mean relation

$$T_{12}^c = (T_{11}^c T_{22}^c)^{\frac{1}{2}}$$

Hudson and McCoubrey[41] and Munn[42] have pointed out that this is valid only when the solute and carrier have similar molecular dimensions and ionization potentials, which is not a likely situation in GC. Therefore eq. 4.45 should be used when B_{12} data are calculated via eq. 4.8. A variety of workers has demonstrated the success of this approach; a detailed consideration of the calculation method by Conder and Langer,[32] for example, showed it to have a likely accuracy of 5—10%, as borne out by the data in Table 4.10, which compares calculated and GLC virial data. One might, in fact, suppose that, provided that the necessary critical-point data are available, the experimental GLC method is somewhat superfluous. Calculated data, however, have so far been found to be accurate within the

Table 4.10 Comparison[23] of GLC and Calculated (Eqs. 4.8, 4.43–4.45) Second-Interaction Virial Coefficients at 35°C

| Solute | $-B_{12}$ (ml/mole) | | | |
| | N$_2$ Carrier | | Ar Carrier | |
	GLC	Calculated	GLC	Calculated
n-Butane	84±9	74	...	...
n-Pentane	85±13	91	...	...
n-Hexane	113±9	102	...	...
n-Heptane	132±9	107	...	...
n-Octane	143±9	119	...	...
iso-Pentane	...	...	78±12	94
2,2-Dimethylbutane	60±20	97	...	...
2,3-Dimethylbutane	...	...	98±12	104
2-Methylpentane	91±9	100	...	...
3-Methylpentane	...	...	95±12	106
2,4-Dimethylpentane	109±9	106	...	...
Cyclohexane	122±13	110	...	...
Benzene	104±9	107	...	...
1-Pentene	64±9	88	84±12	92
1-Hexene	110±9	100	110±12	103
1-Heptene	142±9	109	139±12	112
1-Octene	...	...	136±12	119

stated limits solely for hydrocarbon solutes with permanent carriers and may be presumed to be somewhat less reliable for other chemical types; for example, amines. The chromatographic procedure must in any event be employed in order to measure partition coefficients at zero column pressure drop and B_{12} values are more readily determined from these experimental data than from eq. 4.8.

4.3 SORBED CARRIERS

4.3.1 Effects of Carrier Solubility Two fundamentally similar approaches to a description of sorbed carriers have been presented, —by Cruickshank and co-workers[24] outlined in Section 4.1.2 and the more general treatment of Wicar and Novak[44] reported in 1974. The difference between the two formulations lies in the manner in which carrier solubility is said to affect solute retention and can be expressed as[44]

$$\frac{\exp-\left[(\alpha-\beta)x_2+\beta x_2^2\right]}{1-x_2^L}$$

where x_2^L is the carrier mole fraction in the liquid phase and α and β are coefficients that characterize the three types of binary liquid solution possible, namely, solute + carrier, solute + stationary phase, and carrier + stationary phase. In order to test their ternary solution model, Wicar and Novak studied the elution of tetrachloromethane, iso-octane, and toluene with hydrogen, nitrogen, and carbon dioxide carriers from columns containing Apiezon K at moderate to high pressures. According to the simplest (pseudobinary) model (eq. 4.32), plots of $\ln V_N - \beta p_o J_3^4$ versus $p_o J_3^4$ should for each solute lie on a single straight line (irrespective of the carrier), the slope of which will depend on the partial molar volume of the solute in the stationary phase. On the other hand, if the ternary solution model obtains, different straight lines will be found for each solute with each carrier; this, in fact, was observed in all cases. In addition, Wicar and Novak calculated the effective partial molar volumes of the solutes with each carrier from the plots; the results are shown in Table 4.11. The negative effective molar volumes obtained for iso-octane with CO_2 imply a considerable change in the solute activity coefficient over that to be found with, say, hydrogen carrier which has important implications in light of the work of Sie and co-workers[29] who had earlier concluded that even though CO_2 carrier was appreciably soluble in squalane this behavior had little or no effect on the retention of benzene and methanol, for the pressure dependence of V_N for these solutes was unaltered when the solvent was changed to glycerol (negligible CO_2 solubility). In contrast, Stalkup and Kobayashi[45] showed that at sufficient pressures carrier dissolution may become the predominant factor in retention. Cruickshank and co-workers[24] also found that carrier solubility (CO_2 in squalane) has a nontrivial effect on the pressure dependence of V_N values. These results can be rationalized in terms of the Wicar-Novak model and as postulated some years earlier by Pecsok and Windsor,[46] solubility of the carrier will produce an increase in the volume of stationary phase with a concomitant increase in solute retention. With mobile phases such as CO_2, however, an increase in the solute activity coefficient in the "modified" stationary phase may also occur, in which case solute retention will be decreased. When these two factors compensate, solute retention will appear to be independent of the solubility of the carrier in the liquid phase as found by Sie and co-workers.[29] Cruickshank and co-workers[24] agreed tacitly with this approach but preferred to express carrier solubility effects in terms of the experimentally inaccessible term $(\partial \ln \gamma_1^\infty / \partial x_2)$. Noteworthy in this context is the fact that Wicar and Novak[44] found plots of $\ln V_N$ versus $p_o J_3^4$ to be linear for their three solutes with CO_2 and Apiezon K at pressures up to 35 atm.

Table 4.11 Solute[44] Bulk and "Effective" Partial Molar Volumes in Apiezon K with H_2, N_2, and CO_2 Carriers

Solute	T ($^\circ$C)	v_1 (ml/mole)			
		Bulk	H_2	N_2	CO_2
iso-Octane	50	171.3	195.3	182.2	-67.2
	75	176.9	216.6	181.9	-8.54
Carbon tetrachloride	50	100.2	...	150.0	53.7
	75	103.5	151.5	134.0	101.5
Toluene	50	109.8	150.4	116.9	98.9
	75	113.0	175.3	101.3	94.8

4.3.2 Methane and Ethane Carriers   In light of these interpretations of carrier solubility the GLC technique of virial coefficient measurements may appear to be confined to low column pressure drops and permanent carriers exhibiting limited miscibility with the stationary phase. This is an unwarranted conclusion, however, as indicated, if by nothing else, by continued interest in sorbed carriers in GC[47]; for example, Pecsok and

Windsor[46] employed methane and ethane carriers with hydrocarbon solutes; their B_{12} data are listed in Table 4.12. They used eq. 4.29 to evaluate the virial coefficients and assumed that the term $[1 - (\partial \ln \gamma_1^\infty / \partial x_2)]$ must lie between 0 and 1; that is, 0.5 ± 0.5. Further, λ was expressed as an expansion of x_2:

$$x_2 = \lambda p + \nu p^2 + \cdots \qquad (4.46)$$

where the carrier was assumed to obey ideal solution behavior in the stationary phase. The rhs of eq. 4.29 was then evaluated to be -22 (± 22) and -153 (± 153) ml/mole for methane and ethane, respectively, at 25°C and -10 (± 10) ml/mole for methane at 50°C. The resultant B_{12} data corrected for these effects were found to be in good agreement with static[36,48] values.

Table 4.12   Second-Interaction[46] Virial Coefficients for Hydrocarbon Solutes with Methane and Ethane Carriers at 25 and 50°C

Solute	Carrier	$-B_{12}$ (ml/mole) 25°C	50°C
n-Pentane	Methane	204 ± 42	138 ± 30
n-Hexane	Methane	292 ± 55	280 ± 54
2-Methylpentane	Methane	317 ± 42	144 ± 34
2,2-Dimethylbutane	Methane	216 ± 42	154 ± 29
iso-Pentane	Methane	199 ± 43	123 ± 36
n-Pentane	Ethane	414 ± 171	...

4.3.3   Apparatus for Carrier Generation   Methane and ethane are, of course, readily available in pressurized cylinders which facilitates their use in GC. Other carriers such as benzene require vapor generator apparatus, several devices for which have been described in the literature. The first reported use of such a system appears to be that by Dumazert and Ghiglione[49] in 1959, who employed a vapor entrainment apparatus that, when heated sufficiently, produced a carrier consisting almost entirely of the organic liquid used (here, alcohol). Tsuda and co-workers have considerably improved and simplified earlier equipment in their efforts to increase solute trapping, detection, and collection efficiency. Carbon tetrachloride,[50] ethanol, benzene, and carbon tetrachloride,[51] and tetrachloroethylene[52] have all been used as carriers by these workers (virial coefficients for hydrocarbons with CCl_4 are about five times those with nitrogen[51]) and their method remains the preferred technique: a 0.5-l reservoir is filled with approximately 350 ml solvent, the reservoir is evacuated, then heated until the solvent vapor pressure reaches the desired (column inlet) pressure. The column inlet and outlet pressures may be adjusted with needle valves and the flow rate measured either with a heated rotameter, an electronic flowmeter, or by dividing the column dead volume (determined with a fixed carrier in a separate experiment) by the retention time of a nonsorbed solute eluted with the organic carrier. Clearly this technique is applicable to any material with an appreciable vapor pressure at the desired column temperature; reduced pressure at the column outlet facilitates the use of an even wider range of carriers.

4.4 SUPERCRITICAL FLUID CHROMATOGRAPHY (SFC)

The distinction between GC and LC disappears above the critical point of the carrier, the technique then being termed "fluid chromatography" or "supercritical fluid chromatography" (SFC). Interest in this topic began in 1962 when Klesper and co-workers[53] reported the elution of porphyrins with a freon mobile phase. Since then a number of workers have exploited SFC for a variety of separations (cf. the reviews of Gouw and Jentoft[54]), the most recent being that by Jentoft and Gouw[55] in 1976. The advent of "high-performance" liquid chromatography has, however, brought about a marked decline in studies of this type (even though the speed and efficiency of SFC are inherently superior to those of HPLC; e.g., Ref. 56). In addition, there remain the problems of detection and adequate temperature and pressure control, these having reduced the technique to all but a novelty. Nonetheless, it holds much promise for the physicochemical study of higher order $B, C, \cdots$ coefficients and fluid-stationary phase solubility effects, and even though theory lags far behind experiment in these areas SFC should not be overlooked as a means of elucidating virial and related phenomena.

4.5 REFERENCES

1. E. A. Guggenheim, *Thermodynamics*, North-Holland, Amsterdam, 1967, Chapters 4, 5.
2. H. K. Onnes and W. H. Keesom, *Commun. Phys. Lab. Univ. Leiden*, **11**, Suppl. 23 (1912).
3. D. H. Everett and C. T. H. Stoddart, *Trans. Faraday Soc.*, **57**, 746 (1961).
4. M. L. McGlashan and D. J. B. Potter, *Proc. Roy. Soc. Ser. A*, **267**, 478 (1962).
5. J. A. Beattie and O. C. Bridgeman, *J. Am. Chem. Soc.*, **49**, 1665 (1927).
6. M. L. McGlashan and C. J. Wormald, *Trans. Faraday Soc.*, **60**, 646 (1964).

7. E. A. Guggenheim and C. J. Wormald, *J. Chem. Phys.*, **42**, 3775 (1965).

8. P. Y. Feng and M. Melzer, *J. Chem. Educ.*, **49**, 375 (1972).

9. K. A. Kobe and R. E. Lynn, *Chem. Rev.*, **52**, 117 (1953).

10. R. B. Spertell and G. T. Chang, *J. Chromatogr. Sci.*, **10**, 60 (1972).

11. A. Kwantes and G. W. A. Rijnders, in *Gas Chromatography 1958*, D. H. Desty, Ed., Butterworths, London, 1958, p. 125.

12. D. H. Desty and A. Goldup, in *Gas Chromatography 1960*, R. P. W. Scott, Ed., Butterworths, London, 1960, p. 162.

13. D. H. Desty, A. Goldup, and W. T. Swanton, in *Gas Chromatography*, N. Brenner, J. E. Callen, and M. D. Weiss, Eds., Academic, New York, 1962, p. 105.

14. A. Goldup, G. R. Luckhurst, and W. T. Swanton, *Nature*, **193**, 333 (1962).

15. S. Trestianu, *Stud. Cercet. Chim.*, **17**, 463 (1969).

16. R. J. Laub and R. L. Pecsok, *J. Chromatogr.*, **98**, 511 (1974).

17. P. A. Sewell and R. Stock, *Nature*, **207**, 618 (1965).

18. D. E. Martire and D. C. Locke, *Anal. Chem.*, **37**, 144 (1965).

19. D. H. Desty, A. Goldup, G. R. Luckhurst, and W. T. Swanton, in *Gas Chromatography 1962*, M. van Swaay, Ed., Butterworths, London, 1962, p. 67.

20. D. H. Everett, *Trans. Faraday Soc.*, **61**, 1637 (1965).

21. A. J. B. Cruickshank, M. L. Windsor, and C. L. Young, *Proc. Roy. Soc. Ser. A*, **295**, 259, 271 (1966).

22. D. H. Everett, B. W. Gainey, and C. L. Young, *Trans. Faraday Soc.*, **64**, 2667 (1968).

23. A. J. B. Cruickshank, B. W. Gainey, and C. L. Young, *Trans. Faraday Soc.*, **64**, 337 (1968); in *Gas Chromatography 1968*, C. L. A. Harbourn, Ed., Institute of Petroleum, London, 1969, p. 76.

24. A. J. B. Cruickshank, B. W. Gainey, C. P. Hicks, T. M. Letcher, R. W. Moody, and C. L. Young, *Trans. Faraday Soc.*, **65**, 1014 (1969).

25. S. Chapman and T. G. Cowling, *Mathematical Theory of Non-Uniform Gases*, Cambridge University Press, Cambridge, England, 1939.

26. A. D. Buckingham, *The Laws and Applications of Thermodynamics*, Pergamon, Oxford, England, 1964.

27. B. W. Gainey and R. L. Pecsok, *J. Phys. Chem.*, **74**, 2548 (1970).

28. F. M. Zado and R. S. Juvet, Jr., in *Aspects in Gas Chromatography*, H. G. Struppe, Ed., Akademie Verlag, Berlin, 1971, p. 206.

29. S. T. Sie, W. van Beersum, and G. W. A. Rijnders, *Sep. Sci.*, **1**, 459 (1966).

30. M. Vigdergauz and V. Semkin, *J. Chromatogr.*, **58**, 95 (1971).

31. E. sz. Kovats, *Helv. Chim. Acta*, **41**, 1915 (1958).

32. J. R. Conder and S. H. Langer, *Anal. Chem.*, **39**, 1461 (1967).

33. B. W. Gainey, Ph.D. Dissertation, University of Bristol, England, 1967.

34. C. L. Young, Ph.D. Dissertation, University of Bristol, England, 1967.

35. J. F. Connolly, *Phys. Fluid*, **4**, 1494 (1961); **7**, 1023 (1964).

36. E. M. Dantzler, C. M. Knobler, and M. L. Windsor, *J. Chromatogr.*, **32**, 433 (1968).

37. C. M. Knobler, *Rev. Sci. Instrum.*, **38**, 184 (1967).

38. C. R. Coan and A. D. King, Jr., *J. Chromatogr.*, **44**, 429 (1969).

39. J. M. Prausnitz and P. R. Benson, *AIChE J.*, **5**, 161 (1959).

40. B. W. Gainey and C. L. Young, *Trans. Faraday Soc.*, **64**, 349 (1968).

41. G. H. Hudson and J. C. McCoubrey, *Trans. Faraday Soc.*, **56**, 761 (1960).

42. R. J. Munn, *Trans. Faraday Soc.*, **57**, 187 (1961).

43. E. A. Guggenheim and M. L. McGlashan, *Proc. Roy. Soc. Ser. A*, **206**, 448 (1951).

44. S. Wicar and J. Novak, *J. Chromatogr.*, **95**, 1, 13 (1974).

45. F. I. Stalkup and R. Kobayashi, *AIChE J.*, **9**, 121 (1963).

46. R. L. Pecsok and M. L. Windsor, *Anal. Chem.*, **40**, 1238 (1968).

47. S. Wicar, J. Novak, J. Drozd, and J. Janak, Twelfth International Symposium on Advances in Chromatography, Amsterdam, 1977.

48. Sh. D. Zaalishvili, *Zh. Fiz. Khim.*, **30**, 1891 (1956).

49. C. Dumazert and C. Ghiglione, *Bull. Soc. Chim. Fr.*, 615 (1959).

50. T. Tsuda, H. Mori, and D. Ishii, *Bunseki Kagaku*, **18**, 1328 (1969).

51. T. Tsuda, N. Tokoro, and D. Ishii, *J. Chromatogr.*, **46**, 241 (1970).

52. T. Tsuda and D. Ishii, *Bunseki Kagaku*, **19**, 565 (1970); *J. Chromatogr.*, **47**, 469 (1970).

53. E. Klesper, A. H. Corwin, and D. A. Turner, *J. Org. Chem.*, **27**, 700 (1962).

54. T. H. Gouw and R. E. Jentoft, *J. Chromatogr.*, **68**, 303 (1972); *Adv. Chromatogr.*, **13**, 1 (1975).

55. R. E. Jentoft and T. H. Gouw, *Anal. Chem.*, **48**, 2195 (1976).

56. M. Novotny, W. Bertsch, and A. Zlatkis, *J. Chromatogr.*, **61**, 17 (1971).

CHAPTER 5

Thermodynamic
Properties
of Solution

5.1 THERMODYNAMIC FUNCTIONS DERIVED
FROM INFINITE-DILUTION RETENTION DATA

This topic predominates by far in the use of gas chromatography for physicochemical studies. More than 150 publications have been devoted to the GLC-measurement of infinite-dilution activity coefficients for example, and the list continues to grow. An ever-expanding number of workers in related fields of solution chemistry, including the study of polymers, Lewis acid-base properties, liquid crystals, and gas-liquid interfacial adsorption, have, in addition, found GLC to be a tool of important if not indispensable utility. Chemical engineers have also benefited from the method, and the employment of GC to determine liquid/liquid partition coefficients, isotherms, and the relative merits of entrainment solvents is today not uncommon.

5.1.1 Heats of Solution and Vaporization James[1] was the first to note that plots of log (relative retention volume) versus carbon number were linear for homologous series, and two years later, in 1954, Ray[2] verified the phenomenon for paraffins, alcohols, formates, acetates, propionates, and methyl ketones. Littlewood, Phillips, and Price,[3] however, are generally given credit as the first to take advantage not only of the variation of solute retention with carbon number but, in addition, and of greater thermodynamic importance, plots of $\ln V_g$ versus $1/T$. In 1955 heats of solution were calculated from the slopes of such plots by these workers for esters, alcohols, and aromatic hydrocarbons with the solvents, tritolyl phosphate and silicone oil. Shortly after similar studies were reported by

110

White and Cowen,[4, 5] Anderson and Napier,[6] Herington,[7] Hoare and Purnell,[8, 9] Knox,[10] and others.[11-15]

The equations relating thermodynamic solution properties to retention data are straightforward: first, when the solute activity coefficient is unity, the excess molar enthalpy of mixing $\Delta \bar{H}^e$ is zero and it follows from eq. 2.35 that

$$V_g^0 = \frac{273R}{p_1^0 MW_L} \tag{5.1}$$

The solute vapor pressure is described as a function of temperature by the Clausius-Clapeyron equation:

$$\ln p_1^0 = -\frac{\Delta \bar{H}_v}{RT} + C \tag{5.2}$$

where $\Delta \bar{H}_v$ is the molar heat of vaporization of bulk solute and C is a constant. Combining these two relations,

$$\ln V_g^0 = \ln \left(\frac{273R}{MW_L} \right) + \frac{\Delta \bar{H}_v}{RT} + C \tag{5.3}$$

Because the first term on the rhs is itself a constant and $\Delta \bar{H}_v + \Delta \bar{H}_s = 0$ when $\gamma_1^\infty = 1$, where $\Delta \bar{H}_s$ is the molar heat of solution,

$$\ln V_g^0 = -\frac{\Delta \bar{H}_s}{RT} + C' = \frac{\Delta \bar{H}_v}{RT} + C' \tag{5.4}$$

Thus the slope of a plot of $\ln V_g^0$ versus $1/T$ yields $-\Delta \bar{H}_s/R$ (or $\Delta \bar{H}_v/R$). Because $\Delta \bar{G}_s = \Delta \bar{H}_s - T\Delta \bar{S}_s$, the Gibbs free energy and entropy of solution at the column temperature may also be calculated from these data.

Plots of $\ln V_g^0$ versus $1/T$ have a positive slope, for the heat of solution (generally -1 to -10 kcal/mole) is invariably negative (heat evolved).

Combining eqs. 5.2 and 5.4,

$$\ln V_g^0 = -a \ln p_1^0 + C'' \tag{5.5}$$

where the factor a represents the ratio $\Delta \bar{H}_s/\Delta \bar{H}_v$. Plots of $\ln V_g^0$ versus $\ln p_1^0$ at various temperatures thus provide an interesting test of solution ideality, for these should be linear of negative slope, unity. Hoare and Purnell[8] were the first to use eq. 5.5; their data for a variety of solutes in "liquid paraffin" solvent are given in Table 5.1. As might be expected, the ratio $\Delta \bar{H}_s/\Delta \bar{H}_v$ approaches unity only for n-alkanes (and, fortuitously, ethanol)

Table 5.1   Heats of Solution and Vaporization for Named Solutes in "Liquid Paraffin" Solvent[8]

Solute	$-\Delta\bar{H}_s$ (kcal/mole)	$\Delta\bar{H}_v$ (kcal/mole)	$-\Delta\bar{H}_s/\Delta\bar{H}_v$
n-Pentane	5.99	6.13	0.98
n-Hexane	6.80	6.62	1.01
n-Heptane	7.68	7.32	1.04
Dichloromethane	6.43	6.69	0.96
Chloroform	6.62	7.05	0.94
Carbon tetrachloride	6.62	7.14	0.92
cis-1,2-Dichloroethene	4.92	7.15	0.69
trans-1,2-Dichloroethene	5.48	7.23	0.76
1,1-Dichloroethane	4.44	6.65	0.67
Tetrachloroethylene	10.24	8.30	1.24
Methanol	13.68	8.43	1.61
Ethanol	10.10	9.68	1.04
Acetone	13.83	7.25	1.92

with this liquid phase. Nonetheless, heats of vaporization can be measured (if only approximately) for other solutes from data derived from plots of $\ln V_g^0$ versus $1/T$ and versus $\ln p_1^0$ because these provide, respectively, $-\Delta\bar{H}_s$ and a. Conversely, when $\gamma_1^\infty \neq 1$, the relation $\Delta\bar{H}_s - \Delta\bar{H}^e = \Delta\bar{H}_v$ may be employed. In addition, vapor pressures and (via Trouton's rule) boiling points may, for homologous series at least, also be determined in this manner (the measurement of these and other intrinsic properties is taken up in Chapter 10).

A derivation analogous to the one given in terms of K_R provides

$$\ln K_R = \ln\left(\frac{RT}{\bar{V}_L}\right) - \frac{\Delta\bar{H}_s}{RT} + C \tag{5.6}$$

Plots of $\ln K_R$ versus $1/T$ would therefore be expected to be curved, for the first term on the rhs is temperature-dependent. Further, use of eq. 5.6 refers to a gas-phase standard-state temperature of that of the column, whereas eq. 5.4 refers to 0°C. Equations 5.4 and 5.6 are related by[16]

$$\Delta\bar{G}_s = \Delta\bar{G}_s^T - RT\ln\frac{T\rho_L}{273} \tag{5.7}$$

$$\Delta\bar{H}_s = \Delta\bar{H}_s^T + RT - RT^2\eta \tag{5.8}$$

$$\Delta\bar{S}_s = \Delta\bar{S}_s^T + R - RT\eta + R\ln\frac{T\rho_L}{273} \tag{5.9}$$

Table 5.2 Examples of Thermodynamic Solution Properties Measured by GLC

A. Heats[a] of Solution (40–60°C) for Named Solutes with Substituted Hexadecane Solvents[16]

Solute	$-\Delta\overline{H}_s$ (kcal/mole)					
	n-Hexa-decane	1-Hexa-decene	1-Chlorohexa-decane	1-Bromohexa-decane	1-Cyanohexa-decane	1-Hexadecanol
iso-Pentane	5.24	5.68	5.92	5.60	5.36	6.34
n-Pentane	6.42	6.07	6.30	5.98	5.62	6.55
2,2-Dimethylbutane	6.58	6.35	6.30	6.21	5.72	6.52
2-Methylpentane	6.93	6.71	6.88	6.71	6.43	6.98
3-Methylpentane	7.17	6.95	6.91	6.90	6.46	7.09
n-Hexane	7.43	7.18	7.22	7.14	6.78	7.25
2,4-Dimethylpentane	7.29	7.07	7.22	7.12	6.69	7.08
3-Methylhexane	8.05	7.95	8.04	7.90	7.50	7.86
n-Heptane	8.38	8.33	8.46	8.31	7.78	8.11
1-Pentene	5.98	6.07	5.97	5.88	5.48	6.51
2-Methyl-2-butene	6.29	6.52	6.33	6.40	5.86	6.93
1-Hexene	7.07	7.16	7.18	7.00	6.54	7.09
1-Heptene	8.19	8.13	8.23	8.12	7.67	7.95
Cyclohexane	7.48	7.33	7.42	7.36	6.81	7.30
Benzene	7.36	7.47	7.58	7.50	7.76	7.17
Diethyl ether	6.43	6.07	5.96	6.16	6.26	6.36
Di-n-propyl ether	7.98	7.88	8.21	8.07	7.98	7.76
Methanol	6.67	—	5.55	6.69	7.09	8.96
Ethanol	6.22	—	6.52	6.83	7.71	9.69
n-Propanol	6.66	6.83	6.85	7.00	8.92	10.59
n-Butanol	6.83	7.38	7.96	8.34	9.88	13.78
Methyl formate	4.95	4.50	5.00	5.30	5.57	6.00
Ethyl formate	5.72	5.73	6.22	6.57	6.57	6.04
Ethyl acetate	6.75	5.81	7.22	7.43	7.66	6.77
Methyl ethyl ketone	6.53	6.12	7.22	7.11	7.41	6.63
Propionaldehyde	4.90	5.28	6.33	6.31	6.40	5.35
n-Butyraldehyde	6.59	6.31	7.15	6.55	7.33	6.12
n-Propyl chloride	6.00	6.07	6.55	6.28	6.47	6.26
Ethyl bromide	5.50	5.71	6.02	5.86	6.24	5.86
n-Propyl bromide	6.77	7.02	7.10	7.06	7.28	6.82
sec-Butyl bromide	7.38	7.42	8.05	7.69	7.95	7.25
n-Butyl bromide	7.84	7.87	8.40	8.19	8.40	8.00
Ethyl iodide	6.60	6.60	6.95	6.97	7.04	6.45
Ethyl cyanide	5.85	5.66	7.00	6.78	7.76	5.56
n-Propyl cyanide	6.90	6.85	8.04	8.02	7.76	6.75
Nitroethane	6.98	6.73	7.94	7.48	8.68	6.45
Dichloromethane	5.64	5.83	6.33	6.40	6.83	6.18
Chloroform	6.53	6.87	7.26	7.07	8.45	7.20
Carbon tetrachloride	7.34	7.33	7.50	7.28	7.42	7.36

113

Table 5.2 (*Continued*)

B. Heats of Solution (100–200°C) for SbCl$_3$ and NbCl$_5$ Solutes with Named Solvents[17]

	Solvent	$-\Delta\overline{H}_s$ (kcal/mole)
SbCl$_3$	LiAlCl$_4$	10.7
	NaAlCl$_4$	10.7
	KAlCl$_4$	11.4
	TlAlCl$_4$	11.7
	NaFeCl$_4$	10.8
	KFeCl$_4$	11.0
NbCl$_5$	LiAlCl$_4$	17.3
	NaAlCl$_4$	25.2

C. Heats,[b] Energies, and Entropies of Solution for Named Solutes with Apiezon M Solvent[18]

	T (°C)	$-\Delta\overline{H}_s$ (kcal/mole)	$-\Delta\overline{G}_s$ (kcal/mole)	$-\Delta\overline{S}_s$ (eu)
n-Hexane	125	5.78	2.24	8.9
Chloropentane	125	6.93	3.06	9.8
Capronitrile	125	7.88	3.45	11.2
iso-Butylene	80	4.32	1.06	9.1
Acetone	80	5.20	1.46	10.5

D. Heats,[b] Energies, and Entropies of Solution (76°C) for Named Solutes with n-Tetracosane and Di-n-Nonyl Ketone Solvents[18]

	n-Tetracosane			Di-n-nonyl Ketone		
Solute	$-\Delta\overline{H}_s$ (kcal/mole)	$-\Delta\overline{G}_s$ (kcal/mole)	$-\Delta\overline{S}_s$ (eu)	$-\Delta\overline{H}_s$ (kcal/mole)	$-\Delta\overline{G}_s$ (kcal/mole)	$-\Delta\overline{S}_s$ (eu)
n-Pentane	5.86	−0.56	18.4	5.76	−0.73	18.6
n-Hexane	6.84	0.05	19.4	6.85	−0.12	19.9
n-Heptane	7.96	0.66	20.9	7.90	0.49	21.2
n-Octane	9.00	1.27	22.2	8.98	1.09	22.6
Cyclohexane	7.05	0.45	18.9	7.02	0.29	19.3
Cyclohexene	7.20	0.51	19.2	7.26	0.40	19.6
1,3-Cyclohexadiene	7.00	0.41	18.9	7.29	0.39	19.8
Benzene	6.70	0.33	18.2	7.45	0.43	20.1
Chloropentane	7.65	0.80	19.6	8.47	0.94	21.6
Fluorohexane	7.37	0.39	20.0	8.00	0.51	21.5
1-Hexyne	6.45	−0.04	18.6	7.18	0.06	20.4
Valeronitrile	7.40	0.49	19.8	8.78	1.23	21.6
1-Heptene	7.49	0.56	19.8	7.62	0.46	20.5
Hexaldehyde	8.00	0.86	20.5	8.63	1.24	21.2
Di-n-propyl ether	7.28	0.41	19.7	7.76	0.38	21.1
Methylenecyclohexane	7.70	0.88	19.5	7.84	0.78	20.2
Cyclohexanone	8.31	1.32	20.0	...	...	...
Methylcyclohexane	7.66	0.84	19.5	7.67	0.67	20.0

Table 5.2 (*Continued*)

E. Heats[b] of Solution (25°C) for Named Solutes with Formamide Solvent[19]

Solute	$-\Delta \bar{H}_s$ (kcal/mole)
n-Nonane	8.81
Benzene	6.32
Cyclohexane	5.89

[a] V_g^0 values not corrected for surface adsorption; quoted data are probably accurate to $\pm 10\%$.
[b] V_g^0 values fully corrected for surface adsorption.

where η is the cubical expansion coefficient (approximately 10^{-3} ml/deg) of the liquid phase. Thus plots of $\ln V_g^0$ rather than $\ln K_R$ are preferred because the latter require knowledge of the stationary phase density at several column temperatures in order to derive thermodynamic quantities at the useful standard-state temperature of 0°C.

The examples of GLC-determined thermodynamic solution properties presented in Table 5.2 are indicative of the wide range of compound types to which the technique is applicable.

The accuracy of heats of solution as determined by gas chromatography depends only on the accuracy of the V_g^0 data which, as noted before, can be as good as $\pm 1\%$ even in the presence of gas-liquid interfacial adsorption. Further, provided that helium or hydrogen are used as mobile phases at moderate column pressures, virial corrections are unnecessary, for, as shown in Chapter 4, they are negligible for these carriers; they will, in any event, be approximately self-canceling because B_{12} data do not vary strongly with temperature and because the slopes of the plots (not the intercepts) are used in deriving thermodynamic information.

It may be supposed, given the foregoing, that only three data points (the minimum required for least squares) are necessary to define $\ln V_g^0$ as a function of the reciprocal of temperature. Streuli and co-workers[20] found that the accuracy of heats of solution derived via such plots improved only slightly when 10 points were taken instead of three. They illustrated the number of points used, values for the slope and intercept, and standard deviations for the least squares and experimentally measured V_g^0 data for their systems, which showed that to $\pm 2\%$ they were independent of the number of points taken above three. The scatter was in fact such that the determination of more than five points was not merited, although self-evidently on statistical grounds alone 10 would be preferable. Nonetheless, it is doubtful that the increase in accuracy gained would justify the added work involved nor indeed could it provide data better than the inherent accuracy of the V_g^0 values themselves, namely, 1–2%. Four or five points over a 10–30°C range are therefore adequate for these purposes.

5.1.2 Excess Properties: Activity Coefficients

Although thermodynamic properties of solution have been used in GC in attempts to clarify solute-solvent interactions (e.g., Refs. 21–28), numerous workers have argued that the activity coefficient and its related *excess* thermodynamic quantities provide a more exacting description in the sense that these can, to some degree, be correlated with intrinsic molecular properties and structure.

The fully corrected activity coefficient is determined in GLC via eqs. 4.27 and 4.28, as discussed in Chapter 4, and the molar excess enthalpy of solution $\Delta \bar{H}^e$ from the slope of plots of $\ln \gamma_1^\infty$ versus $1/T$:

$$\ln \gamma_1^\infty = \frac{\Delta \bar{H}^e}{RT} + C \qquad (5.10)$$

The excess energy and entropy are then calculated at each column temperature in the usual manner.

Plots of $\ln \gamma_1^\infty$ versus $1/T$ generally exhibit 1–5% experimental scatter in excess of that of $\ln V_g^0$ because vapor pressure, fugacity, and (except for helium or hydrogen carriers) virial data must be included in calculations of the activity coefficient at each temperature. Standard deviations of 2–3% on the experimental versus "smoothed" γ_1^∞ values are therefore not uncommon.

As mentioned at the outset, γ_1^∞ values have been reported for a wide range of solutes and solvents as exemplified by the data of Harris and Prausnitz[30] given in Table 5.3 (activity coefficients ranging from 0.45 to 30). Locke[31] has recently published a compendium of γ_1^∞ data (106 references) which appears to be more or less complete through 1972; at least 20 additional studies[32–49] have since been reported.

The accuracy of activity coefficients as measured by gas chromatography can be as good as 1% but obviously it depends directly on the reliability of the zero-pressure K_R or V_g values and, in addition, accurate fugacity and vapor pressure data; for example, Locke[31] compared GLC, static, and extrapolated (from finite-concentration GLC) γ_1^∞ data for about 60 systems and found the agreement to be, on average, $\pm 5\%$ (although some data were considerably worse, due most likely to the use of initial retention times, improper measurement of w_L, or incomplete correction for fugacity and virial effects). Recent studies[50, 51] indicate that ± 0.5–3% is a more realistic figure; for example, Martire, Laub, and Purnell[52] have measured the activity coefficients of aliphatic and aromatic hydrocarbons

Table 5.3 Fully Corrected Infinite-Dilution Activity Coefficients[30] at 80°C

	γ_1^∞			
Solute	Squalane	Squalene	n-Octa decane	Benzyl Diphenyl
Acetone	1.93	1.42	2.96	1.44
Methyl ethyl ketone	1.50	1.14	2.20	1.28
Methyl iso-propyl ketone	1.29	1.04	1.89	1.29
Cyclopentanone	1.64	1.13	2.46	0.92
n-Butyraldehyde	1.25	0.99	1.85	1.15
Ethyl acetate	1.27	0.94	1.86	1.26
n-Propyl acetate	1.17	0.89	1.69	1.22
iso-Propyl acetate	1.20	0.93	1.74	1.37
n-Butyl formate	1.18	0.91	1.70	1.27
Dimethyl carbonate	1.87	1.36	2.81	1.39
iso-Propyl alcohol	4.1	2.9	6.0	4.3
Acetonitrile	6.4	3.82	9.7	2.32
Propionitrile	4.4	2.78	6.5	1.86
Methoxyacetonitrile	5.6	3.35	8.5	2.15
Nitroethane	3.85	2.48	5.9	1.92
Trimethyl orthoformate	1.32	0.97	1.87	...
Dimethyl formamide	5.4	3.03	8.2	1.59
Furfural	5.5	3.21	8.1	2.17
Dimethyl sulfoxide	14.	7.	30.	...
Trimethyl phosphite	0.48	0.22	1.86	...
Trimethyl phosphate	7.2	3.3	12.	...
1,2-Dichloroethane	0.89	0.66	1.33	0.81
Carbon disulfide	0.45	0.40	0.65	0.81
Di-iso-propyl ether	0.74	0.72	1.01	1.97
Benzene	0.590	0.499	0.828	0.849
1-Hexene	0.61	0.63	0.84	1.91
n-Hexane	0.61	0.69	0.84	2.36

with squalane and di-nonyl phthalate at 30°C and a comparison of their data with those of Ashworth[53] (derived from static techniques for solvents obtained from the same source) is made in Table 5.4. The high degree of agreement was said to be a consequence of very careful measurement of the weight percent of liquid phase.[54]

In addition to activity coefficients and excess heats, gas chromatographic data may be used to determine excess heat capacities at infinite dilution of the solute C_p^e:

$$\left(\frac{\partial \Delta \bar{H}^e}{\partial T}\right)_p = C_p^e \tag{5.11}$$

Table 5.4   Comparison[52] of Fully Corrected Infinite-Dilution Activity Coefficients for Aliphatic and Aromatic Hydrocarbons with Squalane and Di-nonyl Phthalate at 30°C

| Solute | γ_1^∞, Squalane | | γ_1^∞, Di-nonyl Phthalate | |
	GLC ($\pm 0.9\%$)	Static	GLC ($\pm 0.9\%$)	Static
n-Pentane	0.617	0.620 ± 0.002	1.114	1.115 ± 0.001
n-Hexane	0.642	0.640 ± 0.002	1.203	1.201 ± 0.002
n-Heptane	0.670	0.669 ± 0.004	1.300	1.313 ± 0.007
n-Octane	0.700	0.700 ± 0.007	1.405	1.416 ± 0.007
Cyclohexane	0.515	0.510 ± 0.005	0.932	0.931 ± 0.003
Methylcyclohexane	0.529	...	1.003	...
Benzene	0.701	0.698 ± 0.007	0.552	0.549 ± 0.005
Toluene	0.691	...	0.577	...

Alessi and co-workers[38, 39] found that plots of $\ln\gamma_1^\infty$ versus $1/T$ for aliphatic and aromatic solutes with phthalate ester solvents were curved, indicating that $\Delta\bar{H}^e$ was not constant over the temperature range investigated (25–125°C). By combining eqs. 5.10 and 5.11 the activity coefficient was expressed as

$$\ln\gamma_1^\infty = \left(-\frac{C_p^e}{R} \right) \ln T + \frac{A}{RT} + B \qquad (5.12)$$

and nonlinear least squares were used to fit their data; a portion of the results is given in Table 5.5.

Table 5.5   Excess Heat Capacities for Hydrocarbons with Phthalate Ester Stationary Phases[39]

| Solvent | $-C_p^e$ (cal/mole) | | | |
	n-Heptane	1-Heptene	Methyl-cyclohexane	Toluene
Diethyl phthalate	11.8	10.7	6.24	7.11
Di-n-butyl phthalate	6.88	3.58	4.13	5.76
Di-iso-butyl phthalate	19.2	9.90	9.88	9.78
Dicyclohexyl phthalate	8.78	7.69	7.71	3.58
Di-2-ethylhexyl phthalate	10.4	4.83	6.24	7.21
Di-iso-decyl phthalate	11.3	10.5	9.16	9.36
Butyl benzyl phthalate	10.7	8.60	7.19	5.35
Butyl 2-ethylhexyl phthalate	6.32	3.74	3.22	5.23

5.1.3 Interaction Parameters: Current Topics in Solution Theories

Among the variety of attempts to reconcile deviations from Raoult's law, the most recent and successful approaches arise from the solution theories proposed by Huggins,[55] Flory and co-workers,[56-59] Tompa,[60] Guggenheim,[61] Prigogine,[62] and others.[63-71] Several were developed for straight-chain polymer mixtures and as a result require little modification for solutions of n-alkanes.

In the simplest model, termed quasi-lattice theory, the solute activity coefficient is said to be the sum of two terms:

$$\ln \gamma_1^\infty = \ln \gamma_1^\infty (\text{config}) + \ln \gamma_1^\infty (\text{interact}) \tag{5.13}$$

The configurational (or combinatorial) portion is pictured as due to the solute and solvent occupying sites of a lattice, each site being taken up by equivalent-sized segments. A parameter, r, is defined as the ratio of sites occupied by the solvent and solute. When the coordination number Z of the lattice is allowed to approach infinity, the rigidity of the lattice is destroyed such that the model approaches an unstructured fluid. According to Flory theory, the configurational portion of the infinite-dilution activity coefficient is then given by

$$\ln \gamma_1^\infty (\text{config}) = \ln \left(\frac{1}{r} \right) + (1 - r) \tag{5.14}$$

Ashworth and Everett[72] and Cruickshank and co-workers[73] have argued that for n-alkanes r is best given by the ratio of the solvent: solute molar volumes because Z has been allowed to approach infinity. Others[74] have chosen the ratio of the number of segments in the solvent and solute, equal segments having been designated variously as $-CH_3/-CH_2-$, $-CH_3/$ $-CH_2-CH_2-$, or $-CH_2-CH_3/-CH_2-CH_2-$. An examination of molar volume ratios indicates that the second of these most closely approximates r (however, see below); for species other than n-alkanes molar volumes have been employed.

As for the interactional (structural plus energetic) contribution to γ_1^∞, van Laar[75] originally proposed that van der Waals constants be incorporated into expressions relating the excess Gibbs free energy to size differences between molecules. Dolezalek[76] adopted an alternative view, namely, that mixed solution interactions (not volumes) were responsible for deviations from Raoult's law. Following a somewhat polemical reply from van Laar,[77] Hildebrand[78] introduced the notion that a large number of mixtures which exhibit a positive excess heat of mixing appear to obey a semiempirical "regular" variation of activity coefficient with mole fraction. Heitler[79] later arrived at the same conclusion from what might well be called quasi-lattice considerations. Then, in 1931, Scatchard[80] proposed the following:

(a) Pair-wise intermolecular interactional energies depend only on the relative distance and orientation between molecules.
(b) Molecules are randomly distributed throughout a solution.
(c) In all but extreme cases the excess volume of mixing can be taken to be negligible.

These approximations lead to the relation[80,81]:

$$\Delta \overline{G}^e = A_{12} \overline{V}_s \phi_1 \phi_2 \qquad (5.15)$$

where $\overline{V}_s$ is the molar solution volume, ϕ_1 and ϕ_2 are the volume fractions of components 1 and 2, and A_{12} is defined as

$$A_{12} = \left(\frac{\Delta \overline{E}_{v_2}}{\overline{V}_2} \right)^{\frac{1}{2}} - \left(\frac{\Delta \overline{E}_{v_1}}{\overline{V}_1} \right)^{\frac{1}{2}} \qquad (5.16)$$

where $\Delta \overline{E}_v$ is the molar energy of vaporization (approximately, $\Delta \overline{H}_v - RT$). Equations 5.15 and 5.16 lead, in turn, to an expression for the interactional contribution to the activity coefficient which is of the form

$$\ln \gamma_1 \text{ (interact)} = \frac{v_1^0}{RT} A_{12} \phi_2^2 \qquad (5.17)$$

The terms $(\Delta \overline{E}_v / \overline{V})^{\frac{1}{2}}$ and $v_1^0 A_{12}/RT$ are better known today as the solubility parameter δ and the interaction parameter χ. Equation 5.17 may thus be written in the limit of infinite dilution as

$$\ln \gamma_1^\infty \text{ (interact)} = \chi = \frac{v_1^0 (\delta_2 - \delta_1)^2}{RT} \qquad (5.18)$$

Combination of eqs. 5.14 and 5.18 produces a crude first approximation to the description of solute-solvent interactions:

$$\ln \gamma_1^\infty = \ln \left(\frac{1}{r} \right) + \left(1 - \frac{1}{R} \right) + \chi \qquad (5.19)$$

Cruickshank and co-workers[73] tested eq. 5.19 with several solutes and n-octadecane liquid phase at 35°C; their calculated and experimental χ

values are given in columns 4 and 5 of Table 5.6 and confirm[82] that for n-alkane/n-alkane mixtures "regular" solution theory is of limited utility. Modifications (e.g., Refs. 83, 84) of the method of calculation of solubility parameters have also met with little success.

The quasi-lattice theory has produced more encouraging results[73] when each lattice point is considered to consist of one or more kinds of "contact" points.[85-87] For hydrocarbons the contact points are considered to be hydrogen atoms among which, for n-alkanes, two types can be identified, namely, —CH$_3$ (A) and —CH$_2$— (B). Thus n-alkane/n-alkane interactions are of the type $A, B/A, B$; that is, two types of contact point. The fraction of methyl-group hydrogens in an n-alkane is given by

$$\theta_i' = \frac{3}{n_i + 1} \tag{5.20}$$

and the interaction parameter is then said to be

$$\chi = \frac{c_1(\theta_1' - \theta_2')^2 \omega'}{RT} \tag{5.21}$$

where c_1 is the sum of all contact points in the solute and ω' is the

Table 5.6 Comparison of Experimental and Calculated Interaction Parameters[73]

Solute	$\Delta \bar{H}_v$ (kcal/mole)	v_1^0 (ml/mole)	δ (cal/ml)$^{\frac{1}{2}}$	χ (calculated)	χ (experimental)
n-Butane	4.443	101.43	6.6185	0.332	0.341
n-Pentane	5.724	116.104	7.0215	0.194	0.246
n-Hexane	6.962	131.598	7.2735	0.125	0.177
n-Heptane	8.143	147.456	7.4312	0.088	0.131
n-Octane	9.323	163.530	7.5504	0.040	0.101
iso-Pentane	5.286	117.383	6.7106	0.335	0.269
2,2-Dimethylbutane	6.058	133.712	6.7309	0.371	0.242
2,3-Dimethylbutane	6.392	131.156	6.9811	0.238	0.131
2-Methylpentane	6.567	132.875	7.0300	0.219	0.209
3-Methylpentane	6.662	130.611	7.1418	0.170	0.182
2,4-Dimethylpentane	7.268	149.925	6.9626	0.282	0.196
1-Pentene	5.744	110.384	7.211	0.127	0.228
1-Hexene	6.848	125.892	7.376	0.094	0.216
1-Heptene	7.980	141.744	7.503	0.070	0.150
1-Octene	9.143	157.850	7.609	0.032	0.129
Cyclohexane	7.303	108.744	8.1946	0.045	0.295
Benzene	7.498	89.399	9.1582	0.185	0.553
n-Octadecane	...	...	8.050	...	...

"interchange" energy per contact point; that is, the energy required to transfer a hydrogen from a methylene unit to a methyl end-group. Combining eqs. 5.20 and 5.21, we have

$$\chi = 18(n_1 + 1)\left(\frac{1}{n_1 + 1} - \frac{1}{n_2 + 1}\right)^2 \frac{\omega'}{kT} \tag{5.22}$$

ω' is expected to be a constant for n-alkanes but this was not found to be the case by Cruickshank and co-workers[73]; their experimental data for ω'/kT for n-alkanes with n-octadecane showed a 3% variation which became larger if, instead of n-octadecane, n-hexadecane was employed as the solvent (column 1 of Table 5.7).

More plausible[73] than "contact" points, "segment-contact" theory assumes that an entire unit, for example, $-CH_2-CH_2-$ or $-CH_3$, is involved in intermolecular interactions. The fraction of end segments is,

Table 5.7 Values[73, 88] of ω'/kT ("Contact Point" Theory) and ω/kT ("Segment Contact" Theory) for n-Alkane Systems at 35°C

Solute	Solvent	ω'/KT	ω/kT
n-Butane	n-Octadecane	0.1743 ± 0.0037	0.520 ± 0.011
n-Pentane		0.1755 ± 0.0051	0.511 ± 0.014
n-Hexane		0.1729 ± 0.0071	0.493 ± 0.018
n-Heptane		0.1741 ± 0.0090	0.488 ± 0.023
n-Octane		0.1828 ± 0.0183	0.507 ± 0.034
n-Butane	n-Hexadecane	0.1616 ± 0.0040	0.488 ± 0.013
n-Pentane		0.1870 ± 0.0056	0.551 ± 0.017
n-Hexane		0.1630 ± 0.0059	0.485 ± 0.020
n-Heptane		0.1861 ± 0.0099	0.528 ± 0.030
n-Pentane	n-Heneicosane	0.1686 ± 0.0058	0.482 ± 0.019
n-Hexane		0.1795 ± 0.0067	0.508 ± 0.019
n-Heptane		0.1922 ± 0.0086	0.540 ± 0.029
n-Butane	n-Docosane	0.1569 ± 0.0037	0.480 ± 0.012
n-Pentane		0.1549 ± 0.0054	0.450 ± 0.018
n-Hexane		0.1621 ± 0.0072	0.435 ± 0.021
n-Pentane	n-Octacosane	0.1699 ± 0.0055	0.466 ± 0.020
n-Hexane		0.1693 ± 0.0076	0.437 ± 0.026

for n-alkanes, given by

$$\theta_i = \frac{2}{1 + n_i/2} \tag{5.23}$$

whence

$$\chi = \left(1 + \frac{n_1}{2}\right)(\theta_1 - \theta_2)^2 \frac{\omega}{kT} = 8(n_1 + 2)\left(\frac{1}{n_1 + 2} - \frac{1}{n_2 + 2}\right)^2 \frac{\omega}{kT} \tag{5.24}$$

where ω is now defined as the energy of interchange of "end" and "middle" units. Values of ω/kT are given in column 2 of Table 5.7, where, superficially, the constancy appears to be no better than that for ω'/kT. Refinements to this approach (e.g., Refs. 89–92) have not produced significant improvement. As a result Hicks and Young[88] were led to apply the more comprehensive Prigogine[62] and Flory-Orwoll-Vrij[56-58] theories to n-alkane mixtures. The first of these subdivides $\ln \gamma_1^\infty$ (interact) into further structural and energetic terms: the energetic contribution is described by the molecular parameters

$$\delta = \frac{\varepsilon_{22}^* - \varepsilon_{11}^*}{\varepsilon_{11}^*} \tag{5.25}$$

$$\theta = \left\{ \varepsilon_{11}^* \left[\varepsilon_{12}^* - \tfrac{1}{2}(\varepsilon_{11}^* + \varepsilon_{22}^*) \right] \right\}^{-1} \tag{5.26}$$

$$\rho = \frac{r_{22}^* - r_{11}^*}{r_{11}^*} \tag{5.27}$$

where r^* and ε^* are said to be segment length and interactional scale factors. The structural contribution takes into account the number of external degrees of freedom of each segment $3c$, which is related to the number of segments r by

$$3c = 3 + r \tag{5.28}$$

The number of nearest neighbors Z is given by

$$qZ = rZ - 2r + 2 \tag{5.29}$$

Only the ratio c/q is important, however, and when $\delta = \theta = \rho = 0$

$$\chi = \frac{C_{p_1}}{2R}\left(\frac{c_1}{q_1} - \frac{c_2}{q_2}\right)^2 \tag{5.30}$$

where C_p is the solute isobaric molar heat capacity given, for n-alkanes, by the empirical relation [93]

$$C_{p_i} = \tfrac{1}{2}(\text{const})(n+1) \tag{5.31}$$

where n is the number of carbon atoms in the alkane. Combining eqs. 5.29–5.31 and allowing Z to approach infinity gives

$$\chi = A(n_1+1)\left(\frac{1}{n_1+1} - \frac{1}{n_2+1}\right)^2 \tag{5.32}$$

which is similar to the result obtained for the contact-point theory (eq. 5.22). Hicks and Young[88] found, however, that the heat capacities predicted from eq. 5.32 were small by a factor of 2; application of the more complex Flory-Orwoll treatment by these workers produced similarly disappointing results.[88]

Janini and Martire[94] reported an extension of the Prigogine solution theory in 1974 which appears to be considerably more successful. First they allowed Z to approach infinity as usual, but then noted that, according to eq. 5.29, $q \rightarrow r$, so that terms in mole fraction of either solution component may be replaced by the volume fraction alternative. Further, the average values of c and q, $\bar{c}$ and $\bar{q}$, may be represented as

$$\bar{c} = x_1 c_1 + x_2 c_2 \tag{5.33}$$

$$\bar{q} \equiv \bar{r} = x_1 r_1 + x_2 r_2 \tag{5.34}$$

ρ (eq. 5.27) was then chosen to be zero, thus ensuring equal hard-core volumes per segment (however they are defined) and θ, taken to be $-\delta^2/8$. Combining all the above with Prigogine's expression (eq. 17.6.1 in Ref. 62) for the excess Gibbs free energy produced

$$\chi = \frac{C_{p_1}}{2R}(\lambda + \delta)^2 - \frac{h_1}{4RT}\delta^2 \tag{5.35}$$

where h_1 is the solute configurational energy and

$$\lambda = 1 - \frac{c_2/r_2}{c_1/r_1} \tag{5.36}$$

Equation 5.35 acknowledged[94] as formally similar to a result obtained by Patterson[95] from a reconsideration of Hijman's modification[96,97] of the

Prigogine theory, was shown to be a useful simplification: the first term on the rhs represents structural contributions, χ_S, and the second the energetic effects, χ_E, constituting χ. Evaluation of the parameters r, δ, and λ, was, moreover, comparatively straightforward. The terms ε_{ii}^*, identified as the segment potential-well depth of species i, were known following Kreglewski[98] to be proportional to $T_i^c/(V_i^*)^{1/3}$, the critical temperature and hard-core molar volume of i, so that

$$\delta = \frac{T_2^c/(V_2^*)^{\frac{1}{3}}}{T_1^c/(V_1^*)^{\frac{1}{3}}} - 1 \tag{5.37}$$

Among the variety of methods of calculating V_i^* the simplest appeared to be the use of plots of V_i^* versus carbon number from which it was found for $n_i \geqslant 2$ that

$$V_i^* = 15.52 + 19.68 n_i \tag{5.38}$$

An indication of the accuracy of this relation, for example, was that the ratio V_i^*/V_8^* (i.e., to n-octane) was accurate to $\pm 0.7\%$ over the range $6 \leqslant n_i \leqslant 50$. Analysis of the available data also showed that $T_i^c/(V_i^*)^{\frac{1}{3}}$ ranged from $99.5°K$ cm^{-1} $mole^{\frac{1}{3}}$ for n-hexane to a limiting value at n-tridecane of $104.4°K$ cm^{-1} $mole^{\frac{1}{3}}$, indicating that δ values will fall between zero and a maximum of about 0.05.

Segment assignment was chosen on the basis that $r = 2$ for propane, giving 0.736 segments per —CH_3 group and 0.528 segments per methylene unit. The hard-core segment volume was then calculated to be 37.28 ml/mole from eq. 5.38 and density data. The number of segments for each alkane could as a result be determined from

$$r_i = 2.000 + 0.528(n_i - 3) \tag{5.39}$$

The ratio c_i/r_i could not be predicted a priori and a reiterative ("best-fit") method utilizing experimental γ_1^∞ data was required; the result was

$$\frac{c_i}{r_i} = 0.550 + \frac{1.817}{r_i} - \frac{1.834}{r_i^2} \tag{5.40}$$

Finally, the solute configurational energy h_1 and heat capacity C_{p1} were calculated from latent heats of vaporization and liquid- and vapor-phase heat capacities.[93,99]

Table 5.8 illustrates the success of the Janini-Martire approach: $\ln\gamma_1^\infty$ data calculated via the foregoing relations agree with experimental values (tabulated from other sources) to ± 0.006, on average. Further, the best agreement yet obtained was found for experimental and calculated activity coefficients at finite concentration and excess enthalpies and volumes at equimolar mixtures of solute and solvent, where, perhaps surprisingly, the infinite-dilution relation for c_i/r_i (eq. 5.40) proved to be adequate.

Parcher and Yun[36] reported similarly excellent correlations, using the theory described and a modified version of eq. 5.40, said to be applicable over a wider range of n-alkane mixtures:

$$\frac{c_i}{r_i} = 0.534 + \frac{1.839}{r_i} - \frac{1.813}{r_i^2} \tag{5.41}$$

The differences between their GLC-measured and calculated $\ln\gamma_1^\infty$ data

Table 5.8 Comparison of Experimental and Calculated Activity Coefficients[94]

Solute	Solvent	T (°C)	$-\ln\gamma_1^\infty$ Experimental	Calculated	Calculated − Experimental
n-Hexane	n-Hexadecane	20	0.099	0.092	0.007
		30	0.101	0.096	0.005
		40	0.094	0.099	−0.005
		50	0.110	0.109	0.001
		60	0.117	0.110	0.007
	n-Octadecane	35	0.131	0.135	−0.004
	n-Eicosane	40	0.161	0.175	−0.014
	n-Docosane	60	0.219	0.221	−0.002
	n-Tetracosane	60	0.253	0.255	−0.002
	n-Octacosane	70	0.309	0.307	0.002
n-Heptane	n-Hexadecane	30	0.074	0.071	0.003
	n-Octadecane	35	0.113	0.104	0.009
	n-Eicosane	40	0.134	0.139	−0.005
	n-Tetracosane	60	0.207	0.213	−0.006
	n-Dotriacontane	75	0.364	0.350	0.014
n-Octane	n-Octadecane	35	0.090	0.083	0.007
	n-Tetracosane	60	0.184	0.189	−0.005
	n-Dotriacontane	75	0.325	0.330	−0.005

for six solutes (*n*-butane through *n*-octane; *n*-decane) with seven *n*-alkane solvents at three temperatures is shown in Table 5.9. With the exception of the data for n-$C_{22}H_{46}$, the overall average error is ± 0.008, that is, close to $\pm 1\%$ on $\ln \gamma_1^\infty$, a difference that defies measurement by gas chromatography. Of equal significance, FAGC-determined activity coefficients were also found to be in agreement with those calculated as in Table 5.10.

Table 5.9 Average Error $\left[\dfrac{1}{6} \displaystyle\sum_{i=1}^{6} (\text{Calculated} - \text{Experimental}) \right]$ $\times 100$ for $\ln \gamma_i^\infty$ Data for *n*-Alkane Solutes with *n*-Alkane Solvents[36]

	Error × 100		
Solvent	80°C	100°C	120°C
n-Docosane	4.1	3.9	3.6
n-Tetracosane	1.0	1.0	0.8
n-Octacosane	0.7	0.7	0.7
n-Triacontane	0.9	0.9	1.0
n-Dotriacontane	0.9	0.8	0.8
n-Tetratriacontane	0.2	0.3	0.8
n-Hexatriacontane	0.9	0.8	0.9

Table 5.10 Comparison of Calculated (Eq. 5.41) and Experimental (Finite-Concentration) $\ln \gamma_1$ Data for *n*-Hexane with *n*-Alkane Solvents[36]

Solvent	T (°C)	x_1^L	$-\ln \gamma_1$ Calculated	Experimental (FAGC)	Δ (×100)
n-Tetracosane	60	0.00	0.233	0.219	−1.4
		0.07	0.190	0.180	−1.0
		0.25	0.153	0.137	−1.6
		0.27	0.141	0.164	2.2
		0.29	0.133	0.127	−0.6
		0.34	0.098	0.120	2.3
		0.43	0.079	0.089	1.0
	80	0.00	0.234	0.222	−1.2
		0.04	0.261	0.211	−5.0
		0.12	0.202	0.203	0.1
		0.20	0.171	0.188	1.7
		0.27	0.142	0.137	−0.5
		0.34	0.116	0.122	0.6
		0.41	0.095	0.100	0.5
		0.42	0.090	0.111	2.1

Table 5.10   (*Continued*)

	100	0.00	0.230	0.222	−0.8
		0.03	0.177	0.228	5.1
		0.09	0.185	0.210	2.5
		0.10	0.183	0.192	1.1
		0.21	0.161	0.168	0.7
		0.32	0.120	0.149	2.8
n-Octacosane	80	0.00	0.306	0.307	0.1
		0.04	0.287	0.290	0.3
		0.05	0.286	0.254	−3.2
		0.17	0.249	0.235	−1.1
		0.20	0.219	0.227	0.8
		0.29	0.184	0.180	0.4
		0.33	0.169	0.189	2.0
		0.35	0.164	0.168	0.4
		0.37	0.154	0.144	−1.0
		0.43	0.125	0.145	2.0
n-Dotriacontane	80	0.00	0.375	0.373	−0.2
		0.17	0.299	0.298	−0.1
		0.18	0.280	0.318	3.8
		0.23	0.268	0.229	−3.8
		0.26	0.247	0.255	0.8
		0.39	0.184	0.204	2.0
		0.41	0.179	0.234	5.5
		0.46	0.152	0.105	−4.7

Martire[100] later modified eq. 5.40 to a two-parameter relation

$$\frac{c_i}{r_i} = 0.477 + \frac{1.046}{r_i} \tag{5.42}$$

Calculated and experimental $\ln \gamma_1^\infty$ data based on this expression are given in Table 5.11. (The Janini-Martire method has lately been extended to ternary *n*-alkane mixtures, a discussion of which is deferred to Chapter 6 where its relevance to complexation is of particular significance.)

Table 5.12 presents χ data for systems that at this time are beyond the scope of current theories. Given the success of the Janini-Martire approach, however, anticipation that they, too, may soon prove tractile does not seem to be wholly without foundation.

Table 5.11 Comparison[100] of Calculated (Eq. 5.42) and Experimental γ_1^∞ Data for
n-Alkane Solutes with n-Alkane Solvents at 80°C

	γ_1^∞					
	n-Tetracosane		n-Triacontane		n-Hexatriacontane	
Solute	Calculated	Experimental	Calculated	Experimental	Calculated	Experimental
n-Hexane	0.792	0.792	0.702	0.698	0.628	0.633
n-Heptane	0.812	0.815	0.725	0.720	0.652	0.650
n-Octane	0.834	0.849	0.750	0.736	0.678	0.679
n-Nonane	0.856	0.877	0.777	0.766	0.706	0.716

Table 5.12 Examples of Interaction Parameters χ, Determined by GLC

A. Various Solutes with Alkane and Di-nonyl Phthalate Solvents[101]

Solute	Solvent	T (°C)	χ
Benzene	Squalane	50.0	0.504
	Di-nonyl phthalate	50.0	0.184
	n-Octadecane	50.0	0.486
n-Octane	n-Octadecane	40.0	0.007
	Squalane	65.05	0.182
2,2,4-Trimethylpentane	n-Octadecane	40.00	0.332
	Squalane	65.05	1.142

B. Olefin and Diene Solutes with n-Alkane Solvents[91]

			χ (av)
1-Pentene	n-Octadecane	35	0.287
1-Hexene	n-Octadecane	35	0.215
		40	0.213
		50	0.198
1-Heptene	n-Octadecane	35	0.162
1-Octene	n-Octadecane	35	0.129
1-Hexene	n-Tetracosane	60	0.267
1-Heptene	n-Tetracosane	60	0.217
1,3-Pentadiene	n-Octadecane	35	0.406
1,3-Hexadiene	n-Octadecane	35	0.305

Table 5.12 (*Continued*)

C. Various Solutes[102] with *n*-Eicosane and Squalane Solvents at 93.9°C

| | χ | |
	n-Eicosane	Squalane
Benzene	0.401	0.421
Toluene	0.313	0.335
Cyclopentane	0.162	0.150
Methylcyclopentane	0.151	0.136
Cyclohexane	0.139	0.117
Cyclohexene	0.168	0.159
2-Methyl-2-butene	0.205	0.206
1-Pentene	0.234	0.235
4-Methyl-1-pentene	0.206	0.201
2-Methyl-1-pentene	0.171	0.163
1-Hexene	0.220	0.222
1-Heptene	0.204	0.203
iso-Pentane	0.223	0.215
n-Pentane	0.219	0.219
2,2-Dimethylbutane	0.198	0.187
2,3-Dimethylbutane	0.181	0.165
2-Methylpentane	0.216	0.204
n-Hexane	0.197	0.196
2,2-Dimethylpentane	0.211	0.193
2,4-Dimethylpentane	0.219	0.201
2,3-Dimethylpentane	0.161	0.143
3-Methylhexane	0.180	0.176
n-Heptane	0.194	0.187
Iodomethane	0.739	0.759
Dichloromethane	0.767	0.795
Chloroform	0.455	0.484
Carbon tetrachloride	0.213	0.228
Dibromomethane	0.952	0.977
Bromodichloromethane	0.516	0.556
Bromotrichloromethane	0.266	0.288
cis-1,2-Dichloroethene	0.646	0.652
trans-1,2-Dichloroethene	0.375	0.386
Trichloroethylene	0.300	0.314
1,2-Dichloroethane	0.918	0.906
1,1-Dichloroethane	0.622	0.632
1-Bromo-2-chloroethane	0.973	0.972
1-Chloropropane	0.503	0.493
1-Chlorobutane	0.429	0.423
2-Chlorobutane	0.420	0.417

Table 5.12 (*Continued*)

D. Fluorinated Aromatic Hydrocarbon Solutes with *n*-Octadecane Solvent[103]

	T (°C)	χ
Benzene	32	0.564
	40	0.522
	50	0.484
Fluorobenzene	32	0.733
	40	0.691
	50	0.648
1,2-Difluorobenzene	50	0.901
1,3-Difluorobenzene	50	0.759
1,4-Difluorobenzene	50	0.886
1,3,5-Trifluorobenzene	32	0.954
	40	0.910
	50	0.864
1,2,3,4-Tetrafluorobenzene	50	1.039
1,2,4,5-Tetrafluorobenzene	50	1.146
Pentafluorobenzene	50	1.079
Hexafluorobenzene	32	1.143
	40	1.099
	50	1.050

E. Alkane Solutes with 1-Phenylalkane Solvents[104] at 40°C

	χ	
	1-Phenyldodecane	1-Phenylpentadecane
n-Pentane	0.369	0.389
n-Hexane	0.325	0.320
n-Heptane	0.286	0.292
n-Octane	0.250	—
2-Methylbutane	0.420	0.438
2-Methylpentane	0.358	0.373
3-Methylpentane	0.331	0.341
2,3-Dimethylbutane	0.335	0.348
2,2-Dimethylbutane	0.398	—
2-Methylhexane	0.327	0.371
3-Methylhexane	0.302	0.299
2,3-Dimethylpentane	0.273	0.322
2,4-Dimethylpentane	0.367	0.304
3,3-Dimethylpentane	0.272	0.281
2,2,3-Trimethylbutane	0.302	0.373
2,2,3,3-Tetramethylbutane	—	0.269

Table 5.12 (*Continued*)

F. Halogenated Alkane Solutes with *n*-Alkane Solvents[105] at 82°C

	χ		
	n-Tetracosane	*n*-Triacontane	*n*-Hexatriacontane
1-Chlorobutane	0.489	0.499	0.507
1-Chloropentane	0.422	0.420	0.444
1-Chlorohexane	0.367	0.358	0.382
1-Chloroheptane	0.314	0.306	0.325
1,1-Dichloroethane	0.725	0.736	0.749
1,2-Dichloroethane	1.018	1.023	1.010
1,2-Dichloropropane	0.748	0.751	0.756
1,2,3-Trichloropropane	1.055	1.039	1.036
1-Bromopropane	0.537	0.541	0.541
1,2-Dibromopropane	0.861	0.854	0.854
1,3-Dibromopropane	1.411	1.383	1.377

G. Aliphatic and Aromatic Hydrocarbon Solutes[92]
with *n*-Alkane Solvents at 82°C

n-Hexane	0.229	0.248	0.287
2-Methylpentane	0.254	0.276	0.305
2,2-Dimethylbutane	0.259	0.289	0.317
2,3-Dimethylbutane	0.226	0.244	0.281
3-Methylpentane	0.222	0.245	0.286
n-Heptane	0.185	0.196	0.229
3-Methylhexane	0.192	0.205	0.242
2,4-Dimethylpentane	0.240	0.265	0.304
n-Octane	0.166	0.161	0.199
3-Methylheptane	0.174	0.171	0.214
n-Nonane	0.143	0.138	0.193
1-Hexene	0.253	0.279	0.309
1-Heptene	0.215	0.226	0.263
1-Octene	0.190	0.187	0.237
1-Nonene	0.162	0.157	0.203
Benzene	0.454	0.446	0.471
Toluene	0.345	0.331	0.356
p-Xylene	0.246	0.227	0.242
m-Xylene	0.273	0.255	0.270
o-Xylene	0.304	0.281	0.301
Ethylbenzene	0.324	0.306	0.327
Cumene	0.292	0.283	0.301
Chlorobenzene	0.478	0.460	0.473

5.1.4 Properties of Liquid Crystals The study of liquid crystals has only just begun as an active area of research in GLC but it is one that promises to be fruitful. Three texts by Gray[106, 107] review in detail the physicochemical properties of these materials and Karger,[108] McCrea,[109] and Schroeder[110] have briefly discussed their investigation by gas chromatography. An A.C.S. monograph[111] has become available on the topic and papers presented at a relevant NATO symposium have been published.[112] More recently, and in addition to the determination of phase changes via plots of $\ln V_g^0$ versus $1/T$ (cf. Refs. 109, 113), Martire and co-workers[114–121] have described and discussed plots of retention data versus reciprocal temperature reduced to that of a transition point or region, which produced novel correlations. Kelker and Winterscheidt[122] reported that data for a wide range of liquid crystals could be represented by a common plot of $\ln V_g$ versus $1/\theta$ where θ is a mesomorphic-phase transition temperature. Kelker[123] rationalized this behavior by employing expressions for γ_1^∞ in terms of mass rather than mole fraction. Diffusion and mass transfer effects as a function of heating and cooling temperatures have been studied by Grushka and Solsky[124] and partition isotherms of n-decane with 4,4′-dihexyloxyazoxybenzene (DHAB) have been determined by finite-concentration GC and calorimetry by Bocquet and Pommier.[125]

Of perhaps greater interest, GLC has been used to measure the marked differences in thermodynamic properties of solutes with mesogen phases; for example, in 1968 Martire and co-workers[126] reported excess heats and entropies of solution for aliphatic and aromatic solutes with cholesteryl myristate; the data, given in Table 5.13, show interesting variations on passing through the transitions, smectic → cholesteric → isotropic. Kraus and co-workers[127] found that heats and entropies of solution for a variety of solutes varied in a zig-zag fashion with a homologous series of 4-n-pentylacetophenone-(O-4-n-alkoxybenzoyloxime) liquid crystals. Barry and co-workers[128, 129] measured the thermodynamic properties of (p-ethoxyphenylazo)phenyl heptanoate (EPAPH) and cholesteryl palmitate with aliphatic and aromatic hydrocarbons and esters. A portion of their γ_1^∞ data is shown in Table 5.14.

In view of the orientational effects of solid supports discussed in Section 2.2.5, the transition temperatures of liquid crystals might well be presumed to be altered when coated, for example, onto Chromosorb. Martire and co-workers[126] reported that there were small differences in "bulk" versus "thin-layer" transition temperatures for cholesteryl myristate; Barry and co-workers[128, 129] found that the same was true for their systems, although the discrepancies noted were on the order of two degrees or less. Chow and Martire[130] later showed that the transition temperature of p-azoxyanisole (PAA) and DHAB were accurate to $\pm 1°C$, whether measured with bulk or

Table 5.13 Excess Enthalpies and Entropies for Named Solutes with Cholesteryl Myristate Stationary Phase[126]

Solute	$\Delta \bar{H}^e$ (kcal/mole)			$\Delta \bar{S}^e$ (eu)		
	Smectic (74°C)	Cholesteric (80°C)	Isotropic (86°C)	Smectic (74°C)	Cholesteric (80°C)	Isotropic (86°C)
n-Hexane	1.94	9.66	1.97	4.88	27.04	5.53
n-Heptane	2.72	10.87	1.37	7.02	30.41	3.76
n-Octane	3.77	11.67	2.09	9.98	32.61	5.73
n-Nonane	3.95	9.82	1.52	10.39	27.32	4.04
o-Xylene	3.42	11.90	1.43	9.87	34.28	4.87
m-Xylene	3.57	13.67	1.13	10.29	39.26	4.00
p-Xylene	2.63	13.43	1.29	7.72	38.66	4.51

support-coated materials. They also found a small contribution to retention, due to adsorption at the gas-liquid interface for aliphatic and aromatic hydrocarbons solutes with PAA (none for DHAB), that became negligible above 15% w/w loadings on Chromosorb W.

To confirm unambiguously the validity of the GC technique, Peterson, Martire, and Lindner[131] measured the (finite-concentration) activity coefficient of n-heptane with DHAB at 90.1°C (nematic region) with a McBain balance over a solute mole fraction range of 0.008 to 0.080. The (extrapolated) infinite-dilution value was 3.49±0.02, whereas GLC gave 3.54± 0.05. Thus GLC and nonchromatographic techniques appear to yield data for liquid crystals that are, within the respective error limits, identical.

Table 5.14 Infinite-Dilution Activity Coefficients for Named Solutes with Liquid Crystal Solvents

A. EPAPH Solvent[128]

Solute	γ_I^∞	
	Nematic (106°C)	Isotropic (121°C)
n-Heptane	0.977	0.628
n-Octane	1.917	1.018
n-Nonane	3.710	2.111
Benzene	0.333	0.228
Toluene	0.571	0.387
Ethylbenzene	0.956	0.658

Table 5.14 (*Continued*)

Solute	γ_I^∞	
	Nematic (106°C)	Isotropic (121°C)
o-Xylene	0.994	0.701
p-Xylene	0.891	0.643
cis-4-Methylcyclohexyl acetate	5.035	3.309
trans-4-Methylcyclohexyl acetate	4.263	2.769
cis-4-Methylcyclohexyl propionate	10.91	7.535
trans-4-Methylcyclohexyl propionate	9.390	5.810
cis-4-Methylcyclohexyl butyrate	615.75	315.5
trans-4-Methylcyclohexyl butyrate	447.15	231.1

B. Cholesteryl Palmitate Solvent[129]

	γ_I^∞		
	Smectic (80°C)	Cholesteric (85°C)	Isotropic (150°C)
n-Heptane	0.840	0.750	0.260
n-Octane	1.500	1.420	0.483
n-Nonane	2.232	2.385	0.904
Benzene	0.342	0.343	0.136
Toluene	0.590	0.585	0.235
Ethylbenzene	0.830	0.830	0.418
o-Xylene	0.775	0.770	0.480
p-Xylene	0.750	0.770	0.413
m-Xylene	0.755	0.795	0.401

5.1.5 Properties of Polymer Solutions The use of solutes as molecular probes for the gas-chromatographic determination of properties of polymers has been discussed in comprehensive reviews by Guillet[132, 133] (the originator of the technique); Stevens[134] and Berezkin and co-workers[135, 136] have also published texts on various aspects of the subject. The reader is referred to these publications for details regarding, for example, structure determination, glass point transitions, and degrees of crystallinity by GLC.

More important from a thermodynamics point of view is the use of units other than mole fraction for infinite-dilution activity coefficient data of (probe) solutes with polymer stationary phases. An obvious inconsistency arises, for example, from eq. 2.35 if the molecular weight of the stationary phase is on the order of 10^4 g/mole (as is not uncommon[137–139]). Solute specific retention volumes (or partition coefficients) with such solvents are predicted to tend to zero, contrary to what is, in fact, found in practice (e.g., Ref. 140). This problem, of some importance to polymer chemists, is compounded because number- or weight-averaged polymer molecular weights are generally known only approximately, if at all. Patterson, Guillet, and co-workers[141-144] have circumvented the difficulty by rederiving expressions for the activity coefficient in terms of weight fraction: Henry's law solute vapor pressure in terms of mole fraction x, weight fraction, w, and volume fraction, ϕ, is given by

$$p_1 = {}^x\gamma_1^\infty x_1 p_1^0 = {}^w\gamma_1^\infty w_1 p_1^0 = {}^\phi\gamma_1^\infty \phi_1 p_1^0 \tag{5.43}$$

where the superscripts refer to the appropriate units for γ_1^∞. Redefining the partition coefficient in terms of weight fraction

$$K_R = \frac{n_1^L/V_L}{n_1^M/V_M} = \frac{w_1^L/MW_1 V_L}{p_1/RT} = \frac{w_1^L RT}{{}^w\gamma_1^\infty w_1 p_1^0 MW_1 V_L} \tag{5.44}$$

where w_1^L and w_1 are the weight and weight fraction of solute in the liquid phase. In terms of V_g^0, eq. 5.44 becomes

$$V_g^0 = K_R \frac{273}{\rho_L T} = \frac{w_1^L 273 R}{{}^w\gamma_1^\infty w_1 p_1^0 MW_1 w_L} \tag{5.45}$$

or, because $w_1 \cong w_1^L / w_L$,

$$V_g^0 = \frac{273 R \rho_L}{{}^w\gamma_1^\infty p_1^0 MW_1} \tag{5.46}$$

where the solvent molecular weight has now been eliminated. Following the same procedure, using volume fractions,

$$V_g^0 = \frac{273 R \rho_1}{{}^\phi\gamma_1^\infty p_1^0 MW_1 \rho_L} \tag{5.47}$$

Alternatively,

$$ {}^w\gamma_1^\infty = {}^x\gamma_1^\infty \left(\frac{MW_L}{MW_1} \right) \tag{5.48}$$

$$ {}^\phi\gamma_1^\infty = {}^x\gamma_1^\infty \left(\frac{\overline{V}_L}{\overline{V}_1} \right) \tag{5.49}$$

$$ {}^\phi\gamma_1^\infty = {}^w\gamma_1^\infty \left(\frac{\rho_L}{\rho_1} \right) \tag{5.50}$$

Guillet and co-workers[144] demonstrated that, for n-alkanes, weight-fraction-based solute activity coefficients approached a finite limit as the molecular weight of the solvent was increased; the mole-fraction-based data, on the other hand, approached infinity as shown for n-hexane in Table 5.15.

Table 5.15 Mole Fraction and Weight Fraction Activity Coefficients for n-Hexane as a Function of the Stationary Phase Carbon Number[144]

Solvent	$\ln{}^x\gamma_1^\infty$	$\ln{}^w\gamma_1^\infty$
n-$C_{20}H_{42}$	-0.10	0.90
n-$C_{40}H_{82}$	-0.39	1.25
n-$C_{60}H_{122}$	-0.65	1.39
n-$C_{100}H_{202}$	-1.03	1.50
n-$C_{1000}H_{1002}$	-3.14	1.67
n-$C_\infty H_\infty$	$-\infty$	1.69

Roberts and Hawkes[145] found that activity coefficients for alkanes with SE 30 and SF 96 and for alcohols with Carbowax 400 and 20M could be compared on a common basis only when the data were treated in terms of weight fraction. Kovats and co-workers[146,147] have reported molal-based activity coefficients and Martire[100] showed that the Janini-Martire solution model accurately predicted activity coefficients when this unit was used. Purnell and co-workers[148,150] have employed both molar concentration and volume-fraction units where, when the solute is at infinite dilution,

$$^c\gamma_1^\infty \cong {}^\phi\gamma_1^\infty \tag{5.51}$$

Their data for several aromatic hydrocarbons in di-alkyl phthalate solvents are presented in Table 5.16.

Because interaction parameters are calculated directly from solute activity coefficients, they, too, must be redefined if units other than mole fraction are employed. Roberts and Hawkes[145] and Guillet and Purnell[151] showed that, in terms of weight fraction,

$$^wx = \ln\left(\frac{{}^w\gamma_1^\infty \rho_1}{\rho_L}\right) - 1 \tag{5.52}$$

Guillet and Purnell[151] also demonstrated that because, for polymeric

Table 5.16   Comparison of $^x\gamma_i^\infty$ and $^c\gamma_i^\infty$ Values for Aromatic Hydrocarbon Solutes with Di-alkyl Phthalate Solvents[148] at 60°C

Alkyl Group	Benzene		Toluene		o-Xylene		m-Xylene		p-Xylene	
	$^x\gamma_i^\infty$	$^c\gamma_i^\infty$	$^x\gamma_i^\infty$	$^c\gamma_i^\infty$	$^x\gamma_i^\infty$	$^c\gamma_i^\infty$	$^x\gamma_i^\infty$	$^c\gamma_i^\infty$	$^x\gamma_i^\infty$	$^c\gamma_i^\infty$
n-Propyl	0.837	2.15	0.970	2.10	1.088	2.09	1.151	2.16	1.146	2.14
n-Butyl	0.769	2.25	0.875	2.16	0.969	2.12	1.017	2.18	1.008	2.15
n-Pentyl	0.729	2.41	0.774	2.15	0.840	2.06	0.879	2.12	0.869	2.09
n-Heptyl	0.594	2.40	0.651	2.22	0.695	2.10	0.722	2.13	0.711	2.09
n-Octyl	0.579	2.55	0.632	2.35	0.671	2.21	0.695	2.24	0.686	2.19
iso-Propyl	0.834	2.17	0.977	2.14	1.103	2.14	1.172	2.23	1.160	2.20
iso-Butyl	0.732	2.17	0.829	2.07	0.909	2.01	0.958	2.07	0.946	2.04
iso-Octyl	0.582	2.59	0.639	2.36	0.686	2.24	0.712	2.28	0.692	2.21
iso-Decyl	0.507	2.59	0.548	2.36	0.583	2.22	0.600	2.24	0.594	2.21
3,5,5-Trimethylhexyl	0.588	2.80	0.634	2.61	0.676	2.40	0.695	2.42	0.692	2.40

stationary phases, $r \gg 1$,

$$^\phi x = \ln\left(^\phi\gamma_1^\infty\right) - 1 \tag{5.53}$$

$$^x x = \ln {}^x\gamma_1^\infty + \ln(r) - 1 \tag{5.54}$$

Equation 5.53 is of interest, first, because it does not require density (eq. 5.52) or molar volume (eq. 5.54) ratios and, second, because the volume-fraction excess Gibbs free energy is

$$\Delta^\phi \overline{G}^e = RT \ln {}^\phi\gamma_1^\infty \tag{5.55}$$

the interaction parameter is given by

$$^\phi x = \frac{\Delta^\phi \overline{G}^e - RT}{RT} = \frac{\Delta^\phi \overline{A}^e}{RT} \tag{5.56}$$

where $\Delta^\phi \overline{A}^e$ is the excess volume-fraction-based molar Helmholtz free energy. Further, because

$$\Delta \overline{A}^e = \Delta \overline{E}^e - T\Delta \overline{S}^e \tag{5.57}$$

Guillet and Purnell[151] found

$$^\phi x = \frac{\Delta^\phi \overline{E}^e}{RT} - \frac{\Delta^\phi \overline{S}^e}{R} \tag{5.58}$$

which is a promising result, for solubility parameters (eq. 5.17 et seq.) are also defined in terms of the internal energy.

Assessment of the significance of eq. 5.56 (which, strictly, is valid as derived only for polymer solvents) in terms of macromolecules as well as solvents of more conventional molecular weight is now in progress.[152]

5.1.6 Scales of Relative Lewis Acid-Base Properties Lewis acid-base properties of organic solutes are of commercial importance in extractive distillation, in which use is made of relative activity coefficient data to predict the enhancement of separations.[153-167] Scales of relative Lewis base[163] or acid[167] properties may be compiled from GLC data; for example, Harris and Prausnitz[30] used the relation

$$K_a = \left(\frac{\gamma_1^\infty}{\gamma_1^{\infty'}} \right) - 1 \qquad (5.59)$$

where K_a is the Lewis "acidity constant" of a (Lewis acid) solute with activity coefficients γ_1^∞ and $\gamma_1^{\infty'}$ in inert and Lewis base solvents, respectively. Table 5.17 presents the K_a values of a variety of solutes with the

Table 5.17 Lewis Acidity Constants[30] for Organic Solutes at 80°C

	K_a	
Solute	Squalane/Squalene	n-Octadecane/ Benzyl Diphenyl
Methyl iso-propyl ketone	0.24	0.46
n-Butyraldehyde	0.26	0.61
iso-Propyl acetate	0.29	0.27
n-Butyl formate	0.30	0.34
Methyl ethyl ketone	0.32	0.72
n-Propyl acetate	0.32	0.38
Ethyl acetate	0.35	0.48
1,2-Dichloroethane	0.35	0.64
Acetone	0.36	1.05
Trimethyl orthoformate	0.36	...
Dimethyl carbonate	0.38	1.02
iso-Propyl alcohol	0.41	0.40
Cyclopentanone	0.45	1.7
Nitroethane	0.55	2.1
Propionitrile	0.58	2.5
Methoxyacetonitrile	0.67	2.9
Acetonitrile	0.68	3.2
Furfural	0.71	2.7
Dimethyl formamide	0.78	4.2
Dimethyl sulfoxide	1.0	...
Trimethyl phosphate	1.1	...
Trimethyl phosphite	1.2	...

solvent pairs: squalane/squalene and *n*-octadecane/benzyl diphenyl. GC is of obvious value in generating these data; noteworthy in this regard are the studies by Ono and co-workers[168–171] and Karger and co-workers[172] which relate Hammett σ constants[173] or related parameters[174] to the log of activity coefficients.

5.2 THERMODYNAMIC PROPERTIES AT THE GAS-LIQUID INTERFACE

As mentioned in Section 2.2.4, solute adsorption at the gas-liquid inter-phase may contribute substantially to retention. The phenomenon has been discussed on numerous occasions[51, 175–227] and reviewed elsewhere by Martire[228] (a condensed version of which has been published by Locke[31]). The vast majority of these studies describe various means of determining bulk solution properties in the presence of surface effects (where, when liquid-solid effects are negligible, the method given in Section 2.2.4 suffices to the same extent as any other). Thermodynamic information is also available, however, from retention data of systems exhibiting interfacial adsorption; for example, Eon and Guiochon[210] showed that the surface partition coefficient K_S is given by

$$K_S = \frac{RT}{\overline{A}_L p_1^0 \gamma_1^{\infty,s}} \exp\left[\overline{A}_1 \left(\frac{\sigma_L^0 - \sigma_1^0}{RT} \right) - \frac{RT}{\overline{V}_L p_1^0 \gamma_1^{\infty,b}} \right] \tag{5.60}$$

where $\overline{A}_L$ and $\overline{A}_1$ are molar solvent and partial molar solute surface areas, $\gamma_1^{\infty,s}$ and $\gamma_1^{\infty,b}$ are the solute surface and bulk-solution infinite-dilution activity coefficients, and σ_L^0 and σ_1^0 are the solvent and solute surface tensions; $\overline{A}_1$ is calculated from the relation

$$\overline{A}_1 = \left(\frac{\partial \overline{A}_{\text{tot}}}{\partial x_1^s} \right)_{T,p,\sigma} \tag{5.61}$$

which for pseudospherical molecules may be approximated by the pure solute surface area $\overline{A}_1^0$:

$$\overline{A}_1 \cong \overline{A}_1^0 = \left(v_1^0 \right)^{2/3} N^{1/3} \tag{5.62}$$

where N is Avogadro's number. K_S is obtained from plots of V_N/V_L versus $1/V_L$, thus enabling the surface activity coefficient to be calculated from eq. 5.60. Because the surface and bulk partition coefficients are related by[228]

$$K_S = - \frac{K_R \bar{V}_L}{RT} \left(\frac{\partial \sigma_1}{\partial x_1^b} \right)^\infty \tag{5.63}$$

where the superscript ∞ indicates the limit of infinite dilution; eqs. 5.60 and 5.63 also yield a value for the term $(\partial \sigma_1 / \partial x_1^b)^\infty$. Table 5.18 presents the activity coefficient and surface tension data calculated by Eon and Guiochon[210] for various solutes with thiodipropionitrile and water solvents.

Table 5.18 Surface and Bulk-Solution Properties for Named Solutes with Thiodipropionitrile (TDPN) and Water Solvents[210]

Solute	$\gamma_1^{\infty,s}$ ($\times 10^2$)	$\gamma_1^{\infty,b}$	$-(\partial \sigma_1 / \partial x_1^b)^\infty$
A. TDPN Solvent			
Cyclohexane	19.5	44.6	340
Cyclohexene	19.5	17.8	165
Benzene	16.0	3.4	48
Cyclopentane	15.0	29.0	295
Methylcyclopentane	11.5	49.0	495
2,3-Dimethylbutane	4.1	85.3	1360
2-Methylpentane	4.0	103.	1570
3-Methylpentane	4.8	90.7	1300
n-Hexane	5.8	105.	1400
1-Hexene	3.8	44.0	840
n-Heptane	5.2	156.	1860
2,2,4-Trimethylpentane	2.3	191.	3080
Diethyl ether	2.3	8.8	345
Acetone	3.1	1.5	98
Ethyl acetate	2.8	3.6	200
B. Water Solvent			
Carbon tetrachloride	0.52	7200.	16,400
Chloroform	0.19	640.	6500
Dichloromethane	0.24	197.	2550

Suprynowicz and co-workers[219] claimed that a variety of adsorption phenomena contributes to retention and developed a gas chromatographic method based on the multilayer sorption model by Hill[229] for generating isotherms for each of them. The significance of their experimental curves is questionable, however, for a key feature of the theory was based on the ideal gas-phase behavior of solutes, no account being taken of fugacity or virial effects. Further, Filonenko and Korol[51] reported recently that specific retention volumes (corrected for surface effects) for acetone, alcohols, and benzene with squalane and oxydipropionitrile agreed within 2% with those measured by a static technique. Corrected heats of solution were also in agreement, although, presumably, use of eq. 2.66 would mask multiple sources of adsorption; the dispute concerning this topic has still to be resolved.[230, 231]

5.3 FINITE-CONCENTRATION VAPOR-LIQUID EQUILIBRIA

In addition to their fundamental importance, vapor-liquid data are used in a variety of engineering fields (e.g., Ref. 165); Kobayashi and co-workers have shown that various modes of finite-concentration gas chromatography (cf. Section 2.1.3) can be employed to measure equilibrium solution properties even when the stationary phase is a volatile solvent. Table 5.19 summarizes the systems studied by these workers. Mole-fraction partition coefficients were reported in their work because they are related to the inverse of Henry's constants.

Table 5.19 Multicomponent Finite-Dilution Vapor/Liquid Systems Studied by GLC

Solute/Solvent System	Temperature Range (°C)	Pressure Range (psia)	Ref.
n-Butane, 1-butene, cis-2-butene, trans-2-butene, 1,3-butadiene/n-decane, furfural	−100 to 100	20 to 1500	232
Ethane, propane, n-butane/methane + n-decane, methane + n-hexadecane	−110 to 130	20 to 2000	233
Methane, propane/n-decane, methane + propane + n-heptane, methane + propane + n-decane	−40 to 20	20 to 1000	234, 235

Table 5.19 (*Continued*)

Solute/Solvent System	Temperature Range (°C)	Pressure Range (psia)	Ref.
Methane/n-heptane	−70 to −20	up to critical point	236
Methane, ethane, propane/ n-heptane, toluene	−70 to −20	100 to 1600	237
Methane/toluene, methylcyclohexane	−70 to −20	100 to 3500	238
Carbon dioxide, hydrogen sulfide/ methane + n-octane	−40 to 20	100 to 1500	239–241
Methane/propane; methane/ethane; methane/ethane + propane	−140 to −60	25 to 950	242
Nitrogen/methane, ethane	−160 to −80	20 to 1000	243

Various forms of head-space analysis[51, 226, 244–250] have also been used to a considerable extent for the measurement of vapor pressure-composition data; gas chromatography, applied in these cases primarily as an analytical tool, is nonetheless an indispensible feature of studies of this kind.

5.4 REFERENCES

1. A. T. James, *Biochem. J.*, **52**, 242 (1952).
2. N. H. Ray, *J. Appl. Chem.*, **4**, 21 (1954).
3. A. B. Littlewood, C. S. G. Phillips, and D. T. Price, *J. Chem. Soc.*, 1480 (1955).
4. D. White, *Nature*, **179**, 1075 (1957).
5. D. White and C. T. Cowen, *Trans. Faraday Soc.*, **54**, 557 (1958); in *Gas Chromatography 1958*, D. H. Desty, Ed., Butterworths, London, 1958, p. 116.
6. J. R. Anderson and K. H. Napier, *Aust. J. Chem.*, **10**, 250 (1957).
7. E. F. G. Herington, *Analyst*, **81**, 52 (1956); in *Vapour Phase Chromatography*, D. H. Desty, Ed., Butterworths, London, 1957, p. 52.
8. M. R. Hoare and J. H. Purnell, *Research*, **8**, S41 (1955); *Trans. Faraday Soc.*, **52**, 222 (1956).
9. J. H. Purnell, in *Vapour Phase Chromatography*, D. H. Desty, Ed., Butterworths, London, 1957, p. 52; *J. Roy. Inst. Chem.*, **82**, 586 (1958).

10. J. H. Knox, *Sci. Progr.*, **45**, 227 (1957).

11. A. I. M. Keulemans, A. Kwantes, and P. Zaal, *Anal. Chim. Acta*, **13**, 257 (1955).

12. G. J. Pierrotti, C. H. Deal, E. L. Derr, and P. E. Porter, *J. Am. Chem. Soc.*, **78**, 2989 (1956).

13. J. R. Young, *Chem. Ind.*, 594 (1958).

14. D. Jentzsch and G. Bergmann, *Z. Anal. Chem.*, **165**, 401 (1959).

15. B. H. Pike and E. S. Swinbourne, *Aust. J. Chem.*, **12**, 104 (1959).

16. A. B. Littlewood, *Anal. Chem.*, **36**, 1441 (1964).

17. R. S. Juvet, Jr., V. R. Shaw, and M. A. Khan, *J. Am. Chem. Soc.*, **91**, 3788 (1969).

18. E. F. Meyer and R. A. Ross, *J. Phys. Chem.*, **75**, 831 (1971).

19. R. C. Castells, *J. Chromatogr.*, **111**, 1 (1975).

20. C. A. Streuli, W. H. Muller, and M. Orloff, *J. Chromatogr.*, **101**, 17 (1974).

21. E. F. Meyer, K. S. Steck, and R. D. Hotz, *J. Phys. Chem.*, **77**, 2140 (1973).

22. E. B. Molnar, P. Mority, and J. Takacs, *J. Chromatogr.*, **66**, 205 (1972).

23. J. Takacs, *J. Chromatogr. Sci.*, **11**, 210 (1973).

24. G. D. Mitra and N. C. Saka, *Chromatographia*, **6**, 93 (1973).

25. A. W. London and S. Sandler, *Anal. Chem.*, **45**, 921 (1973).

26. C. A. Streuli and M. Orloff, *J. Chromatogr.*, **101**, 23 (1974).

27. P. H. Weiner, H.-L. Liao, and B. L. Karger, *Anal. Chem.*, **46**, 2182 (1974).

28. B. L. Reinbold and T. H. Risby, *J. Chromatogr. Sci.*, **13**, 372 (1975).

29. C. E. Figgins, T. H. Risby, and P. C. Jurs, *J. Chromatogr. Sci.*, **14**, 453 (1976).

30. H. G. Harris and J. M. Prausnitz, *J. Chromatogr. Sci.*, **7**, 685 (1969).

31. D. C. Locke, *Adv. Chromatogr.*, **14**, 87 (1976).

32. C. Eon, C. Pommier, and G. Guiochon, *Bull. Soc. Chim. Fr.*, 1277 (1974).

33. V. Brandani, *Ind. Eng. Chem. Fundam.*, **13**, 154 (1974).

34. M. H. Abraham, P. L. Grellier, and J. Mana, *J. Chem. Thermodyn.*, **6**, 1175 (1974).

35. J. W. King and P. R. Quinney, *J. Phys. Chem.*, **78**, 2635 (1974).

36. J. F. Parcher and K. W. Yun, *J. Chromatogr.*, **99**, 193 (1974).

37. T. M. Letcher and F. Marsicano, *J. Chem. Thermodyn.*, **6**, 501 (1974).

38. I. Kikic and P. Alessi, *J. Chromatogr.*, **100**, 202 (1974).

39. P. Alessi, I. Kikic, and G. Torriano, *J. Chromatogr.*, **105**, 257 (1975).

40. J. F. Parcher, P. H. Weiner, C. L. Hussey, and T. N. Westlake, *J. Chem. Eng. Data*, **20**, 145 (1975).

41. V. M. Nabovach and L. A. Venger, in *Advances in Chromatography*, Vol. IV, Pt. 1, M. S. Vigdergauz, Ed., Akad. Nauk SSSR, Kazanskii Filial, Kazan, Russia, 1975, p. 133.

42. A. N. Korol, *Chromatographia*, **8**, 385 (1975).

43. V. A. Chernoplekova, A. N. Korol, K. I. Sakodynskii, and K. A. Kocheshkov, *Izv. Akad. Nauk SSSR, Ser. Khim.*, 834 (1975).

44. M. A. Pais, M. F. Bondarenko, Z. I. Abramovich, and E. A. Kruglov, *Neftekhim.*, **15**, 626 (1975).

45. G. Castello and G. D'Amato, *J. Chromatogr.*, **107**, 1 (1975).

46. A. B. Sund and P. E. Barker, *J. Chromatogr. Sci.*, **13**, 541 (1975).

47. B. S. Rawat, K. L. Mallik, and I. B. Gulati, *J. Appl. Chem. Biotechnol.*, **22**, 1001 (1972).

48. R. K. Kuchhal and K. L. Mallik, *J. Chem. Eng. Data.* **17**, 49 (1972); *J. Appl. Chem. Biotechnol.*, **26**, 67 (1976).

49. R. K. Kuchhal, K. L. Mallik, and P. L. Gupta, *Can. J. Chem.*, **55**, 1273 (1977).

50. A. S. Bogeatzes and D. P. Tassios, *Ind. Eng. Chem. Proc. Des. Dev.*, **12**, 274 (1973).

51. G. V. Filonenko and A. N. Korol, *J. Chromatogr.*, **119**, 157 (1976).

52. R. J. Laub, D. E. Martire, and J. H. Purnell, *J. Chem. Soc. Faraday Trans. II*, **74**, 213 (1978).

53. A. J. Ashworth, *J. Chem. Soc. Faraday Trans. I*, **69**, 459 (1973).

54. R. J. Laub, J. H. Purnell, P. S. Williams, M. W. M. Harbison, and D. E. Martire, *J. Chromatogr.*, in press.

55. M. L. Huggins, *J. Chem. Phys.*, **9**, 440 (1941); *Ann. N.Y. Acad. Sci.*, **1**, 43 (1942).

56. P. J. Flory, *J. Chem. Phys.*, **10**, 51 (1942); *J. Am. Chem. Soc.*, **87**, 1833 (1965); *Discuss. Faraday Soc.*, **49**, 7 (1970).

57. P. J. Flory, R. A. Orwoll, and A. Vrij, *J. Am. Chem. Soc.*, **86**, 3507, 3515 (1964).

58. R. A. Orwoll and P. J. Flory, *J. Am. Chem. Soc.*, **89**, 6814, 6822 (1967).

59. B. E. Eichinger and P. J. Flory, *Trans. Faraday Soc.*, **64**, 2035, 2053, 2061, 2066 (1968).

60. H. Tompa, *Trans. Faraday Soc.*, **45**, 107 (1949); *Polymer Solutions*, Butterworths, London, 1956, Chapter 4.

61. E. A. Guggenheim, *Mixtures*, Oxford University Press, Oxford, England, 1952, Chapters X and XI.

62. I. Prigogine, *The Molecular Theory of Solutions*, Interscience, New York, 1957, Chapters 16 and 17.

63. W. G. McMillan and J. E. Mayer, *J. Chem. Phys.*, **13**, 276 (1945).

64. J. H. van der Waals and J. J. Hermans, *Rec. Trav. Chim. Pays-Bas*, **69**, 949 (1950).

65. J. H. van der Waals, *Rec. Trav. Chim. Pays-Bas*, **70**, 101 (1951).

66. G. S. Rushbrooke, H. I. Scoins, and A. J. Wakefield, *Discuss. Faraday Soc.*, **15**, 57 (1953).

67. H. C. Longuet-Higgins, *Discuss. Faraday Soc.*, **15**, 73 (1953).

68. J. L. Copp and D. H. Everett, *Discuss. Faraday Soc.*, **15**, 268 (1953).

69. T. L. Hill, *J. Am. Chem. Soc.*, **79**, 4885 (1957); *J. Chem. Phys.*, **30**, 93 (1959).

70. Th. Holleman and J. Hijmans, *Physica*, **28**, 604 (1962).

71. D. F. G. Pusey, *Proc. Chem. Soc.*, **11**, 108 (1974).

72. A. J. Ashworth and D. H. Everett, *Trans. Faraday Soc.*, **56**, 1609 (1960).

73. A. J. B. Cruickshank, B. W. Gainey, and C. L. Young, *Trans. Faraday Soc.*, **64**, 337 (1968).

74. C. L. Young, *Trans. Faraday Soc.*, **64**, 1537 (1968).

75. J. J. van Laar, *Z. Phys. Chem.*, **72**, 723 (1910).

76. F. Dolezalek, *Z. Phys. Chem.*, **64**, 727 (1908).

77. J. J. van Laar, *Z. Phys. Chem.*, **83**, 599 (1913).

78. J. H. Hildebrand, *J. Am. Chem. Soc.*, **38**, 1452 (1916); **41**, 1067 (1919); **42**, 2180 (1920); *Proc. Nat. Acad. Sci.*, **13**, 267 (1927).

79. W. Heitler, *Ann. Phys.*, **80**, 30 (1926).

80. G. Scatchard, *Chem. Rev.*, **8**, 321 (1931).

81. J. H. Hildebrand and S. E. Wood, *J. Chem. Phys.*, **1**, 817 (1933).

82. R. L. Scott, *Ann. Rev. Phys. Chem.*, **7**, 43 (1956).

83. T. M. Reed, *J. Am. Chem. Soc.*, **59**, 429 (1955).

84. A. G. Williamson, *Ann. Rev. Phys. Chem.*, **15**, 63 (1964).

85. J. A. Barker, *J. Chem. Phys.*, **20**, 1526 (1952).

86. J. B. Ott, J. R. Goates, and R. L. Snow, *J. Phys. Chem.*, **67**, 515 (1963).

87. D. H. Everett and R. J. Munn, *Trans. Faraday Soc.*, **60**, 1951 (1964).

88. C. P. Hicks and C. L. Young, *Trans. Faraday Soc.*, **64**, 2675 (1968).

89. D. E. Martire, in *Gas Chromatography*, L. Fowler, Ed., Academic, New York, 1963, p. 33.

90. D. E. Martire and L. Z. Pollara, *Adv. Chromatogr.*, **1**, 335 (1965).

91. A. J. B. Cruickshank, B. W. Gainey, C. P. Hicks, T. M. Letcher, and C. L. Young, *Trans. Faraday Soc.*, **65**, 2356 (1969).

92. Y. B. Tewari, D. E. Martire, and J. P. Sheridan, *J. Phys. Chem.*, **74**, 2345 (1970).

93. J. F. Messerly, G. B. Guthrie, S. S. Todd, and H. L. Finke, *J. Chem. Eng. Data*, **12**, 338 (1967).

94. G. M. Janini and D. E. Martire, *J. Chem. Soc. Faraday Trans. II*, **70**, 837 (1974).

95. D. Patterson, *Rubber Chem. Tech.*, **40**, 1 (1967); *Macromolecules*, **1**, 279 (1969); *J. Polym. Sci., Pt. C*, **16**, 3379 (1969).

96. J. Hijmans, *Physica*, **27**, 433 (1961).

97. J. Hijmans and Th. Holleman, *Adv. Chem. Phys.*, **16**, 223 (1969).

98. A. Kreglewski, *J. Phys. Chem.*, **71**, 3860 (1967); **72**, 1879, 2280 (1968); **73**, 3359 (1969).

99. R. R. Dreisbach, *Physical Properties of Chemical Compounds*, Vol. II, American Chemical Society Publications, Washington, D.C., 1959.

100. D. E. Martire, *Anal. Chem.*, **46**, 626 (1974).

101. D. H. Everett, B. W. Gainey, and C. L. Young, *Trans. Faraday Soc.*, **64**, 2667 (1968).

102. D. E. Martire, in *Gas Chromatography 1966*, A. B. Littlewood, Ed., Institute of Petroleum, London, 1967, p. 21.

103. A. J. B. Cruickshank, B. W. Gainey, and C. L. Young, in *Gas Chromatography 1968*, C. L. A. Harbourn, Ed., Institute of Petroleum, London, 1969, p. 76.

104. B. W. Gainey and R. L. Pecsok, *J. Phys. Chem.*, **74**, 2548 (1970).

105. Y. B. Tewari, J. P. Sheridan, and D. E. Martire, *J. Phys. Chem.*, **74**, 3263 (1970).

106. G. W. Gray, *Molecular Structure and Properties of Liquid Crystals*, Academic, New York, 1962.

107. G. W. Gray and P. A. Winsor, Eds., *Liquid Crystals and Plastic Crystals*, Vols. 1 and 2, Ellis Horwood, London, 1974.

108. B. L. Karger, *Anal. Chem.*, **39**(8), 24A (1967).

109. P. F. McCrea, *Adv. Anal. Chem. Instrum.*, **11**, 87 (1973).

110. J. P. Schroeder, in *Liquid Crystals and Plastic Crystals*, Vol. 1, G. W. Gray and P. A. Winsor, Eds., Ellis Horwood, London, 1974, p. 356.

111. R. S. Porter and J. F. Johnson, Eds., *Ordered Fluids and Liquid Crystals*, A. C. S. Publications, Washington, D.C., 1967.

112. "NATO Advanced Study Institute: The Molecular Physics of Liquid Crystals," Cambridge, England, 1978.

113. V. Pacakova, H. Ullmannova, and E. Smolkova, *Chromatographia*, **7**, 171 (1974).

114. L. C. Chow and D. E. Martire, *Mol. Cryst. Liq. Cryst.*, **14**, 293 (1971).

115. H. T. Peterson and D. E. Martire, *Mol. Cryst. Liq. Cryst.*, **25**, 89 (1974).

116. H. T. Peterson, D. E. Martire, and M. A. Colter, *J. Chem. Phys.*, **61**, 3547 (1974).

117. G. I. Agren and D. E. Martire, *J. Chem. Phys.*, **61**, 3959 (1974).

118. G. I. Agren and D. E. Martire, *J. Phys. Colloq.*, **36**, Cl-141 (1975).

119. D. E. Martire, G. A. Oweimreen, G. I. Agren, S. G. Ryan, and H. T. Peterson, *J. Chem. Phys.*, **64**, 1456 (1976).

120. F. Dowell and D. E. Martire, *J. Chem. Phys., in press.*

121. D. E. Martire, papers presented in Ref. 112.

122. H. Kelker and H. Winterscheidt, *Z. Anal. Chem.*, **220**, 1 (1966).

123. H. Kelker, *J. Chromatogr.*, **112**, 165 (1975).

124. E. Grushka and J. F. Solsky, *Anal. Chem.*, **45**, 1836 (1973); *J. Chromatogr.*, **99**, 134 (1974); **112**, 145 (1975).

125. J. F. Bocquet and C. Pommier, *J. Chromatogr.*, **117**, 315 (1976).

126. D. E. Martire, P. A. Blasco, P. F. Carone, L. C. Chow, and H. Vicini, *J. Phys. Chem.*, **72**, 3489 (1968).

127. G. Kraus, K. Seifert, and H. Schubert, *J. Chromatogr.*, **100**, 101 (1974).

128. A. A. Jeknavorian and E. F. Barry, *J. Chromatogr.*, **101**, 299 (1974).

129. A. A. Jeknavorian, P. Barrett, A. C. Watterson, and E. F. Barry, *J. Chromatogr.*, **107**, 317 (1975).

130. L. C. Chow and D. E. Martire, *J. Phys. Chem.*, **73**, 1127 (1969).

131. H. T. Peterson, D. E. Martire, and W. Lindner, *J. Phys. Chem.*, **76**, 596 (1972).

132. J. E. Guiller, *Adv. Anal. Chem. Instrum.*, **11**, 187 (1973).

133. J.-M. Braun and J. E. Guillet, *Adv. Polym. Sci.*, **21**, 108 (1976).

134. M. P. Stevens, *Characterization and Analysis of Polymers by Gas Chromatography*, Marcel Dekker, New York, 1969.

135. V. G. Berezkin, *Analytical Reaction Gas Chromatography*, Plenum, New York, 1968.

136. V. G. Berezkin, V. R. Alishoyev, and I. B. Nemirovskaya, *Gas Chromatography in Polymer Chemistry*, Nauka, Moscow, 1972; Elsevier, New York, 1977.

137. H. Rotzsche, *Plaste Kautsch.*, **15**, 477 (1968).

138. C. R. Trash, *J. Chromatogr. Sci.*, **11**, 196 (1973).

139. A. E. Coleman, *J. Chromatogr. Sci.*, **11**, 198 (1973).

140. A. R. Cooper, C. W. P. Crowne, and P. G. Farrell, *Trans. Faraday Soc.*, **62**, 2725 (1966); **63**, 447 (1967).

141. O. Smidrod and J. E. Guillet, *Macromolecules*, **2**, 272 (1969).

142. A. Lavoie and J. E. Guillet, *Macromolecules*, **2**, 443 (1969).

143. J. E. Guillet and A. N. Stein, *Macromolecules*, **3**, 102 (1970).

144. D. Patterson, Y. B. Tewari, H. P. Schreiber, and J. E. Guillet, *Macromolecules*, **4**, 356 (1971).

145. G. L. Roberts and S. J. Hawkes, *J. Chromatogr. Sci.*, **11**, 16 (1973).

146. G. A. Huber and E. sz. Kovats, *Anal. Chem.*, **45**, 1155 (1973).

147. D. F. Fritz and E. sz. Kovats, *Anal. Chem.*, **45**, 1175 (1973).

148. D. L. Meen, F. Morris, J. H. Purnell, and O. P. Srivastava, *J. Chem. Soc. Faraday Trans. I*, **69**, 2080 (1973).

149. J. M. Vargas de Andrade, Ph.D. Dissertation, University of Wales, 1975.

150. C. S. Allen, Ph.D. Dissertation, University of Wales, 1977.

151. J. E. Guillet and J. H. Purnell, *Macromolecules, in press.*

152. J. E. Guillet, *private communication.*

153. H. Röcke, *Chem.-Ing.-Tech.,* **28**, 489 (1956).

154. G. W. Warren, R. R. Warren, and V. A. Yarborough, *Ind. Eng. Chem.,* **51**, 1475 (1959).

155. J. A. Gerster, J. A. Gorton, and R. B. Eklund, *J. Chem. Eng. Data,* **5**, 423 (1960).

156. C. Döring, *Z. Chem.,* **11**, 347 (1961).

157. J. M. Prausnitz and R. Anderson, *AIChE J.,* **7**, 96 (1961).

158. A. N. Genkin, S. K. Ogorodnikov, and M. S. Memtsov, *Neftekhim.,* **2**, 837 (1962).

159. M. R. Sheets and J. M. Marcello, *Hydroc. Proc. Pet. Refin.,* **42**(12), 99 (1963).

160. R. V. Orye, R. F. Weiner, and J. M. Prausnitz, *Science,* **148**, 74 (1965).

161. B. G. Kyle and D. E. Leng, *Ind. Eng. Chem.,* **57**, 43 (1965).

162. J. Bonastre and P. Grenier, *Bull. Soc. Chim. Fr.,* **118**, 1292 (1968).

163. I. Brown, I. L. Chapman, and G. J. Nicholson, *Aust. J. Chem.,* **21**, 1125 (1968).

164. P. Vernier, C. Raimbault, and H. Renon, *J. Chim. Phys.,* **66**, 429 (1969).

165. D. Tassios, *Chem. Eng. J.,* **76**(3), 118 (1969); *Hydroc. Proc. Pet. Refin.,* **49**(7), 114 (1970); *Ind. Eng. Chem. Process Des. Dev.,* **15**, 574 (1976).

166. R. S. Sheehan and S. H. Langer, *Ind. Eng. Chem. Process Res. Dev.,* **10**, 44 (1971).

167. P. M. Cukor and J. M. Prausnitz, *J. Phys. Chem.,* **76**, 598 (1972).

168. R. Goto, T. Araki, and A. Ono, *Nippon Kagaku Zasshi,* **81**, 1318 (1960).

169. T. Araki, R. Goto, and A. Ono, *Nippon Kagaku Zasshi,* **82**, 1081 (1961).

170. T. Araki, *Bull. Chem. Soc. Jap.,* **36**, 879 (1963).

171. A. Ono, *Nippon Kagaku Zasshi,* **91**, 72 (1970); **92**, 429, 986 (1971); *J. Chromatogr.,* **110**, 233 (1975).

172. B. L. Karger, Y. Elmehrik, and R. L. Stern, *Anal. Chem.,* **40**, 1227 (1968).

173. L. P. Hammett, *Chem. Rev.,* **17**, 125 (1935); *Physical Organic Chemistry,* McGraw-Hill, New York, 1940, pp. 184–199.

174. A. Vetere, R. DeSimone, and A. Ginnassi, *Ind. Eng. Chem. Process Des. Dev.,* **14**, 141 (1975).

175. F. T. Eggertsen and H. S. Knight, *Anal. Chem.,* **30**, 15 (1958).

176. R. L. Martin and J. C. Winters, *Anal. Chem.,* **31**, 1954 (1959).

177. R. L. Martin, *Anal. Chem.,* **33**, 347 (1961); **35**, 116 (1963).

178. R. L. Pecsok, A. deYllana, and A. Abdul-Karim, *Anal. Chem.,* **36**, 452 (1964).

179. D. H. Everett, *Trans. Faraday Soc.,* **60**, 1803 (1964); **61**, 2478 (1965).

180. D. E. Martire, R. L. Pecsok, and J. H. Purnell, *Nature,* **203**, 1279 (1964); *Trans. Faraday Soc.,* **61**, 2496 (1965).

181. M. R. James, J. C. Giddings, and H. Eyring, *J. Phys. Chem.,* **69**, 2351 (1961).

182. P. Urone and J. F. Parcher, *Anal. Chem.,* **38**, 270 (1966).

183. D. E. Martire, *Anal. Chem.,* **38**, 244 (1966).

184. J. H. Purnell, S. P. Wasik, and R. S. Juvet, Jr., *Acta Chim. Acad. Sci. Hung.,* **50**, 201 (1966).

185. P. Urone, J. F. Parcher, and E. N. Baylor, *Sep. Sci.,* **1**, 595 (1966).

186. R. L. Pecsok and B. H. Gump, *J. Phys. Chem.*, **71**, 2202 (1967).

187. J. Bonastre, P. Grenier, and P. Cazenave, *Bull. Soc. Chim. Fr.*, 3885 (1968).

188. B. L. Karger and A. Hartkopf, *Anal. Chem.*, **40**, 215 (1968).

189. P. Urone, Y. Takahashi, and G. H. Kennedy, *Anal. Chem.*, **40**, 1130 (1968); *J. Phys. Chem.*, **74**, 2326 (1970).

190. D. E. Martire and P. Riedl, *J. Phys. Chem.*, **72**, 3478 (1968).

191. B. L. Karger, A. Hartkopf, and H. Postmanter, *J. Chromatogr. Sci.*, **7**, 315 (1969).

192. J. R. Conder, D. C. Locke, and J. H. Purnell, *J. Phys. Chem.*, **73**, 700 (1969).

193. D. F. Cadogan, J. R. Conder, D. C. Locke, and J. H. Purnell, *J. Phys. Chem.*, **73**, 708 (1969).

194. D. F. Cadogan and J. H. Purnell, *J. Phys. Chem.*, **73**, 3849 (1969).

195. J. R. Conder, *J. Chromatogr.*, **39**, 273 (1969).

196. V. G. Berezkin and V. M. Fateeva, *Zh. Anal. Khim.*, **25**, 2023 (1970); **29**, 2453 (1974); *J. Chromatogr.*, **58**, 73 (1971); *Chromatographia*, **4**, 19 (1971).

197. A. A. Zhukhovitskii, M. L. Sazanov, M. K. Lunskii, and V. Yusfin, *J. Chromatogr.*, **58**, 87 (1971).

198. B. L. Karger, P. A. Sewell, R. C. Castells, and A. Hartkopf, *J. Colloid Interface Sci.*, **35**, 328 (1971).

199. C. W. P. Crowne, M. Harper, and P. G. Farrell, *J. Chromatogr.*, **61**, 1 (1971).

200. N. D. Petsev, *Dokl. Bolg. Akad. Nauk*, **24**, 1043 (1971).

201. B. L. Karger, C. Castells, P. A. Sewell, and A. Hartkopf, *J. Phys. Chem.*, **75**, 3870 (1971).

202. A. K. Chatterjee, J. W. King, and B. L. Karger, *J. Colloid Interface Sci.*, **41**, 71 (1972).

203. J. W. King, A. K. Chatterjee, and B. L. Karger, *J. Phys. Chem.*, **76**, 2769 (1972).

204. H.-L. Liao and D. E. Martire, *Anal. Chem.*, **44**, 498 (1972).

205. C. Eon, A. K. Chatterjee, and B. L. Karger, *Chromatographia*, **5**, 28 (1972).

206. V. G. Berezkin, V. S. Gavichev, and G. P. Zorina, *Izv. Akad. Nauk SSSR, Ser. Khim.*, 1975 (1972).

207. V. G. Berezkin, *J. Chromatogr.*, **65**, 227 (1972); **98**, 477 (1974); *Usp. Khromatogr.*, 215 (1972).

208. Z. Suprynowicz, A. Waksmundzki, and W. Rudzinski, *J. Chromatogr.*, **67**, 21 (1972); **72**, 5 (1972).

209. J. F. Parcher and C. L. Hussey, *Anal. Chem.*, **45**, 188 (1973).

210. C. Eon and G. Guiochon, *J. Colloid Interface Sci.*, **45**, 521 (1973).

211. M. O. Burova, M. Kh. Lunskii, and A. A. Zhukhovitskii, *Zavod. Lab.*, **39**, 1194 (1973); in *Khromatografiya*, Vol. I, A. A. Zhukhovitskii, Ed., VINITI, Moscow, 1974, p. 7.

212. D. W. Connell and P. J. Malcom, *J. Chromatogr.*, **78**, 251 (1973).

213. V. G. Berezkin, N. S. Nikitina, and V. M. Fateeva, *Dokl. Akad. Nauk SSSR*, **209**, 1131 (1973).

214. A. Hartkopf and B. L. Karger, *Acc. Chem. Res.*, **6**, 209 (1973).

215. R. N. Nikolov, *Chromatographia*, **6**, 451 (1973).

216. F. Bruner, G. Ciccioli, G. Crescentini, and M. T. Pistolesi, *Anal. Chem.*, **45**, 1851 (1973).

217. T. P. Khobotova, M. Kh. Lunskii, B. I. Anvayer, and A. A. Zhukhovitskii, *Zh. Anal. Khim.*, **29**, 551 (1974).

218. L. V. Semenchenko and L. T. Lantsova, *Zh. Anal. Khim.*, **29**, 1805 (1974).

219. Z. Suprynowicz, A. Waksmundzki, W. Rudzinski, and J. Rayss, *J. Chromatogr.*, **91**, 67 (1974).

220. P. H. Weiner, H.-L. Liao, and B. L. Karger, *Anal. Chem.*, **47**, 288 (1974).

221. R. Fontaine, C. Pommier, C. E. Guiochon, and G. Guiochon, *J. Chromatogr.*, **104**, 1 (1975).

222. A. A. Zhukhovitskii, V. G. Berezkin, M. Kh. Lunskii, M. O. Burova, and T. P. Khobotova, *J. Chromatogr.*, **104**, 241 (1975).

223. T. Fukuda, *Jap. Anal.*, **8**, 627 (1959).

224. E. Soczewinski, W. Golkiewicz, and A. Markowski, *Chromatographia*, **8**, 13 (1975).

225. S. M. Yanovskii and T. P. Khobotova, *Zh. Fiz. Khim.*, **49**, 2124 (1975).

226. A. G. Vitenberg, B. V. Ioffe, Z. St. Dimitrova, and I. L. Butaeva, *J. Chromatogr.*, **112**, 319 (1975).

227. T. Komaita, K. Naito, and S. Takei, *J. Chromatogr.*, **114**, 1 (1975).

228. D. E. Martire, *Adv. Anal. Chem. Instrum.*, **6**, 93 (1968).

229. T. L. Hill, *J. Chem. Phys.*, **14**, 263 (1964).

230. J. R. Conder, *Anal. Chem.*, **48**, 917 (1976).

231. J. Serpinet, *Anal. Chem.*, **48**, 2264 (1976).

232. G. Lopez-Mellado and R. Kobayashi, *Pet. Refin.*, **39** (2), 125 (1960).

233. F. I. Stalkup and R. Kobayashi, *AIChE J.*, **9**, 121 (1963); *J. Chem. Eng. Data*, **8**, 564 (1963).

234. K. T. Koonce and R. Kobayashi, *J. Chem. Eng. Data*, **9**, 490, 494, (1964).

235. K. T. Koonce, H. A. Deans, and R. Kobayashi, *AIChE J.*, **11**, 259 (1965).

236. H. L. Chang, L. J. Hurt, and R. Kobayashi, *AIChE J.*, **12**, 1212 (1966).

237. L. D. van Horn and R. Kobayashi, *J. Chem. Eng. Data*, **12**, 294 (1967).

238. H. L. Chang and R. Kobayashi, *J. Chem. Eng. Data*, **12**, 517, 520 (1967).

239. P. S. Chappelear, K. Asano, T. Nakahara, and R. Kobayashi, *Proc. Annu. Conv. Nat. Gas Process. Assoc.*, 23 (1970).

240. K. Asano, T. Nakahara, and R. Kobayashi, *J. Chem. Eng. Data*, **16**, 16 (1971).

241. Y. Arai, P. S. Chappelear, and R. Kobayashi, *Bull. Jap. Pet. Inst.*, **15**, 156 (1973).

242. I. Wichterle and R. Kobayashi, *J. Chem. Eng. Data*, **17**, 4, 9, 13 (1972).

243. P. S. Chappelear, R. Stryjek, and R. Kobayashi, *Proc. Conf. Nat. Gas Res. Technol.*, **3**, 1 (1972).

244. H. J. Arnikar, T. S. Lao, and A. A. Bodhe, *J. Chem. Educ.*, **47**, 826 (1970).

245. R. Barrett and T. Stewart, *J. Chem. Educ.*, **49**, 492 (1972).

246. A. G. Vitenberg, B. V. Ioffe, and V. N. Borisov, *Chromatographia*, **7**, 610 (1974).

247. A. G. Vitenberg, I. L. Butaeva, and Z. St. Dimitrova, *Chromatographia*, **8**, 693 (1975).

248. A. G. Vitenberg, B. V. Ioffe, Z. St. Dimitrova, and T. P. Strukova, *J. Chromatogr.*, **126**, 205 (1976).

249. S. P. Wasik, *J. Chromatogr. Sci.*, **12**, 845 (1974).

250. H. Hachenberg and A. P. Schmidt, *Gas Chromatographic Headspace Analysis*, Heydon, London, 1977.

CHAPTER 6

Complexation

6.1 INTRODUCTION

In 1948 Benesi and Hildebrand[1] reported the discovery of a UV absorption band for a solution of iodine and benzene. The band occurred at 297 nm and had a molar absorptivity (based on the iodine concentration) of 9600. Because the band height varied linearly with the concentration of either component, the evidence seemed to suggest a 1 : 1 stoichiometric complex: benzene was taken to be a Lewis base, iodine, a Lewis acid, and the resultant complex, a Lewis acid-base adduct. Strongly colored complexes had been known for at least a hundred years[2-6] but in terms of the state of chemistry at the time were inexplicable. The benzene-iodine complex however, gave new impetus to the study of weak intermolecular interactions and at least 10 books[7-12] that review various aspects of the subject, adequate testimony to its importance, have appeared since 1950. Nonetheless, intermolecular complexation remains, by and large, an enigma (as might well be said of solution interactions in general) and is still a controversial area of chemistry. Gas chromatography has from many points of view contributed substantially to the understanding of these phenomena but, from the standpoint of conventional theories, recent developments seem only to have added to the present state of confusion.

Chromatography has a well-founded tradition in the investigation of complexation dating back, in fact, at least 35 years, when, in much of the early work, charge transfer "derivatives" of aromatic hydrocarbons were utilized for the LC separation of mixtures or purification of individual solutes: for example, in 1940 Sinomiya[13] resolved complexes of nitrotoluene and substituted naphthalenes on an alumina column; Jones and Neuworth[14] derivatized polycyclic aromatic hydrocarbons with *sym*-trinitrobenzene which were subsequently purified on alumina and Super-Cel; Orchin and Woolfolk[15] employed 2,4,7-trinitrofluorenone (TNF) in a similar manner; Ayres and Mann[16] and Smets and co-workers[17] nitrated polystyrene resins for the column chromatography of aromatic hydro-carbons; Parihar and co-workers[18] used explosives (such as 2,4,6,2',4',6'-hexanitrodiphenyl sulfide) as acceptors impregnated in silica gel for the thin-layer chromatographic separation of vitamin-A-related mixtures; and Gil-Av and co-workers[19] and Heath and co-workers[20] employed rhodium and silver salts for the "high-performance" liquid chromatographic separation of olefins.

Complexation interactions have also been employed in analytical gas chromatography. In 1955 Bradford, Harvey, and Chalkley[21] reported the enhanced separation of ethylene from ethane with a column containing silver nitrate. Norman[22] used TNF in 1958 to resolve the three nitrotoluene isomers and Langer and co-workers[23] found in 1960 that di-alkyl tetrahalophthalate esters were selective stationary phases for aromatic hydrocarbon mixtures. Cooper and co-workers[24] studied a variety of aromatic hydrocarbons and amines with TNF, the remarkable selectivity of which is illustrated in Table 6.1. Martire and co-workers[25] noted similar results for aromatic hydrocarbons with tetra-n-butyl pyromellitate. Steric hindrance[26,27] to charge transfer is thought to play a major role in the utility of stationary phases of this kind.[28]

Multicomponent liquid phases have also been used in efforts to enhance separations[29-65] in which, in most cases, some form of the solute partition coefficient has been found to vary linearly with the composition of the

Table 6.1 Vapor Pressures and Specific Retention Volumes of Aromatic Amines[24] at 180°C

| | | V_g^0 (ml/g) | |
| | p_1^0 | | |
Solute	(torr)	Silicone Oil	TNF
Aniline	675.6	24.6	168
p-Toluidine	438.7	35.3	292
N-Methylaniline	495.5	36.6	200
o-Toluidine	447.9	37.7	273
N,N-Dimethyl-o-toluidine	667.8	37.9	28.8
m-Toluidine	409.4	38.7	298
N,N-Dimethylaniline	530.9	41.3	141
N,N-Dimethyl-2,6-xylidine	511.1	50.7	28.8
p-Ethylaniline	393.6	57.1	324
2,4-Xylidine	309.0	57.1	476
N,N-Dimethyl-2,4-xylidine	447.0	57.4	41.8
N,N-Dimethyl-p-toluidine	340.4	62.0	241
N,N-Diethylaniline	288.4	73.2	108
N,N-Dimethyl-p-$tert$-butylaniline	110.7	156	204

phase, albeit not in conformity with conventional solution theories.[66] Laub and Purnell[67, 68] rationalized this behavior only recently and have placed the use of mixed phases on a quantitiative basis; the significance of these findings is discussed later.

In general terms charge transfer reagents and interactions have been found to be useful in analytical chromatography because of their selectivity either as derivatizing agents or stationary phases. However, an equal (if not greater) number of studies have been conducted by GC with regard to the fundamental solution interactions involved in complexation in which (as here) much of the emphasis is placed on the determination of the stoichiometric complexation equilibrium (formation) constant K_1.

6.2 CLASSIFICATION OF TYPES OF COMPLEXATION INTERACTION

Donors and acceptors are generally classified according to the type of electron transfer thought to occur:[11]

Donors (D)

Electron taken from	Type	Example
Nonbonding lone pair	n	$R_3N:$, RO:
Bonding π orbital	$b\pi$	Benzene

Acceptors (A)

Electron goes to	Type	Example
Vacant orbital	v	BCl_3, Ag^+
Antibonding π orbital	$a\pi$	TNF
Antibonding σ orbital	$a\sigma$	I_2, RH

On this basis silver ion-olefin complexes are v-$b\pi$ interactions, TNF–benzene is an $a\pi$-$b\pi$ complex, and hydrogen bonding is classified as n-$a\sigma$. Donors and acceptors may also be characterized as one-electron or two-electron Lewis acids and bases:

A charge transfer acceptor shares (gains) part (or all) of the donor electron that is placed in the lowest unoccupied acceptor molecular orbital (LUMO); it is a Lewis acid.

A charge transfer donor shares (loses) part (or all) of one of its electrons from the highest occupied molecular orbital (HOMO); it is a Lewis base.

Charge transfer interactions may be further classifed according to the degree of electron sharing (cf. Ref. 11, Chapter 16). *Inner* complexes result from close approach of donor and acceptor species with resultant strong interaction. Donors and acceptors may in addition form ions (D^+, A^-) such that bonding becomes appreciably ionic in character; for example, crystals of the tetramethyl-*p*-phenylenediamine (TMPPD)/*p*-chloranil complex exhibit photoconduction, semiconduction, and paramagnetism,[11] although the latter is smaller than would be expected if the compounds were totally ionic. *Outer* complexes are bound together by a relatively weak covalently shared electron at a greater internuclear distance and exhibit the properties of inner complexes to a lesser extent. Each of these types is strongly influenced by solvents; for example, TMPPD–chloranil forms outer complexes in cyclohexane but inner complexes in even moderately polar solvents that appear to solvate the organic ions, thus stabilizing the inner state.[69]

Inner and outer complexes are easily differentiated by changes in absorption spectra; for example, outer-complex crystals show approximately the same infrared spectra as the individual components but inner complexes yield very different IR spectra from either the donor or acceptor.[70] Visible spectra of some *b*π-*a*π outer complexes show a band attributable to charge transfer but the spectra for inner complexes consist of two bands, presumably due to the ionic species D^+ and A^-, in which transfer of an electron is virtually complete.[71, 72] Thus inner and outer complexes differ substantially, the former consisting esssentially of an ion pair, D^+–A^-, the latter being a loosely held covalent complex in which electron transfer is minimal.

Finally, *contact* charge transfer is said to result from a random collision of donor and acceptor species. In a solution containing both A and D in appreciable quantities the number of collisions would be expected to be large on statistical grounds alone, giving rise to discrete contact charge transfer spectra; this hypothesis has been used to explain the apparent "complexation" of iodine with heptane and other solvents. Contact pairs differ from the above mentioned types in that the intermolecular orientation is almost certainly random.

Hydrogen-bonded complexes are a special case of charge transfer, being *n-a*σ interactions in which two electrons are involved. The "normal" hydrogen bond $D \cdots H\text{-}R$ is an outer complex in which both D and A are polarized slightly but certainly not ionized. Inner complexes[73] are also known [e.g., pyridine-methyl iodide and normal alcohols with donors such as $:N(C_2H_5)_3$], the existence of which serves to justify the classification of hydrogen bonding as merely a specific type of charge transfer.

Purnell[74] has further described the various solution interactions possible in gas chromatography; here, solutes are identified as donors (D) with acceptor (A) additives in solvents (S):

CLASS A.   Solute interacts with additive to give complexes of the type

    1. $A-D_n$ 2. A_m-D 3. A_m-D_n

where $m, n \geqslant 1$. In Class A types of interaction the donor solute reacts only with the acceptor additive that is dissolved in an inert solvent, an example being the elution of benzene from a column containing Celite coated with a solution of TNF dissolved in di-n-butyl phthalate (DNBP). Because the solute is effectively at infinite dilution in the stationary phase higher (greater than $1:1$), nonionic complex formation is rare.

CLASS B.   Solute interacts with pure (complexing) solvent to give complexes of the type

    1. $S-D_n$ 2. S_p-D 3. S_p-D_n

where $n, p \geqslant 1$. Here the stationary phase might be (liquid) TNF.

CLASS C.   Solute reacts with itself:

    1. Polymerizes in solution   2. Depolymerizes in solution

An example of such a solute is acetic acid, which as a vapor is a dimer at moderate temperatures but exists as a monomer in solvents such as water.

CLASS D.   Additive reacts with solvent to give complexes of the type

    1. $S-A_m$ 2. S_p-A 3. S_p-A_m

In Class D interactions between the additive and solvent are examined (inferentially) with inert solutes; for example, discrete complexes between barium and diglycerol were identified when their mixtures were used as stationary phases and the retention volumes of inert solutes measured as a function of the normality of the solution.[75]

The two types of interaction of greatest interest in GC are Classes A and B and various methods have been proposed for determining formation constants for each; although Classes C and D have been little used, they should not be overlooked as useful possibilities for the study of complexation.

6.3 MEASUREMENT OF COMPLEXATION EQUILIBRIUM CONSTANTS BY SPECTROSCOPY

Rose[10] has listed more than 20 methods by which charge transfer equilibrium constants have been measured. The UV/Vis and NMR spectroscopic techniques have by far been used most frequently; they have a direct bearing on matters discussed later; therefore, they are briefly recounted here.

6.3.1 UV/Vis Method The Benesi-Hildebrand[1] method makes use of the variation of the charge transfer absorption band height with the concentration of donor or acceptor (one or the other being held at fixed concentration), both being dissolved in a third supposedly inert solvent:

$$\frac{C_A^t l}{A_{DA}} = \frac{1}{\varepsilon_{DA} K_1 C_D^t} + \frac{1}{\varepsilon_{DA}} \tag{6.1}$$

where C_A^t and C_D^t are the total concentrations of acceptor and donor, A_{DA} and ε_{DA} are the absorbance and absorptivity of the charge transfer band, and l is the cell path length. By holding C_A^t constant A_{DA} is measured as a function of C_D^t (which is in excess over C_A^t); a plot of the lhs of eq. 6.1 versus $1/C_D^t$ then yields K_1 from the intercept/slope quotient. Precautions concerning the use of this relation have been discussed thoroughly elsewhere.[11]

Scott[76] has advocated a modified version of the form

$$\frac{C_A^t C_D^t l}{A_{DA}} = \frac{1}{\varepsilon_{DA} K_1} + \frac{C_D^t}{\varepsilon_{DA}} \tag{6.2}$$

which has the advantage of extrapolation through decreasing concentration, for the lhs is plotted versus C_D^t.

6.3.2 NMR Method Foster[12] has shown that the NMR analogue of the Benesi-Hildebrand equation is

$$\frac{1}{\Delta} = \frac{1}{\Delta^0 K_1 C_D^t} + \frac{1}{\Delta^0} \tag{6.3}$$

where Δ^0 is the difference between the NMR shift of pure acceptor and complex ($= \delta_A - \delta_{DA}$) and Δ is the observed NMR shift of a given mixture at some value of C_D^t ($= \delta_{obs} - \delta_A$); $1/\Delta$ is plotted versus $1/C_D^t$ and K_1 is determined as above from the intercept/slope quotient. The rearranged form of eq. 6.3 analogous to the Scott modification is

$$\frac{\Delta}{C_D^t} = K_1 \Delta^0 - K_1 \Delta \qquad (6.4)$$

where Δ / C_D^t is plotted versus Δ. Extrapolation to the ordinate thus avoids passing through a region of infinite donor concentration.

6.3.3 Comparison of Spectroscopic Results Comparison of spectroscopically derived formation constants yields conflicting results. Table 6.2 lists the concentration formation constants for several complexes determined at

Table 6.2 Comparison of Optical and NMR Formation Constants[12, 77]

Solvent	Donor	Acceptor	K_1 (l/mole) Optical	NMR
Chloroform	N,N-Dimethylaniline	sym-Trinitro-benzene	0.68	0.41
1,2-Dichloroethane	N,N,N',N'-tetramethyl-p-phenylenediamine	sym-Trinitro-benzene	1.45	1.01
1,2-Dichloroethane	Hexamethylbenzene	Fluoranil	3.30	2.66
Carbon tetrachloride	Hexamethylbenzene	Fluoranil	16.5	9.70

Table 6.3 Comparison of Optical and NMR Formation Constants[78] with TNF Acceptor at 40°C

Solvent	Donor	$\left(\dfrac{K_1}{1/\text{mole}}\right)$ UV	NMR	$\left(\dfrac{K_1}{\text{mole fraction}}\right)$ UV	NMR
Di-n-butyl succinate	Toluene	0.116	−0.019	−0.045	−0.624
	m-Xylene	0.210	0.072	0.210	−0.173
	o-Xylene	0.167	0.105	0.241	−0.033
Di-n-butyl adipate	Toluene	−0.030	−0.010	−0.710	−0.571
	m-Xylene	0.082	−0.096	−0.246	−0.198
	o-Xylene	...	...	...	...
Di-n-butyl sebacate	Toluene	−0.008	0.053	−0.730	−0.519
	m-Xylene	0.065	0.041	−0.448	−0.448
	o-Xylene	0.145	0.098	−0.180	−0.356

33.5°C by UV and NMR from the data of Emslie and co-workers[77] and unpublished work (cf. Ref. 12, p. 158) by Fyfe and Matheson; in all cases $C_A^t \ll C_D^t$. A more recent comparison has been presented by Purnell and Srivastava,[78] whose UV and NMR data for aromatic hydrocarbons with TNF at 40°C are given in Table 6.3. Many of the formation constants were found to be negative, which is physically meaningless. Laub and Pecsok,[79] in a recent review, also found that UV/Vis and NMR data are rarely in agreement.

These results indicate that optical and NMR charge transfer formation constants must be regarded with some suspicion and agreement between these and other methods, for example, calorimetry,[80] is of doubtful value (see, however, Section 6.7).

6.4 MEASUREMENT OF COMPLEXATION EQUILIBRIUM CONSTANTS BY GC

6.4.1 Method of Gil-Av and Herling Before 1962 several groups had employed silver nitrate-ethylene glycol mixtures for the separation of olefins[21, 81-86] and in 1959 du Plessis and Spong[87] used this phase to determine the dissociation constants of the silver-diammine complex. Gil-Av and Herling[88] and Muhs and Weiss[89] were then able to deduce the following: if silver ion is added to a notionally-inert (noncomplexing) solvent such as ethylene glycol, the enhanced solubility of an olefinic donor solute (e.g., ethylene) can be expressed as

$$C_D = C_D^0 + C_{DA} \qquad (6.5)$$

where C_D is the total molar concentration of solute in the stationary phase, C_D^0 is the concentration of solute in pure (sans additive) solvent, and C_{DA} is the concentration of complex in the additive-solvent solution. [An important presumption of eq. 6.5 is that the solubility of solutes (even in the absence of complexation) is not *decreased* by the presence of additive; i.e., $C_D \geqslant C_D^0$ is assumed always to be true.] The solute liquid/vapor partition coefficient K_R therefore becomes

$$K_R = \frac{C_D^0 + C_{DA}}{C_D^M} \qquad (6.6)$$

where C_D^M is the solute concentration in the mobile phase. Multiplying the rhs of eq. 6.6 by $C_D^0 C_A / C_D^0 C_A$, where C_A is the concentration of additive in the stationary phase, and recognizing that because the complex is at infinite dilution the complex formation constant is given by

$$K_1 = \frac{C_{DA}}{C_D^0 C_A} \tag{6.7}$$

provides

$$K_R = K_{R(S)}^0 + K_{R(S)}^0 K_1 C_A \tag{6.8}$$

where $K_{R(S)}^0$ is the partition coefficient of solute with pure (inert) solvent. A plot of K_R versus C_A should therefore yield a straight line of slope, $K_{R(S)}^0 K_1$, and intercept, $K_{R(S)}^0$, from which K_1 can be calculated. [Note that in GLC it is possible to measure $K_{R(S)}^0$ directly, whereas the analogous spectroscopic equations of Benesi and Hildebrand (or Scott) yield $(\epsilon_{DA} K_1)^{-1}$ and $(\epsilon_{DA})^{-1}$ as the slope and intercept, respectively, but separation of the product terms is by no means straightforward.] Muhs and Weiss[89] determined K_1 values for 131 solutes with silver nitrate/ethylene glycol, including alkenes, alkynes, dienes, trienes, and aromatic hydrocarbons; the formation constants ranged from 0.1 to > 1000 l/mole (measured with columns of 0.00, 0.85, 1.59, 2.38, 3.13, and 4.18 M Ag$^+$), which indicated that the GLC technique was useful for very weak to very strong interactions.

Many workers have employed this and related methods for the study of a wide variety of ionic[90-121] and nonionic[122-156] systems; these have been reviewed comprehensively by Wellington,[157] Laub and Pecsok,[79] Laub and Wellington,[158] and Vigdergauz and co-workers.[159, 160] Wellington[157] has published an extensive list of K_1 data.

There are, however, implied limitations to the use of eq. 6.8 as well as practical difficulties encountered with its application. First, in rearranged form,

$$K_R - K_{R(S)}^0 = K_{R(S)}^0 K_1 C_A \tag{6.9}$$

Letting $K_1 = 0.1$ l/mole and $K_{R(S)}^0 = 100$, $K_R - K_{R(S)} = 5$ at $C_A = 0.5$ M, a relatively small difference between the two partition coefficients; that is, the equipment must be capable of high precision. If K_1 were 0.01 l/mole and $C_A = 0.2$ M, $K_R - K_{R(S)} = 0.10$. At the other extreme, that is, strong interactions, C_A will have to be made small. Let $K_1 = 10^6$ l/mole; for a 10% change in K_R, C_A must be 10^{-7} M and injections of even trivial amounts of solute may overload the additive; that is, produce higher than 1:1 equilibria if the donor and acceptor are so inclined (as with silver ion and benzene). Second, a serious problem arises when suitable "inert" solvents are sought. With the exceptions of DDQ in di-alkyl phthalates and inorganic salts in glycols, most complexing agents are only slightly soluble in common GC stationary phases. TNF, for example, is soluble only to the extent of about 0.2 M in DNBP and for all practical purposes, insoluble in less "polar" liquids. No general remedy has yet been found and it is therefore possible to use only low additive concentrations which, for weak interactions, decrease proportionately the accuracy of K_1 measurements.

6.4.2   Method of Martire and Riedl   Martire and Riedl,[161] following Langer and co-workers,[23] factorized the solute activity coefficient into two components: in terms of the specific retention volume V_g^0,

$$V_g^0 = \frac{273R}{\gamma_1^\infty (1-c)p_1^0 MW_L} \tag{6.10}$$

where c is the fraction of solute that is complexed and γ_1^∞ refers to dissolved but uncomplexed solute. In other words, it is assumed that the complex has zero vapor pressure, and although the solute is complexed it does not move down the column (if $c=1$, $V_g^0 = \infty$). Thus eq. 6.10 divides the behavior of the complexing solute into two portions: γ_1^∞, the free solute activity coefficient, and $(1-c)$, the fractional amount of free solute. The problem now reduces to one of relating c to K_1.

The thermodynamic equilibrium constant is given by

$$K_{eq} = \frac{a_{DA}}{a_D a_A} = K_1 \frac{\gamma_{DA}}{\gamma_D \gamma_A} \tag{6.11}$$

Adopting the following conventions,

$$\gamma_{DA}^\infty \to 1 \quad \text{and} \quad a_{DA} \to C_{DA} \quad \text{as} \quad C_{DA} \to 0$$

$$\gamma_D^\infty \to 1 \quad \text{and} \quad a_D \to C_D^0 \quad \text{as} \quad C_D^0 \to 0$$

gives

$$K_{eq} \to \frac{C_{DA}}{C_D^0 a_A} \tag{6.12}$$

A new constant, K', is now defined:

$$K' = \frac{C_{DA}}{C_D^0} = a_A K_{eq} \tag{6.13}$$

The fraction of complexed solute c is thus given by

$$c = \frac{C_{DA}}{C_{DA} + C_D^0} = \frac{C_{DA}}{C_{DA}/K' + C_{DA}} = \frac{K'}{1 + K'} \tag{6.14}$$

or

$$\frac{1}{1 - c} = 1 + K' \tag{6.15}$$

which, when substituted into eq. 6.10, provides

$$V_g^0 = \frac{273\,R}{\gamma_1^\infty p_1^0 MW_L}(1 + K') \tag{6.16}$$

K' cannot be calculated from a single V_g^0 measurement because it is not possible at this point to separate the quotient terms $(1 + K')$ and γ_1^∞. To do so four additional specific retention volumes must be defined:

$$V_g^{0(A)} = \frac{273\,R}{\gamma_A^\infty p_N^0 MW_N} \tag{6.17}$$

$$V_g^{0(B)} = \frac{273\,R}{\gamma_B^\infty p_N^0 MW_C} \tag{6.18}$$

$$V_g^{0(C)} = \frac{273\,R}{\gamma_C^\infty p_C^0 MW_C}(1 + K') \tag{6.19}$$

$$V_g^{0(D)} = \frac{273\,R}{\gamma_D^\infty p_C^0 MW_N} \tag{6.20}$$

$V_g^{0(A)}$ is the specific retention volume of a *noncomplexing* solute with vapor pressure p_N^0 and a *noncomplexing* stationary phase of molecular weight MW_N; $V_g^{0(B)}$ is that for the same solute with a pure *complexing* stationary phase of molecular weight MW_C; $V_g^{0(C)}$ relates to a *complexing* solute with vapor pressure p_C^0 in the *complexing* phase and $V_g^{0(D)}$, to the *complexing* solute in the *noncomplexing* phase. If the stationary phases have similar physical properties (except for complexation behavior), then[161]

$$\frac{\gamma_D^\infty}{\gamma_C^\infty} = \frac{\gamma_A^\infty}{\gamma_B^\infty} \tag{6.21}$$

that is, the ratio of infinite-dilution activity coefficients for (free) *complexing* solute in inert and complexing phases is equal to the ratio of *noncomplexing* solute activity coefficients in the same phases. This seems to be a reasonable assumption because *free* complexing solute present in the complexing stationary phase is by definition not complexed and should therefore elute in a manner similar to the noncomplexing solute.

Taking the ratios of eqs. 6.17/6.18 and 6.19/6.20 followed by rearrangement yields

$$K' = \frac{V_g^{0(A)} V_g^{0(C)}}{V_g^{0(B)} V_g^{0(D)}} - 1 \tag{6.22}$$

To calculate K_{eq} from eq. 6.22 the pure stationary phase activity a_A or activity coefficient γ_A must be known. Martire and Riedl[161] reasoned that γ_A is given by

$$\gamma_A = \frac{\gamma_A^\infty}{\gamma_B^\infty} = \frac{V_g^{0(B)} MW_C}{V_g^{0(A)} MW_N} \tag{6.23}$$

where the underlying assumption is that differences between two closely related stationary phases can be measured by the ratio of activity coefficients of an inert (noncomplexing) solute. For two such phases the charge transfer equilibrium constant then becomes

$$K_{eq} = \frac{K'}{a_A} = \frac{K' \overline{V}_A}{\gamma_A} \tag{6.24}$$

where $\overline{V}_A$ is the molar volume of the pure complexing phase. The desired relation is finally arrived at by combining eqs. 6.22–6.24:

$$K_{eq} = \left(\frac{\overline{V}_A}{\gamma_A} \right) \left(\frac{V_g^{0(A)} V_g^{0(C)}}{V_g^{0(B)} V_g^{0(D)}} - 1 \right) \tag{6.25}$$

Thus K_{eq} values are said to be calculable from the ratio of four specific retention volumes (or corrected retention times) and $\overline{V}_A$, MW_C, and MW_N.

Equation 6.25 is an interesting development in the study of charge transfer by GLC because apparently it enables the determination of true thermodynamic equilibrium constants. Several questions are immediately raised, however, about the assumptions that were used. First, the ratio, $V_g^{0(B)}/V_g^{0(A)}$, was said to be a measure of the correspondence of the

Table 6.4 Specific Retention Volumes for Noncomplexing Alkanes with Closely Related Phases[161]

Solute	V_g^0, ml/g, at T, °C						
	22.5	30	40	50	60	70	80

A. *n*-Heptadecane Stationary Phase

n-Pentane	173.6	133.4	95.24	69.54	52.52	31.38	
iso-Pentane	125.2	97.21	71.23	53.08	40.94	25.08	
n-Hexane	575.6	418.3	286.7	197.1	140.5	75.83	
2-Methylpentane	396.4	293.4	204.8	144.6	104.9	58.59	
3-Methylpentane	460.4	343.2	236.2	165.5	119.7	66.20	
2,2-Dimethylbutane	260.9	198.3	141.9	102.3	75.95	44.43	
2,3-Dimethylbutane	372.5	279.4	195.0	138.9	101.5	57.53	
n-Heptane	1860.	1303.	831.9	539.4	366.0	179.9	
3-Methylhexane	1375.	975.2	634.4	419.7	290.1	147.0	
2,4-Dimethylpentane	811.2	583.6	391.0	266.2	187.9	99.49	

B. Di-*n*-octyl Ether Stationary Phase

n-Pentane	175.5	132.8	93.38	67.31			
iso-Pentane	127.1	97.26	69.98	52.14			
n-Hexane	576.4	412.6	279.0	191.4			
2-Methylpentane	398.7	290.6	200.5	139.9			
3-Methylpentane	462.9	338.7	230.4	160.1			
2,2-Dimethylbutane	261.2	195.9	136.5	98.61			
2,3-Dimethylbutane	372.9	276.4	189.7	133.8			
n-Heptane	1863.	1278.	800.7	514.7			
3-Methylhexane	1376.	962.3	614.3	401.9			
2,4-Dimethylpentane	812.5	580.5	382.4	254.8			

C. Di-*n*-Octyl Ketone Stationary Phase

n-Pentane					55.91	42.23	32.70	25.78
iso-Pentane					42.92	33.06	26.01	20.71
n-Hexane					155.0	111.0	81.86	61.50
2-Methylpentane					113.8	83.06	62.42	47.69
3-Methylpentane					130.6	95.06	71.24	54.09
2,2-Dimethylbutane					81.19	60.81	46.72	36.23
2,3-Dimethylbutane					111.7	81.09	61.14	47.09
n-Heptane					414.7	282.5	199.0	142.8
3-Methylhexane					326.3	225.5	162.2	117.7
2,4-Dimethylpentane					208.7	148.4	108.5	80.68

noncomplexing and complexing phases; that is, their identity in all respects except that one forms complexes, whereas the other does not. Martire and Riedl[161] initially examined di-n-octyl ether/n-heptadecane and di-n-octyl ketone/n-heptadecane as pairs of phases that are similar in all properties except that the first member of each forms hydrogen bonds with alcohols. If this is the case, the noncomplexing (alkane) solute ratio $V_g^{0(B)}/V_g^{0(A)}$ should be a constant for a particular solvent pair at a given column temperature and should also be independent of the alkane solute. Table 6.4 lists the specific retention volumes of 10 alkanes with n-heptadecane, di-n-octyl ether, and di-n-octyl ketone at several temperatures; $V_g^{0(B)}/V_g^{0(A)}$ ratios are given in Table 6.5 and the average values, in Table 6.6. The standard deviations for the latter data are approximately 1%. Clearly $V_g^{0(B)}/V_g^{0(A)}$ is independent of the choice of noncomplexing alkane solute.

Table 6.5 $V_g^{0(B)}/V_g^{0(A)}$ Values for Alkanes[161]

Solute	$V_g^{0(B)}/V_g^{0(A)}$ at T, °C						
	22.5	30	40	50	60	70	80

A. Di-n-octyl Ether/n-Heptadecane Stationary Phases

Solute	22.5	30	40	50	60	70	80
n-Pentane	1.001	0.996	0.981	0.968			
iso-Pentane	1.015	1.001	0.983	0.982			
n-Hexane	1.001	0.986	0.973	0.971			
2-Methylpentane	1.006	0.991	0.980	0.968			
3-Methylpentane	1.005	0.987	0.976	0.967			
2,2-Dimethylpentane	1.001	0.988	0.962	0.964			
2,3-Dimethylbutane	1.001	0.989	0.973	0.963			
n-Heptane	1.002	0.981	0.963	0.954			
3-Methylhexane	1.000	0.987	0.968	0.958			
2,4-Dimethylpentane	1.002	0.995	0.978	0.957			

B. Di-n-octyl Ketone/n-Heptadecane Stationary Phases

Solute	22.5	30	40	50	60	70	80
n-Pentane					0.804	0.804	0.812	0.822
iso-Pentane					0.809	0.808	0.815	0.826
n-Hexane					0.786	0.790	0.799	0.811
2-Methylpentane					0.787	0.792	0.803	0.814
3-Methylpentane					0.789	0.794	0.808	0.817
2,2-Dimethylbutane					0.794	0.801	0.810	0.815
2,3-Dimethylbutane					0.804	0.799	0.807	0.869
n-Heptane					0.769	0.772	0.780	0.794
3-Methylhexane					0.778	0.778	0.791	0.817
2,4-Dimethylpentane					0.784	0.790	0.800	0.811

Table 6.6   Average $V_g^{0(B)}/V_g^{0(A)}$ Values[161]

Solvent Pair	Av $V_g^{0(B)}/V_g^{0(A)}$ at T, °C						
	22.5	30	40	50	60	70	80
Di-n-octyl ether/n-heptadecane	1.005	0.990	0.973	0.965			
Di-n-octyl ketone/n-heptadecane					0.790	0.793	0.802	0.813

Next, the validity of eq. 6.23 must be questioned; namely the accuracy of the γ_A values. It can be demonstrated[161] that the relation will be valid only if the phases have similar molar volumes; the values for n-heptadecane, di-n-octyl ether, and di-n-octyl ketone are 317.4, 308.3, and 313.0 ml/mole, respectively, at 50°C; therefore this condition is adequately met. The polarizabilities of each phase (314, 303, and 314×10^{-25} ml/molecule) are also nearly identical, so that London dispersion forces, van der Waals interactions, and all other solute/solvent behavior (except complexation) will be about the same for each of the three phases.

So far Martire and co-workers have applied their method to di-n-octyl ether and di-n-octyl ketone,[161] di-n-octyl thioether,[162] di-n-octyl methylamine and tri-n-hexylamine,[163] and di-n-octyl methylamine again.[164] This technique may be employed to a greater extent in the future if a larger number of suitable complexor/reference phase pairs can be found.

6.4.3   Method of Langer, Johnson, and Conder

An alternative method of determining K_1 values has been reported by Langer and co-workers[165]: solute partition coefficients are determined for several homologous complexing solvents K_R plotted versus the molar concentration (inverse molar volume) of each phase, and K_1 is found from the slope and intercept of the line; for example, these workers used di-n-propyl tetrachlorophthalate, di-n-butyl tetrachlorophthalate, di-n-amyl tetrachlorophthalate, and di-n-octyl tetrachlorophthalate and plotted the partition coefficients of 36 solutes against the inverse molar volumes of the phases; several lines were curved, perhaps indicating that higher complexes were formed or other interactions contributed to retention. This method may still prove useful for the investigation of modifications to complexing solvents (in this case, increased alkyl substitution in the acceptor).

6.4.4   Method of Cadogan and Sawyer

Cadogan and Sawyer[166] noted that the Gil-Av/Herling equation should be applicable to gas-solid chromatography, where the concentration of complexing agent in the stationary phase is replaced by the surface concentration of an acceptor salt in a mixed-solid coating (e.g., $LaCl_3/NaCl$; units are moles/m^2) and where the adsorbent surface area (m^2/g) and the column packing weight give the total surface area of stationary (adsorbent) phase. It was assumed that a 10% w/w salt mixture would cover the support surface to the extent of only one monolayer or less and that C_A could be calculated by dividing the number of moles of additive by the surface area in the column. Their data for aromatic hydrocarbon solutes with $LaCl_3$ at 200°C are given in Table 6.7. They also pointed out that such studies could prove useful in a comparison of various transition metal salts as complexing agents and in the investigation of gas-solid catalysis.

Table 6.7   Formation Constants[166] for Aromatic Hydrocarbons-LaCl₃ at 200°C

Solute	K_1 (m²/mole, $\times 10^{-5}$)
Benzene	2.09
Toluene	4.37
Ethylbenzene	5.06
iso-Propylbenzene	4.43
Fluorobenzene	3.00
Chlorobenzene	2.17
Bromobenzene	2.84
Iodobenzene	2.75
1-Hexene	1.62

6.5   COMPARISON OF GLC AND SPECTROSCOPIC DATA

In contrast to the spectroscopic data given in Table 6.3, Purnell and Srivastava[78] found that the GLC K_1 values for the same donors, solvents, and TNF were all positive and decreased as the temperature was increased (see Table 6.8). Martire and co-workers[162, 163, 167] compared GLC and NMR K_1 data for haloform (hydrogen-bonding) acceptors with the donors, di-n-octyl ether, di-n-octyl thioether, and di-n-octylmethylamine at 10–60°C. The GLC K_1 values were consistently larger than those derived from NMR, the differences being attributed to contact charge transfer. Table 6.9 lists a portion of their data for these systems.

Chamberlain and Drago[168] recently found that heats of complex formation for hydrogen-bonding systems determined by calorimetry or infrared frequency-shift correlations[169] compared moderately well with those of the Martire-Riedl GLC method; the data are shown in Table 6.10.

Table 6.8 GLC K_1 Values for Aromatic Hydrocarbons with TNF[78]

Solvent	Donor	K_1 (l/mole, at T, °C) 40	50	60
Di-*n*-butyl succinate	Benzene	0.590	0.464	0.423
	Toluene	0.702	0.521	0.449
	m-Xylene	0.825	0.638	0.628
	o-Xylene	0.871	0.671	0.603
	p-Xylene	0.764	0.649	0.613
	Ethylbenzene	0.615	0.444	0.317
Di-*n*-butyl adipate	Benzene	0.481	0.416	0.374
	Toluene	0.491	0.451	0.417
	m-Xylene	0.615	0.561	0.470
	o-Xylene	0.606	0.581	0.506
	p-Xylene	0.624	0.550	0.498
	Ethylbenzene	0.448	0.338	0.288
Di-*n*-butyl sebacate	Benzene	0.353	0.309	0.262
	Toluene	0.332	0.281	0.240
	m-Xylene	0.401	0.343	0.329
	o-Xylene	0.393	0.371	0.327
	p-Xylene	0.425	0.381	0.322
	Ethylbenzene	0.355	0.251	0.208

Table 6.9 GLC and NMR K_1 (l/mole) Data[167] for Chloroform and Bromoform Acceptors with DOE, DOTE, and DOMA Donors at 20–60°C

System	20°C K_1^{GLC}	K_1^{NMR}	40°C K_1^{GLC}	K_1^{NMR}	60°C K_1^{GLC}	K_1^{NMR}
$CHCl_3$/DOE	0.458	0.327	0.348	0.225	0.278	0.162
$CHBr_3$/DOE	0.467	0.280	0.368	0.212	0.299	0.166
$CHCl_3$/DOTE	0.469	0.317	0.379	0.250	0.314	0.203
$CHBr_3$/DOTE	0.830	0.337	0.649	0.276	0.524	0.232
$CHCl_3$/DOMA	0.573	0.374	0.390	0.237	0.281	0.159
$CHBr_3$/DOMA	0.773	0.417	0.542	0.276	0.399	0.193

Table 6.10 Comparison[168] of GLC and Calorimetric/Infrared Frequency-Shift Data for Enthalpies of Complexation of Named Acceptors with DOMA

Acceptor	ΔH^{static} (kcal/mole)	ΔH^{GLC} (kcal/mole)
2,2,2-Trifluoroethyl alcohol	5.9	6.0
tert-Butyl alcohol	3.0	3.1
iso-Propyl alcohol	3.2	3.2
n-Propyl alcohol	3.5	3.8
n-Butyl alcohol	3.5	3.8
iso-Butyl alcohol	3.4	3.9
sec-Butyl alcohol	3.2	3.4

The only comparison presented so far that tests the internal consistency of the GLC techniques was made by Liao, Martire, and Sheridan.[164] The (concentration) formation constant K_1 should approach the equilibrium constant K_{eq} as the relevant activity coefficients approach unity. Because the concentration of (donor) solute and complex are at infinite dilution ($\gamma_D^\infty \to 1$ and $a_D \to C_D^0$ as $C_D \to 0$),

$$K_{eq} \cong \frac{C_{DA}}{C_D^0 a_A} \cong \frac{K_1}{\gamma_A} \qquad (6.26)$$

Martire and co-workers[164] chose to examine di-*n*-octylmethylamine and *n*-octadecane as the complexing and reference phases, whose molecular weights and molar volumes are 255.5 g/mole, 254.5 g/mole, 325.4 ml/mole, and 331.2 ml/mole, respectively, at 40°C. As expected, the ratio of specific retention volumes of noncomplexing alkanes was constant, irrespective of the choice of solute (the average $V_g^{0(B)}/V_g^{0(A)}$ value was 0.9876 ± 0.0020). It was therefore assumed that the only difference between the phases was that di-*n*-octyl methylamine formed complexes, whereas

Table 6.11 Comparison of GLC K_1 and K_{eq} Data[164] for DOMA at 40°C

Solute	K_1 [l/mole (eq. 6.8)]	K_{eq} (eq. 6.25)
$CHCl_3$	0.405 ± 0.019	0.403 ± 0.006
CH_2Cl_2	0.179 ± 0.014	0.187 ± 0.004
CH_2Br_2	0.222 ± 0.004	0.219 ± 0.004

n-octadecane did not. They also found (eq. 6.23) that γ_D for the former was 0.993 ± 0.002 1/mole so that $K_1 \cong K_{eq}$. They next determined K_1 and K_{eq} values via eqs. 6.8 and 6.25 for the solutes $CHCl_3$, CH_2Cl_2, and CH_2Br_2; the results are presented in Table 6.11, where the agreement is sufficient to indicate that the GLC technique offers an internally consistent alternative to spectroscopy for the measurement of charge transfer interactions.

6.6 DIACHORIC SOLUTIONS

6.6.1 Experimental Anomalies Of the many anomalies in spectroscopic charge transfer studies, perhaps the most striking is the effect of different solvents on the formation constant K_1, regardless of the units in which K_1 is cast.[79] Solvent effects (as well as zero or negative formation constants) have also been reported in GLC. Figures 6.1 and 6.2 show K_R versus C_A plots[170] for various donors with tetracyanoethylene (TCNE) in di-n-butyl adipate (DBA) and di-n-decyl sebacate (DDSEB); a cursory inspection of these figures indicates that TCNE apparently does not form complexes

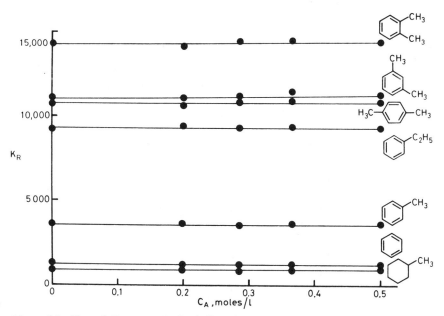

Figure 6.1 Plots of K_R versus C_A for indicated solutes with tetracyanoethylene/di-n-butyl adipate stationary phase mixtures at 25°C. Data of Laub, Meen, and Purnell.[170]

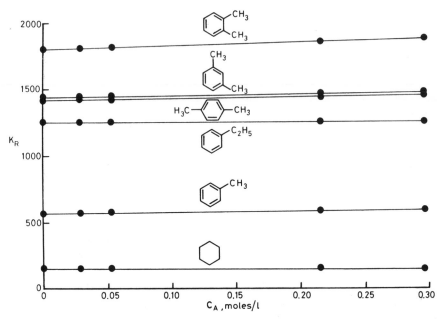

Figure 6.2 Plots of K_R versus C_A for indicated solutes with tetracyanoethylene/di-n-decyl sebacate stationary phase mixtures at 60°C. Data of Laub, Meen, and Purnell.[170]

(zero slopes). Further anomalies are found in the data of Castells,[171] whose K_R versus C_A plots for aromatic hydrocarbon donors with TNB acceptor in di-nonyl phthalate (DNP) solvent are given in Figs. 6.3 and 6.4.

Several of these systems also show zero or negative K_1 values by UV and NMR; for example, two representative Benesi-Hildebrand type plots[170] are shown in Figs. 6.5 and 6.6 for o-oxylene-TCNE in DBA at 25°C and toluene/TCNE in DDSEB at 22°C, respectively. The zero intercepts indicate that in both cases $K_1 = 0$. These findings are somewhat surprising because TCNE is claimed to be second only to tetracyanoquinodimethane (TCNQ) as a accepter as far as strength of interaction is concerned. No explanation can, of course, be given for the negative K_1 data presented so far in terms of traditional charge transfer theory.

Mulliken and Person[11] and numerous others have discussed the solvent-dependent nature of K_1 and the relation between these and vapor-phase spectroscopic data, but an adequate explanation for the discrepancies has not yet been presented; for example, Carter and co-workers[172] and others[173-176] have used semiempirical relations that attempt to redefine the Benesi-Hildebrand equation in terms of strong solvation of D and A by S

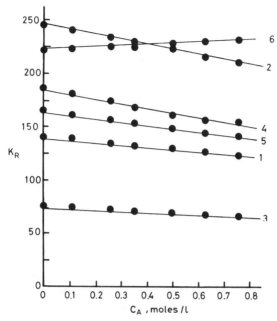

Figure 6.3 Plots of K_R versus C_A for (1) cyclohexane, (2) methylcyclohexane, (3) *n*-hexane, (4) *n*-heptane, (5) 2,2,4-trimethylpentane, (6) benzene, (7) toluene, (8) ethylbenzene, (9) iso-propylbenzene, (10) *o*-xylene, (11) *m*-xylene, and (12) *p*-xylene with *sym*-trinitrobenzene/di-nonyl phthalate stationary phase mixtures at 60°C. Data of Castells.[171]

such that

$$AS_n + DS_m \rightleftharpoons (DA)S_p + qS$$

where the subscripts indicate the number of solvent molecules involved with each species. Carter and co-workers,[172] however, were forced to set $q(m+1)=9$ for iodine, 30 for TNB, and 6 for chloranil with methylbenzene donors in order that $\varepsilon_{DA}^{gas \; phase} \cong \varepsilon_{DA}^{solution}$. These appear to be unrealistic numbers for it would have to be supposed that iodine is more heavily solvated than chloranil, whereas TNB (said to be a weaker acceptor than the latter) is solvated at least three times as heavily as either.

Elaborate theories developed in GLC by several workers, discussed by Laub and Wellington,[157, 158] purport to explain these phenomena, but no coherent theory seems to account for all GLC data heretofore reported; for example, Castells[171] originally plotted $K_R \overline{V}_{(A,S)}/K_{R_{(S)}}^0 \overline{V}_s$ against the acceptor mole fraction in accordance with the theory of Eon and co-

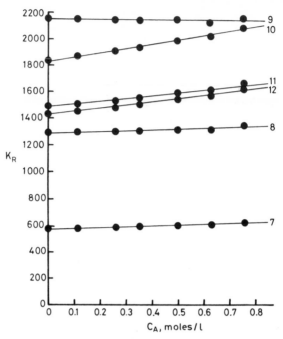

Figure 6.4   Cf. Fig. 6.3.

workers[134, 140] and obtained curved lines. When K_R is plotted versus C_A, as in Figs. 6.3 and 6.4, however, straight lines are found. Similarly, the original data of Eon and Guiochon[134] are plotted in Fig. 6.7 as K_R versus C_A which, when plotted versus x_A, are curved.

Exceptions to the additive-solvent solubility difficulty noted are those systems found by Eon and co-workers[134, 140] and Martire and co-workers[161–164, 177] in which solubility is extant over the entire range, pure solvent to pure acceptor. The interesting aspect of these systems is that K_R versus C_A plots are also found to be linear over the entire solubility range $C_A = 0$ to V_A^{-1}. Spectroscopic Benesi-Hildebrand and Foster-type plots exhibit similar linearity over extended concentration ranges. It appears that few curved plots have been reported in the literature to date for nonionic GLC, UV, and NMR charge transfer data when concentration units have been employed. This, in itself, is anomalous because one would normally suspect that the activity of an acceptor dissolved in a solvent would be greatly affected over a concentration range as narrow as 0.05 M. Yet K_R versus C_A (or analogous spectroscopic) plots are found to be linear and in the cases investigated by Eon and Martire are linear over the *entire* concentration range, pure solvent to pure acceptor.

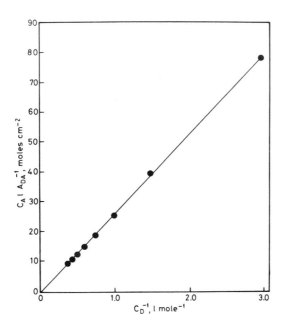

Figure 6.5 Benesi-Hildebrand plot of $C_A l A_{DA}^{-1}$ versus C_D^{-1} for *o*-xylene with tetracyanoethylene/di-*n*-butyl adipate mixtures at 25°C. Data of Laub, Meen, and Purnell.[170]

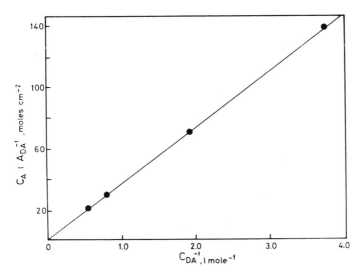

Figure 6.6 Benesi-Hildebrand plot of $C_A l A_{DA}^{-1}$ versus C_D^{-1} for toluene with tetracyanoethylene/di-*n*-decyl sebacate mixtures at 22°C. Data of Laub, Meen, and Purnell.[170]

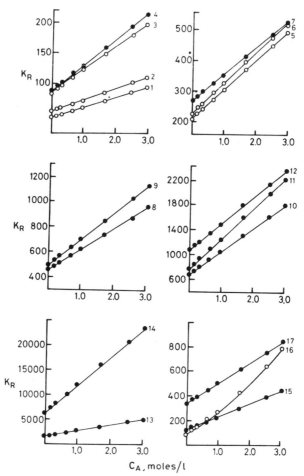

Figure 6.7 Plots of K_R versus C_A for (1) 2-methylfuran, (2) 2,5-dihydrofuran, (3) benzene, (4) thiophene, (5) 2-methylthiophene, (6) 3-methylthiophene, (7) tetrahydrothiophene, (8) 2-ethyl-thiophene, (9) 2,5-dimethylthiophene, (10) 2-bromothiophene, (11) 3-bromothiophene, (12) 2,5-dichlorothiophene, (13) 2-iodothiophene, (14) 3,4-dibromothiophene, (15) 1-methylpyrrole, (16) pyrrole, and (17) 2-chlorothiophene with di-*n*-butyl tetrachlorophthalate/squalane stationary phase mixtures at 80.3°C. Data of Eon, Pommier, and Guiochon.[135]

6.6.2 Diachoric Solutions Approximation Purnell and Vargas de Andrade[178] first attempted to explain the above-noted linear behavior by reexamining the activity coefficients of donor solutes, $\gamma^\infty_{D_{(A,S)}}$, in $A + S$ stationary phase mixtures. If it is assumed that the activity coefficient of D in $A + S$ is a linear function of that of D in A and D in S [i.e., $\gamma^\infty_{D_{(S)}}$ (pure solvent) and $\gamma^\infty_{D_{(A)}}$ (pure additive) are directly related to $\gamma^\infty_{D_{(A,S)}}$], then from

Henry's law

$$\frac{p_D}{p_D^0} = \gamma_{D\,(A)}^{\infty} x_{D(A)} = \gamma_{D\,(S)}^{\infty} x_{D(S)} = \gamma_{D\,(A,S)}^{\infty} x_{D(A,S)} \tag{6.27}$$

where p_D and p_D^0 are the donor vapor pressure and saturation vapor pressure and $x_{D(A)}$ and $x_{D(S)}$ are the donor mole fractions in the additive and solvent; but because D is at infinite dilution

$$x_{D(A)} = \frac{n_{D(A)}}{n_A} \tag{6.28}$$

$$x_{D(S)} = \frac{n_{D(S)}}{n_S} \tag{6.29}$$

$$x_{D(A,S)} = \frac{n_{D(A)} + n_{D(S)}}{n_A + n_S} \tag{6.30}$$

where $n_{i(j)}$ is the mole number of i in environment j. Substituting these relations into eq. 6.27 yields

$$\frac{1}{\gamma_{D\,(A,S)}^{\infty}} = \frac{p_D^0}{p_D} \left(\frac{n_{D(A)}}{n_A + n_S} + \frac{n_{D(S)}}{n_A + n_S} \right) \tag{6.31}$$

or

$$\frac{1}{\gamma_{D\,(A,S)}^{\infty}} = \frac{x_A}{\gamma_{D\,(A)}^{\infty}} + \frac{x_S}{\gamma_{D\,(S)}^{\infty}} \tag{6.32}$$

Putting this result into the partition coefficient expression (eq. 2.34) and recognizing that in the absence of mixing volumes

$$\frac{\overline{V}_i x_i}{\overline{V}_{i,j}} = \phi_i \tag{6.33}$$

produces

$$K_R = \phi_A K_{R(A)}^0 + \phi_S K_{R(S)}^0 \tag{6.34}$$

which is the equation that would be derived if A+S mixtures were either perfect or totally immiscible. The former is highly unlikely, however, because activity coefficients of third components (D) in such mixtures range over an order of magnitude around unity. Furthermore, it would have to be supposed that limited-miscibility additives such as TNF and TCNE form ideal solutions up to a limiting concentration, yet precipitate on further addition.

Equation 6.34 can also be derived by examining those systems for which miscibility extends over the entire volume fraction range $\phi_A = 0$ to 1. At the last point of such K_R versus C_A plots K_R becomes $K_{R_{(A)}}^0$ and C_A is identical to $\bar{V}_A^{-1}$; solving for K_1,

$$K_1 = \frac{\bar{V}_A \Delta K_R^0}{K_{R_{(S)}}^0} \tag{6.35}$$

where $\Delta K_R^0 = K_{R_{(A)}}^0 - K_{R_{(S)}}^0$. Solving eq. 6.8 for K_1 and equating the result to eq. 6.35, we obtain

$$K_R = \bar{V}_A C_A \Delta K_R^0 + K_{R_{(S)}}^0 \tag{6.36}$$

Because $\bar{V}_A C_A = \phi_A$ and the slope at the end-point is the same as that at any other point on a straight line, eqs. 6.34 and 6.36 must be identical. Further, because $\bar{V}_A$ is a constant, plots of K_R versus C_A or versus ϕ_A would be expected to be linear. Note, however, that the volume fraction is not linearly related to the mole fraction

$$\phi_A = \frac{x_A \bar{V}_A}{x_A (\bar{V}_A - \bar{V}_S) + \bar{V}_S} \tag{6.37}$$

Plots of K_R versus x_A would therefore be expected to be curved.

To test these relations Purnell and Vargas de Andrade[178] calculated K_1 values from Martire's data by using only $K_{R_{(S)}}^0$, $K_{R_{(A)}}^0$, and $\bar{V}_A$ and compared the results with those Martire found from the slopes of K_R versus C_A plots; the data are given in Table 6.12 in which the agreement is within the experimental error. Next, K_R was plotted versus ϕ_A for several other systems; linearity was observed over the entire volume fraction range $\phi_A = 0$ to 1. Finally, calculated and experimental K_1 data for the systems in these latter plots were compared as shown in Table 6.13. Again, excellent agreement between K_1 and $(\bar{V}_A \Delta K_R^0 / K_{R_{(S)}}^0)$ is observed. Purnell and Vargas de Andrade further noted that for all systems examined plots of $1/\gamma_{D_{(A,S)}}^\infty$ versus x_A were also linear, thus verifying eq. 6.32; for example, the data of Castells[171] are plotted in this fashion in Fig. 6.8.

Table 6.12 Comparison of Experimental (Eq. 6.8) and Calculated (Eq. 6.35) K_1 Data[178]

Solute	30°C		40°C		50°C	
	Experimental	Calculated	Experimental	Calculated	Experimental	Calculated
A. DOE/n-Heptadecane Stationary Phases						
Methyl alcohol	1.52	1.56	1.30	1.29	0.94	0.91
Ethyl alcohol	1.02	1.05	0.86	0.85	0.69	0.67
n-Propyl alcohol	0.98	1.00	0.81	0.81	0.67	0.65
iso-Propyl alcohol	0.78	0.80	0.62	0.62	0.52	0.51
n-Butyl alcohol	0.99	1.02	0.77	0.77	0.61	0.60
iso-Butyl alcohol	0.99	1.02	0.77	0.76	0.64	0.62
sec-Butyl alcohol	0.69	0.71	0.53	0.52	0.44	0.43
tert-Butyl alcohol	0.53	0.56	0.42	0.42	0.35	0.34
B. DOE/n-Octadecane Stationary Phases						
n-Propyl alcohol	1.20	1.22	0.97	0.95	0.80	0.76
iso-Propyl alcohol	0.90	0.92	0.75	0.74	0.64	0.60
n-Butyl alcohol	1.15	1.17	0.93	0.91	0.77	0.73
iso-Butyl alcohol	1.15	1.17	0.93	0.91	0.76	0.72
sec-Butyl alcohol	0.77	0.79	0.64	0.63	0.54	0.51
tert-Butyl alcohol	0.62	0.64	0.52	0.51	0.44	0.42
Dichloromethane	0.28	0.30	0.26	0.26	0.24	0.23
Bromochloromethane	0.29	0.31	0.26	0.26	0.24	0.23

K_1
(1/mole)

Table 6.12 (*Continued*)

Solute	K_1 (l/mole)					
	30°C		40°C		50°C	
	Experimental	Calculated	Experimental	Calculated	Experimental	Calculated
B. DOE/n-Octadecane Stationary Phases						
Dibromomethane	0.30	0.31	0.27	0.27	0.25	0.24
Chloroform	0.39	0.41	0.35	0.35	0.31	0.29
Bromodichloromethane	0.41	0.42	0.36	0.36	0.32	0.31
Chlorodibromomethane	0.42	0.43	0.36	0.36	0.32	0.30
Bromoform	0.41	0.43	0.37	0.37	0.33	0.31
Carbon tetrachloride	0.07	0.08	0.06	0.07	0.06	0.06
Bromotrichloromethane	0.11	0.12	0.09	0.10	0.08	0.08
Carbon tetrabromide	0.12	0.14	0.11	0.12	0.09	0.09
1,1,1-Trichloroethane	0.10	0.12	0.10	0.10	0.09	0.09
C. DOTE/n-Octadecane Stationary Phases						
n-Propyl alcohol	0.72	0.66	0.63	0.59	0.56	0.52
iso-Propyl alcohol	0.60	0.55	0.53	0.49	0.48	0.44
n-Butyl alcohol	0.72	0.66	0.63	0.58	0.55	0.51
iso-Butyl alcohol	0.74	0.68	0.62	0.57	0.53	0.49
sec-Butyl alcohol	0.54	0.50	0.49	0.45	0.43	0.40
tert-Butyl alcohol	0.48	0.44	0.42	0.39	0.38	0.35
Dichloromethane	0.36	0.33	0.33	0.31	0.29	0.27

Bromochloromethane	0.41	0.37	0.38	0.35	0.33	0.30
Dibromomethane	0.45	0.42	0.42	0.40	0.37	0.35
Chloroform	0.42	0.38	0.38	0.35	0.31	0.29
Bromodichloromethane	0.50	0.46	0.45	0.42	0.37	0.35
Chlorodibromomethane	0.61	0.56	0.55	0.51	0.46	0.43
Bromoform	0.73	0.67	0.65	0.60	0.52	0.49
Carbon tetrachloride	0.14	0.12	0.13	0.12	0.11	0.10
Bromotrichloromethane	0.29	0.27	0.27	0.25	0.23	0.21
Carbon tetrabromide	0.93	0.85	0.80	0.74	0.59	0.55
1,1,1-Trichloroethane	0.16	0.15	0.15	0.14	0.13	0.12

D. DOMA/n-Octadecane Stationary Phases

n-Propyl alcohol	2.47	2.51	1.91	1.92	1.48	1.47
iso-Propyl alcohol	1.53	1.55	1.23	1.23	0.98	0.97
n-Butyl alcohol	2.41	2.45	1.85	1.86	1.43	1.41
iso-Butyl alcohol	2.53	2.57	1.92	1.93	1.46	1.44
sec-Butyl alcohol	1.34	1.37	1.07	1.07	0.84	0.84
tert-Butyl alcohol	0.89	0.91	0.73	0.74	0.60	0.59
Dichloromethane	0.21	0.21	0.18	0.18	0.16	0.16
Bromochloromethane	0.22	0.23	0.20	0.20	0.18	0.17
Dibromomethane	0.25	0.26	0.22	0.23	0.20	0.20
Chloroform	0.47	0.48	0.39	0.40	0.33	0.33
Bromodichloromethane	0.51	0.53	0.43	0.43	0.36	0.35
Chlorodibromomethane	0.58	0.60	0.50	0.49	0.43	0.41
Bromoform	0.64	0.66	0.54	0.55	0.46	0.45
Carbon tetrachloride	0.06	0.07	0.06	0.06	0.06	0.05
Bromotrichloromethane	0.21	0.22	0.18	0.18	0.15	0.15
1,1,1-Trichloroethane	0.07	0.08	0.06	0.06	0.06	0.06

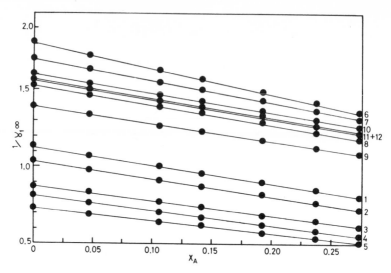

Figure 6.8 Plots of $1/\gamma_1^\infty$ versus x_A for solutes and stationary phases of Fig. 6.3. Data of Castells.[171]

Table 6.13 Comparison of K_1^{exptl} and $\overline{V}_A \Delta K_R / K_{R_{(S)}}^0$ Data[178] for Named Solutes with DOE and DBTC at 30°C

Solute	$K_{R_{(A)}}^0$	$K_{R_{(S)}}^0$	$\overline{V}_A \Delta K_R / K_{R_{(S)}}^0$ (l/mole)	K_1^{exptl} (l/mole)
A. DOE/n-Heptadecane Stationary Phases				
Chloroform	579	250	0.398	0.407
1,2-Dichloroethane	535	317	0.208	0.215
Benzene	663	509	0.092	0.093
Toluene	2189	1718	0.083	0.083
Ethylbenzene	5787	4619	0.077	0.077
o-Xylene	8508	6816	0.075	0.075
m-Xylene	6930	5804	0.059	0.059
p-Xylene	6735	5595	0.062	0.062
B. DBTC/Squalane Stationary Phases				
Chloroform	718	232	0.664	0.640
1,2-Dichloroethane	984	294	0.744	0.745
Benzene	1218	442	0.557	0.548
Toluene	4141	1446	0.591	0.572
Ethylbenzene	8805	3688	0.440	0.410
o-Xylene	18034	5405	0.741	0.698
m-Xylene	12118	4408	0.555	0.521
p-Xylene	12896	4379	0.609	0.580

Altogether, more than 100 solute/solvent/additive systems had to this point been shown to obey one form or another of eq. 6.34. Because each system constituted an average of six data points, more than 600 binary solutions of varying composition (each containing an infinite-dilution third component) obeyed the relations.

To test eq. 634 further, Laub and Purnell[179] examined those systems described in the literature for which no further experiments (other than density measurements) were required; that is, those for which partition coefficients and volume fractions could be calculated. The systems examined are listed in Table 6.14 for which linearity of K_R as a function of ϕ_A was observed within, at worst, an experimental error of $\pm 10\%$. As noted by these workers, many other systems, such as sterols with silicone oils, also obeyed a form of eq. 6.34. Several thousand solutions had thus

Table 6.14   Solute/Additive/Solvent Systems[179] Said to Obey Eq. 6.34

Solute	$K_{R_{(S)}}^0$	$K_{R_{(A)}}^0$
A. 1-Dodecanol/Squalane (56°C)		
Methyl alcohol	12.5	110
Ethyl alcohol	23.9	213
n-Propyl alcohol	50.5	512
iso-Propyl alcohol	26.2	291
n-Butyl alcohol	134	1190
iso-Butyl alcohol	87.0	1019
sec-Butyl alcohol	72.6	751
tert-Butyl alcohol	38.5	393
n-Pentyl alcohol	321	4240
2-Pentyl alcohol	191	1926
3-Pentyl alcohol	198	1795
2-Methyl-1-butyl alcohol	259	2925
3-Methyl-2-butyl alcohol	164	1394
tert-Pentyl alcohol	119	1006
neo-Pentyl alcohol	136	1289
B. Lauronitrile/Squalane (56°C)		
Nitromethane	29.3	370
Nitroethane	75.8	759
Nitropropane	191	1582
Ethyl cyanide	39.0	367
n-Propyl cyanide	97.6	844

Table 6.14 (*Continued*)

Solute	$K^0_{R_{(S)}}$	$K^0_{R_{(A)}}$
C. DPTC/Squalane (90°C)		
Benzene	71.8	156
Toluene	163	360
Ethylbenzene	329	649
p-Xylene	368	809
m-Xylene	373	780
o-Xylene	437	1044
iso-Propylbenzene	524	934
n-Propylbenzene	658	1215
1-Methyl-3-ethylbenzene	729	1333
1-Methyl-4-ethylbenzene	742	1397
1,3,5-Trimethylbenzene	849	1566
1-Methyl-2-ethylbenzene	826	1735
tert-Butylbenzene	882	1499
1,2,4-Trimethylbenzene	978	2203
iso-Butylbenzene	994	1731
sec-Butylbenzene	1003	1669
1-Methyl-3-iso-propylbenzene	1113	1802
1,2,3-Trimethylbenzene	1192	3051
1-Methyl-4-iso-propylbenzene	1178	1915
1-Methyl-2-iso-propylbenzene	1239	2261
1,3-Diethylbenzene	1364	2259
1-Methyl-3-propylbenzene	1416	2423
n-Butylbenzene	1441	2603
1-Methyl-4-propylbenzene	1478	2583
1,2-Diethylbenzene	1484	2769
1,3-Dimethyl-5-ethylbenzene	1595	2539
1,4-Diethylbenzene	1487	2469
1-Methyl-2-propylbenzene	1560	3053
1,4-Dimethyl-2-ethylbenzene	1748	3463
1,3-Dimethyl-4-ethylbenzene	1833	3485
1,2-Dimethyl-4-ethylbenzene	1901	3581
1,2-Dimethyl-3-ethylbenzene	2164	4703
1,2,4,5-Tetramethylbenzene	2508	5759
1,2,3,5-Tetramethylbenzene	2618	5869
1,2,3,4-Tetramethylbenzene	3169	8692

Table 6.14 *(Continued)*

Solute	$K^0_{R_{(S)}}$	$K^0_{R_{(A)}}$
D. DNP/Squalane (100°C)		
Benzaldehyde	281	884
Acetophenone	618	1906
Benzyl alcohol	462	2233
Phenol	227	3068
Ethoxybenzene	416	745
Benzene	41.2	58.6
Ethylbenzene	200	254
n-Butylbenzene	830	980
n-Hexane	33.7	24.5
n-Heptane	67.2	49.0
n-Octane	140	97.1
n-Nonane	301	202
n-Decane	630	414
n-Undecane	1315	847
n-Dodecane	...	1734
E. Quinoline/Diethyl Maleate (35°C)		
iso-Pentane	37.4	39.0
n-Pentane	48.1	57.1
3-Methylpentane	116	131
Cyclopentane	116	167
n-Hexane	131	172
Furan	302	227
Cyclohexane	294	404
n-Heptane	352	508
Cyclohexene	505	713
Benzene	1239	1321
Thiophene	1704	1998

been described by the relation. Noteworthy is that it appears to be equally as valid for those systems for which no conceivable manner of complexing can be envisaged as for those in which charge transfer is traditionally considered to obtain.

Because eq. 6.34 appeared to describe a large number of systems, Laub and Purnell[179] proposed that such mixtures be called *diachoric* (partitioned-volume).

6.6.3 Extension of the Diachoric Solutions Approximation to NMR Spectroscopy Purnell and Vargas de Andrade[178] examined the complexation of several solutes with the systems di-n-octyl ether (DOE)/n-heptadecane and di-n-butyl tetrachlorophthalate (DBTC)/squalane by GLC and NMR and correlated the results. The chemical shift of a nucleus in two (competing) environments was written as

$$\delta_D^{A,S} = \sum P_D^i \delta_D^{0,i} \tag{6.38}$$

where P_D^i and $\delta_D^{0,i}$ are the fractional population and chemical shift, respectively, of D in the (pure) ith medium. (Note that, as in the GLC case, eq. 6.38 implies that A and S form perfect or totally immiscible solutions.) Because

$$P_D^A + P_D^S = 1 \tag{6.39}$$

must be true, it follows that

$$P_D^A = \frac{\delta_D^{A,S} - \delta_D^{0,S}}{\delta_D^{0,A} - \delta_D^{0,S}} = \frac{\Delta}{\Delta^0} \tag{6.40}$$

In terms of moles of donor n_D

$$P_D^A = \frac{n_D^A}{n_D^A + n_D^S} \tag{6.41}$$

where the subscript D denotes the donor and the superscripts, the solvent, additive, or a mixture of the two in which the donor is found. Dividing the numerator and denominator by $(V_A + V_S)$, the total solution volume, and noting that

$$\frac{1}{V_A + V_S} = \frac{\phi_A}{V_A} = \frac{\phi_S}{V_S} \tag{6.42}$$

yields

$$P_D^A = \frac{\phi_A n_D^A / V_A}{(\phi_A n_D^A / V_A) + (\phi_S n_D^S / V_S)} \tag{6.43}$$

Because the solute is partitioned in GLC between the stationary and mobile phases, the top and bottom of eq. 6.43 must be divided by n_D^M / V_M, the number of moles of D solute per unit volume mobile phase:

$$P_D^A = \frac{\phi_A n_D^A V_M / V_A n_D^M}{\left(\phi_A n_D^A V_M / V_A n_D^M\right) + \left(\phi_S n_D^S V_M / V_S n_D^M\right)} \tag{6.44}$$

but

$$\frac{n_D^A / V_A}{n_D^M / V_M} = K_{R_{(A)}}^0 \tag{6.45}$$

and

$$\frac{n_D^S / V_S}{n_D^M / V_M} = K_{R_{(S)}}^0 \tag{6.46}$$

Equation 6.44 therefore becomes

$$P_D^A = \frac{\Delta}{\Delta^0} = \frac{\phi_A K_{R_{(A)}}^0}{\phi_A K_{R_{(A)}}^0 + \phi_S K_{R_{(S)}}^0} = \frac{\phi_A K_{R_{(A)}}^0}{K_R} \tag{6.47}$$

Equation 6.47 relates GLC and NMR data explicitly and was tested by Purnell and Vargas de Andrade[178] by plotting Δ versus ϕ_A / K_R for the solvent systems, DOE/n-heptadecane and DBTC/squalane with 10 solutes; with one possible exception (1,2-dichloroethane solute), eq. 6.47 appeared to be obeyed. Further, the slopes of such plots were calculated a priori from the product, $K_{R_{(A)}}^0 \Delta^0$; good agreement was obtained. Thus, if nothing else, eq. 6.47 permits the calculation of NMR shifts from purely GLC data.

6.7 MICROSCOPIC PARTITION (MP) MODEL OF SOLUTION

Development of the diachoric solutions approximation seemed to leave the matter of complexation in perplexity: a simple equation had been found which was obeyed to within a reasonable estimate of experimental error by many gas-liquid systems reported in the literature for which sufficient data were available. GLC and NMR data were also shown to be interrelated. None of the equations, however, required the invocation of any charge transfer or hydrogen bonding theory or terms; K_1, in fact, was explained as a mixed-solvent effect even for such systems as aromatic hydrocarbons and TCNE. Yet there is strong (although inferential) spectroscopic evidence that charge transfer does take place; some complexes can be isolated as solids. To rationalize the diachoric solutions relations with charge transfer interactions Laub and Purnell[179] postulated a new model of solution, called the microscopic partition (MP) theory, which is now described.

The equations already developed begin with the *model* of microscopically immiscible solvents. In the case of charge transfer, however, third-component solutes may interact with the additive and indeed may also interact with the solvent. To account for all possible interactions Laub and Purnell[179] first defined the following terms (cf. Fig. 6.9):

$$\text{A.} \qquad K^0_{R_{(A)}} = \frac{n\, C^A_{DA} + C^A_D}{C^M_D}$$

Solute D partitions between the (pure) additive portion of the stationary phase and the gas phase, whereas in the additive phase D may form complexes:

$$mA + nD \underset{}{\overset{K^A_1}{\rightleftarrows}} A_m D_n$$

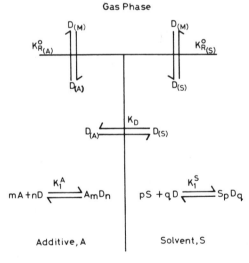

Figure 6.9   Schematic diagram of microscopic partition (MP) model of solution. After Laub and Purnell.[179]

Where m and n are not necessarily integral values. Thus the *measured* partition coefficient in pure A is given by the *total* concentration of D in A [i.e., $(nC_{DA}^A + C_D^A)$, where C_{DA}^A is the concentration of *complex* and C_D^A is the concentration of *free*, uncomplexed, donor, both in pure A] divided by the concentration of donor in the gas phase C_D^M.

$$\text{B.} \qquad K_{R_{(S)}}^0 = \frac{qC_{DS}^S + C_D^S}{C_D^M}$$

Similarly, D may interact with S in the pure solvent portion of the binary stationary phase such that

$$pS + qD \overset{K_i^S}{\rightleftharpoons} S_p D_q$$

and the *measured* partition coefficient of D in S will be the quotient of *complexed* plus *free* donor and donor in the gas phase.

$$\text{C.} \qquad K_{R_{(A)}}^{0,t} = \frac{C_D^A}{C_D^M}$$

The *true* partition coefficient of D in A, $K_{R_{(A)}}^{0,t}$, is the concentration of *free* donor in A, C_D^A, divided by C_D^M.

$$\text{D.} \qquad K_{R_{(S)}}^{0,t} = \frac{C_D^S}{C_D^M}$$

$K_{R_{(S)}}^{0,t}$ is the *true* partition coefficient of *free* D in S, C_D^S, divided by C_D^M.

$$\text{E.} \qquad K_D = \frac{C_D^A + nC_{DA}^A}{C_D^S + qC_{DS}^S}$$

K_D represents the distribution of *total* D across the (hypothetical) phase boundary that separates A and S.

$$\text{F.} \qquad K_D^t = \frac{C_D^A}{C_D^S} = \frac{K_{R_{(A)}}^{0,t}}{K_{R_{(S)}}^{0,t}}$$

K_D^t is the *true* distribution coefficient of (free) D between A and S.

$$\text{G.} \qquad K_i^{t,A} = \frac{C_{DA}^A}{\left(C_A^A\right)^m \left(C_D^A\right)^n}$$

The *true* stability constant of the interaction between D and A in pure A, $K_1^{t,A}$, is given in terms of *free* A and *free* D. Note that, in GLC, $C_A^A \gg C_D^A$ because the solute concentration is maintained at infinite dilution.

$$\text{H.} \qquad K_1^{t,S} = \frac{C_{DS}^S}{(C_S^S)^p (C_D^S)^q}$$

As in definition G, the *true* stability constant of the interaction between D and S in pure S, $K_1^{t,S}$, is formulated in terms of *free* S and *free* D. Once again, $C_S^S \gg C_D^S$.

Figure 6.9 illustrates the microscopic partition model. Altogether there are five equilibria that must be considered, and a careful distinction between true distributions or interactions and the respective apparent or measured quantities is made for each. In order to simplify the following derivation, only 1:1 interactions are considered, and it is assumed that the solvent is noncomplexing ($K_1^S, K_1^{t,S} = 0$; the case in which this is not true is examined later).

From definition G

$$C_{DA}^A = K_1^{t,A} C_A^A C_D^A \qquad (6.48)$$

The term C_{DA}^A represents the number of moles of complex per liter of A; multiplying by $V_A/(V_A + V_S)$ will yield $C_{DA}^{A,S}$, the concentration of complex in the entire solution; but this quotient is just ϕ_A so that

$$C_{DA}^{A,S} = \phi_A K_1^{t,A} C_A^A C_D^A \qquad (6.49)$$

The number of moles of D throughout the system is

$$n_D^{A,S} = n_D^A + n_D^S + n_{DA}^A \qquad (6.50)$$

which, when divided by $(V_A + V_S)$, yields

$$C_D^{A,S} = \phi_A C_D^A + \phi_S C_D^S + C_{DA}^{A,S} \qquad (6.51)$$

The partition coefficient K_R of D with the total binary stationary phase is, in the present nomenclature, given by

$$K_R = \frac{C_D^{A,S}}{C_D^M} \qquad (6.52)$$

Combining eqs. 6.49–6.52 (recall that $C_A^A = \overline{V}_A^{-1}$ and $C_S^S = \overline{V}_S^{-1}$)

$$K_R = \phi_A K_{R(A)}^{0,t}\left(1 + \frac{K_1^{t,A}}{\overline{V}_A}\right) + \phi_S K_{R(S)}^0 \tag{6.53}$$

The generalized form of the Gil-Av/Herling complexation equation is

$$K_{R(i)} = K_{R(i)}^{0,t}\left(1 + \frac{K_1^{t,i}}{\overline{V}_i}\right) \tag{6.54}$$

so that eq. 6.53 reduces immediately to the diachoric solutions relation and it appears that all possible solution equilibria are contained within it. Thus not only does the MP model offer a theory that is consistent with observed solution behavior but, in addition, it allows for the incorporation of multiple stoichiometric interactions or reactions in A or S. Suppose, for example, that the possibility of solute-solvent interactions were to be included; according to eq. 6.54,

$$K_{R(S)}^0 = K_{R(S)}^{0,t}\left(1 + \frac{K_1^{t,S}}{\overline{V}_S}\right) \tag{6.55}$$

Equation 6.53 therefore becomes

$$K_R = \phi_A\left[K_{R(A)}^{0,t}\left(1 + \frac{K_1^{t,A}}{\overline{V}_A}\right) - K_{R(S)}^{0,t}\left(1 + \frac{K_1^{t,S}}{\overline{V}_S}\right)\right] + K_{R(S)}^{0,t}\left(1 + \frac{K_1^{t,S}}{\overline{V}_S}\right) \tag{6.56}$$

Turning now to the matter of the stability constant, the conventional GLC equation in the nomenclature of the MP model is

$$K_R = K_{R(S)}^0\left(1 + K_1^A C_A^{A,S}\right) \tag{6.57}$$

By considering only 1:1 interactions, letting $K_1^S, K_1^{t,S} = 0$, and equating eq. 6.53 with the conventional relation above

$$K_1^{GLC} = \frac{\Delta K_R^{0,t}\overline{V}_A}{K_{R(S)}^0} + \frac{K_1^{t,A}K_{R(A)}^{0,t}}{K_{R(S)}^0} \tag{6.58}$$

where

$$\Delta K_R^{0,t} = K_{R_{(A)}}^{0,t} - K_{R_{(S)}}^{0} \tag{6.59}$$

and a superscript GLC has been added to indicate the apparent (experimentally measured) GLC K_1 value. Note that even when $K_1^{t,A} = 0$ eq. 6.58 reduces to the conventionally employed GLC equation; K_1^{GLC} may therefore appear to be finite even in the absence of complexation interactions.

If solvent effects are important, eq. 6.58 becomes

$$K_1^{GLC} = \frac{K_{R_{(A)}}^{0,t} \bar{V}_A - \bar{V}_A K_{R_{(S)}}^{0,t} \left(1 + K_1^{t,S}/\bar{V}_S\right)}{K_{R_{(S)}}^{0,t} \left(1 + K_1^{t,S}/\bar{V}_S\right)} + \frac{K_1^{t,A} K_{R_{(A)}}^{0,t}}{K_{R_{(S)}}^{0,t} \left(1 + K_1^{t,S}/\bar{V}_S\right)}$$

$$= \bar{V}_S \frac{K_{R_{(A)}}^{0,t} \left(\bar{V}_A + K_1^{t,A}\right)}{K_{R_{(S)}}^{0,t} \left(\bar{V}_S + K_1^{t,S}\right)} - \bar{V}_A \tag{6.60}$$

Equations 6.58 and 6.60 allow for negative, zero, or positive K_1^{GLC} values, depending on the magnitudes of $\bar{V}_S$, $\bar{V}_A$, $K_1^{t,S}$, and $K_1^{t,A}$. Solvent effects are also accounted for and consequently it becomes possible, in principle, to determine *true* values of $K_1^{t,A}$. This point is discussed later, for spectroscopic K_1 data must be examined first in light of the MP solution model.

6.7.1 Extension of MP Theory to UV/Vis Spectroscopy To extend the MP model to UV definitions F and G and eq. 6.51 are used where, for convenience, interactions in the solvent S are taken to be negligible:

$$C_D^{A,S} - C_{DA}^{A,S} = \frac{C_{DA}^{A}}{K_1^{t,A} C_A^{A}} \left(\phi_A + \frac{\phi_S}{K_D^t}\right) \tag{6.61}$$

but $C_{DA}^{A}/C_A^{A} = C_{DA}^{A,S}/C_A^{A,S}$ when $K_1^{t,S} = 0$ so that

$$C_{DA}^{A,S} = \frac{K_1^{t,A} C_A^{A,S} C_D^{A,S} K_D^t}{\phi_S + \phi_A K_D^t + K_1^{t,A} C_A^{A,S} K_D^t} \tag{6.62}$$

The Beer-Lambert law is

$$A_{DA} = \varepsilon_{DA} l C_{DA}^{A,S} \tag{6.63}$$

By substituting this relation into eq. 6.62 and recognizing that $\phi_A/C_A^{A,S} =$

$\overline{V}_A$ and $\phi_S = 1 - \phi_A$

$$\frac{C_D^{A,S}l}{A_{DA}} = \frac{1}{K_1^{t,A}K_D^t C_A^{A,S}\varepsilon_{DA}} + \frac{1}{\varepsilon_{DA}}\left(1 + \frac{\overline{V}_A}{K_1^{t,A}} - \frac{\overline{V}_A}{K_1^{t,A}K_D^t}\right) \quad (6.64)$$

The Benesi-Hildebrand method of obtaining K_1^{UV} values requires that $C_D^{A,S}l/A_{DA}$ be plotted versus $1/C_A^{A,S}$ from which K_1^{UV} is found from the intercept/slope quotient, as noted in Section 6.3.1. To apply this procedure to eq. 6.64 definition F is used to eliminate K_D^t:

$$\frac{C_D^{A,S}l}{A_{DA}} = \frac{K_{R_{(S)}}^0}{K_{R_{(A)}}^{0,t}K_1^{t,A}C_A^{A,S}\varepsilon_{DA}} + \frac{1}{\varepsilon_{DA}}\left(1 + \frac{\overline{V}_A}{K_1^{t,A}} - \frac{\overline{V}_A K_{R_{(S)}}^0}{K_{R_{(A)}}^{0,t}K_1^{t,A}}\right) \quad (6.65)$$

The intercept/slope quotient of eq. 6.65 is

$$K_1^{UV} = \left(1 + \frac{\overline{V}_A}{K_1^{t,A}} - \frac{\overline{V}_A K_{R_{(S)}}^0}{K_{R_{(A)}}^{0,t}K_1^{t,A}}\right) \bigg/ \left(\frac{K_{R_{(S)}}^0}{K_{R_{(A)}}^{0,t}K_1^{t,A}}\right)$$

$$= \frac{K_{R_{(A)}}^{0,t}K_1^{t,A}}{K_{R_{(S)}}^0} + \frac{\Delta K_R^{0,t}\overline{V}_A}{K_{R_{(S)}}^0}$$

$$= K_1^{GLC} \quad (6.66)$$

Thus UV and GLC should yield identical *experimentally measured* K_1 values; this has previously been postulated but never proved nor, indeed, found in practice.

6.7.2 Extension of MP Theory for NMR Spectroscopy

The term P_D^A in eq. 6.38 is now expanded to include the free donor in A:

$$\delta_D^{A,S} = P_D^{t,A}\delta_D^{t,A} + P_{DA}^A\delta_{DA}^A + P_D^S\delta_D^{0,S} \quad (6.67)$$

but because $\quad P_D^S = 1 - P_D^{t,A} - P_{DA}^A$

$$\delta_D^{A,S} - \delta_D^{0,S} = P_D^{t,A}\left(\delta_D^{t,A} - \delta_D^{0,S}\right) + P_{DA}^A\left(\delta_{DA}^A - \delta_D^{0,S}\right) \quad (6.68)$$

In terms of moles

$$P_{DA}^A = \frac{n_{DA}^A}{n_{DA}^A + n_D^{t,A} + n_D^S} \quad (6.69)$$

Dividing by $C_D^M(V_A + V_S)$ and using definitions A–C,

$$P_{DA}^A = \frac{\phi_A\left(K_{R_{(A)}}^0 - K_{R_{(A)}}^{0,t}\right)}{\phi_A K_{R_{(A)}}^0 + \phi_S K_{R_{(S)}}^0} \tag{6.70}$$

Following the same procedure, $P_D^{t,A}$ is found to be

$$P_D^{t,A} = \frac{\phi_A K_{R_{(A)}}^{0,t}}{\phi_A K_{R_{(A)}}^0 + \phi_S K_{R_{(S)}}^0} \tag{6.71}$$

Substituting these relations into eq. 6.68 and inverting gives

$$\frac{1}{\delta_D^{A,S} - \delta_D^{0,S}} = \frac{1}{C_A^{A,S}} \frac{K_{R_{(S)}}^0}{V_A K_{R_{(A)}}^{0,t} \Delta^{0,t}} + \left(\frac{\Delta K_R^{0,t}}{K_{R_{(A)}}^{0,t}} + \frac{K_1^{t,A}}{V_A}\right)\frac{1}{\Delta^{0,t}} \tag{6.72}$$

where

$$\Delta^{0,t} = \left(\delta_D^{t,A} - \delta_D^{0,S}\right) + \frac{K_1^{t,A}\left(\delta_{DA}^A - \delta_D^{0,S}\right)}{\overline{V}_A}$$

Because K_1^{NMR} is found by dividing the intercept of eq. 6.72 by the slope, as outlined in Section 6.3.2,

$$K_1^{NMR} = \frac{K_1^{t,A} K_{R_{(A)}}^{0,t}}{K_{R_{(S)}}^0} + \frac{\Delta K_R^{0,t} \overline{V}_A}{K_{R_{(S)}}^0}$$

$$= K_1^{GLC} = K_1^{UV} \tag{6.73}$$

6.7.3   Disparity Between GLC, UV, and NMR K_1^{exptl} Data   The preceding Sections have argued that K_1^{exptl} data should be identical, whether measured by GLC, UV, or NMR. In practice, this has not been found to be true. First, in GLC D is maintained at infinite dilution, whereas A is generally varied between 0 and 0.5 M in S. In UV and NMR S and D are usually in great excess over A, which even so is not at infinite dilution. Finally, even trivial errors in UV render the accurate determination of K_1^{UV} and ε_{DA} as separate quantities virtually impossible. On the other hand, in the GLC-NMR comparison by Purnell and Vargas de Andrade[178] cited earlier both A and S were liquids which enabled the accurate measurement of Δ^0.

One might suppose in light of the foregoing that UV and NMR could be fairly contrasted, according to simplified MP theory,

$$\frac{1}{\Delta} = \frac{1}{\Delta^0 K_1^{NMR} C_A^{A,S}} + \frac{1}{\Delta^0} \tag{6.74}$$

and

$$\frac{C_D^{A,S} l}{A_{DA}} = \frac{1}{\varepsilon_{DA} K_1^{UV} C_A^{A,S}} + \frac{1}{\varepsilon_{DA}} \tag{6.75}$$

Thus at equal values of $C_A^{A,S}$

$$\frac{C_D^{A,S} l \varepsilon_{DA}}{A_{DA}} = \frac{\Delta^0}{\Delta} \tag{6.76}$$

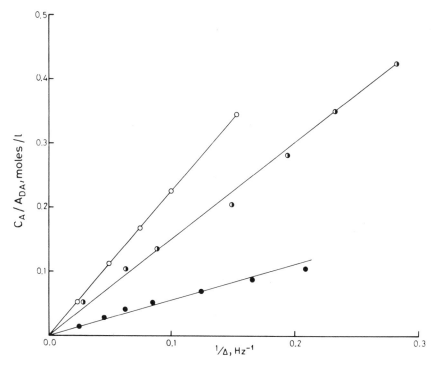

Figure 6.10 Plots of C_A / A_{DA}^{-1} versus Δ^{-1} for toluene with 2,4,7-trinitrofluorenone and the solvents, di-n-butyl adipate ($\bigcirc$), di-n-butyl succinate ($\ominus$), and di-n-butyl sebacate ($\bullet$) at 40°C. Data of Laub and Purnell.[179]

Laub and Purnell[179] tested eq. 6.76 by using the data of Purnell and Srivastava[78]; a representative plot of $C_A^{A,S} l / A_{DA}$ versus $1/\Delta$ is shown in Fig. 6.10 for toluene with TNF in di-n-butyl succinate (DBSUCC), DBA, and DDSEB ($C_A^{A,S}$ is used here because D was held constant and A varied). Equation 6.76 appears to be obeyed, the lines passing through the origin as demanded. Laub and Purnell[179] also noted that values of $\varepsilon_{DA}/\Delta^0$ differed up to factor of three when compared with the UV/NMR Benesi-Hildebrand and Foster-plot data, which again emphasizes that evaluation of ε_{DA} from the intercept of UV plots is by no means straightforward.

6.7.4 Solvent Effects The MP solution model indicates that K_1^{exptl} may be different from $K_1^{t,A}$, irrespective of the technique used to measure it; the relations discussed allow the quantification of these differences; for example, it was pointed out earlier that, even if $K_1^{t,A}=0$, K_1^{exptl} may still be finite, since

$$K_1^{exptl} = \frac{K_{R_{(A)}}^0 \overline{V}_A}{K_{R_{(S)}}^0} - \overline{V}_A \tag{6.77}$$

Thus for K_1^{exptl} to be positive $K_{R_{(A)}}^0 / K_{R_{(S)}}^0 > 1$ must be true. Conversely, of course, K_1^{exptl} may be negative if $K_{R_{(A)}}^0 / K_{R_{(S)}}^0 < 1$; that is, if the solute is less soluble in the A + S mixture than in pure S, which is not an inconceivable notion.

Consider now the case in which $K_1^{t,A}$ is not zero; the following three inequalities are possible:

$$\text{A.} \qquad K_{R_{(A)}}^{0,t} > K_{R_{(S)}}^0$$

In this instance K_1^{exptl} will be greater than $K_1^{t,A}$, for $K_1^{t,A}$ and $\overline{V}_A$ cannot be negative. If complexation is strong, that is, $K_1^{t,A} \gg \overline{V}_A$,

$$\frac{K_1^{exptl}}{K_1^{t,A}} \simeq \frac{K_{R_{(A)}}^0}{K_{R_{(S)}}^0} \tag{6.78}$$

and the magnitude of the error in taking $K_1^{exptl} = K_1^{t,A}$ will be given directly by the ratio $K_{R_{(A)}}^{0,t} / K_{R_{(S)}}^0$. If $K_{R_{(A)}}^{0,t} = K_{R_{(S)}}^0$, $K_1^{exptl} = K_1^{t,A}$. However, with the possible exception of the systems studied by Martire, noted earlier, in which the additive and solvent molecular weights, dimensions, polarizabilities, and molar volumes were virtually indentical, this appears to be an unlikely situation. It is difficult to imagine that the relation will hold for other systems for which these criteria are not even approximately met.

$$\text{B.} \quad K_{R_{(A)}}^{0,t} < K_{R_{(S)}}^{0}$$

$K_1^{t,A}$ will in these cases be larger than K_1^{exptl} and the apparent strength of interaction will be less than that which actually occurs.

$$\text{C.} \quad K_1^{t,A} < \bar{V}_A \left(1 - \frac{K_{R_{(S)}}^{0}}{K_{R_{(A)}}^{0,t}}\right)$$

Here K_1^{exptl} will be negative because $K_{R_{(S)}}^{0}$ and $K_{R_{(A)}}^{0,t}$ must be zero or positive.

$$\text{D.} \quad K_1^{exptl} = 0$$

which will obtain when

$$K_1^{t,A} = -\frac{\Delta K_R^{0,t} \bar{V}_A}{K_{R_{(A)}}^{0,t}} \tag{6.79}$$

$K_1^{t,A}$ cannot be less than zero, however, and $K_{R_{(S)}}^{0} \geqslant K_{R_{(A)}}^{0,t}$ must also be true; for example, if $\bar{V}_A = 0.1$ to 0.5 1/mole and $K_1^{t,A} = 0.1$ 1/mole, $K_{R_{(S)}}^{0}/K_{R_{(A)}}^{0,t}$ need be only 2 to 1.2 in order that $K_1^{\text{exptl}} = 0$, an often encountered situation in weak complexation. Conversely, when complexation is strong, say, $K_1^{t,A} = 4.5$ 1/mole, $K_{R_{(S)}}^{0}/K_{R_{(A)}}^{0,t}$ must fall between 50 and 10 if K_1^{exptl} is to be zero. These cases are less likely to be encountered but are nevertheless not impossible; K_1^{exptl} might therefore be zero, even for strong complexation, as found for the TCNE/aromatic hydrocarbon systems discussed earlier.

Thus it is possible via the MP model to quantify variations of K_1^{exptl} with solvents on the basis of the solubility of D in A and in S. Note that no effort was made to detail the differences in the solubility of D in various solvents; the model instead rationalizes solvent effects in terms of the *relative* solubility of D in A and S. In other words, D is used as a molecular probe to obtain information about A + S mixtures.

6.7.5 Partially Miscible Solvent Systems The diachoric solutions approximation indicates that K_R should be linear in ϕ_A, irrespective of the *macroscopic* miscibility of A in S; that is, because the equations are based on the postulate that A and S are microscopically immiscible they should also apply to A + S mixtures that exhibit partial macroscopic miscibility. This proposition forms an interesting test of MP theory, and Purnell and co-workers[180] chose to investigate the system tributyl phosphate (TBP) in ethylene glycol (EG). The consolute temperature curve for these solvents is shown in Fig. 6.11, in which the critical temperature appears to be about 53°C. A plot of specific volume versus weight fraction is given in Fig. 6.12. Thus, although the system exhibits only partial macroscopic miscibility, there is no excess volume of mixing. This enables the volume fraction to be calculated over the entire concentration range from the relation

$$\phi_A = \frac{\rho_{A,S} - \rho_S^0}{\rho_A^0 - \rho_S^0}. \tag{6.80}$$

Figures 6.13 to 6.15 show plots of K_R versus ϕ_{TBP} for the solutes, furan, diethyl ether, and methyl formate at 21°C, where linearity is extant over

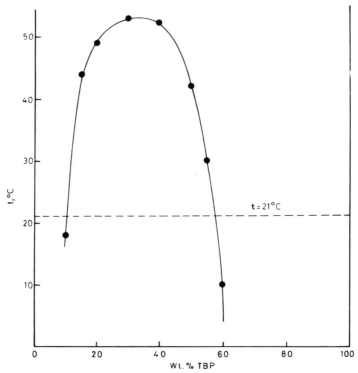

Figure 6.11 Consolute-temperature curve for tributyl phosphate/ethylene glycol. Data of Laub, Purnell, and Summers.[180]

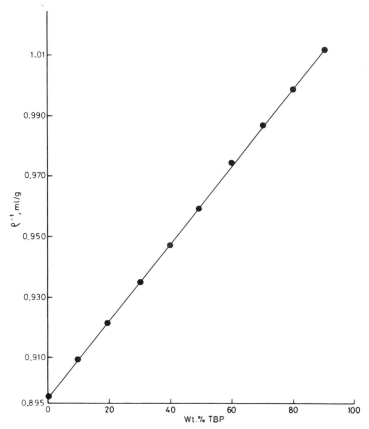

Figure 6.12 Plot of ρ^{-1} versus weight fraction tributyl phosphate in mixtures with ethylene glycol at 21°C. Data of Laub, Purnell, and Summers.[180]

the entire volume fraction range $\phi_{TBP} = 0$ to 1. The MP model therefore applies to several systems at least for which there is no question of ideal solution behavior.

6.7.6 Polymer Solutions Short-range ordering of polymers in solutions has long been recognized[181-190] (although some doubt remains about long-range ordering[191]); presumably they, too, should conform precisely to the diachoric solutions relations. Patterson and co-workers[192] determined $K^0_{R_{(S)}}$ and $K^0_{R_{(A)}}$ for alkanes, cyclohexane, and carbon tetrachloride with the solvents n-tetracosane and *poly*(dimethylsiloxane) (PDMS). Only one intermediate mixture of the stationary phases was attempted and, significantly, the solute retention data at this value were higher (by about 5–10%) than

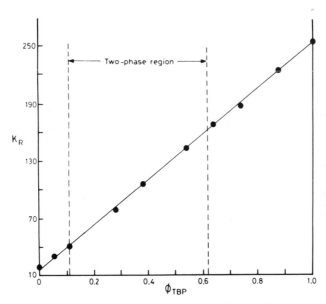

Figure 6.13 Plot of K_R versus ϕ_A for furan with tributyl phosphate/ethylene glycol stationary phase mixtures at 21°C. Data of Laub, Purnell, and Summers.[180]

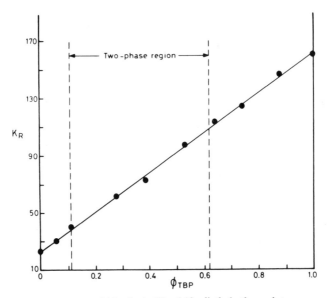

Figure 6.14 As in Fig. 6.13; diethyl ether solute.

198

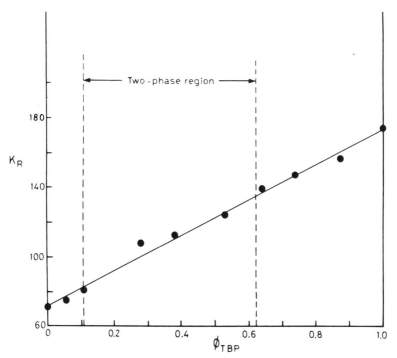

Figure 6.15 As in Fig. 6.13; methyl formate solute.

expected on the basis of linearity of K_R versus ϕ_A. Martynyuk and Vigdergauz,[193] on the other hand, showed that V_g^0 varied linearly with weight fraction for a variety of solutes with colloidal polymer stationary phases. Recognizing that

$$\frac{\phi_i \rho_i^0}{\rho_{i,j}} = w_i \tag{6.81}$$

where w_i is the weight fraction of component i of the i,j (mixed) stationary phase, eq. 6.34 becomes[194, 195]

$$V_g^0 = w_A V_{g\,(A)}^0 + w_S V_{g\,(S)}^0 \tag{6.82}$$

which accounts for plots linear in w_A. Most recently, Klein and Widdecke[195] employed eq. 6.82 to determine polymer compositions. These workers also found no difference in the behavior of mixtures of polystyrene and polybutadiene, no matter whether they were block, homopolymer, graft, statistical, or graft + homopolymer sequence structures. There was, moreover, no difference in solute retention behavior when stationary phases composed of 1:1 mixtures of the polymers were constructed from mixed-bed packings or packings coated with intimately mixed polymer solutions, partial block copolymer solutions, or graft polymer solutions. Noteworthy in this connection is that Lynch and co-workers[196] demonstrated that the properties of methylphenylsilicone copolymer stationary phases could be duplicated within ±0.5% with mechanical mixtures of dimethylsilicone and methylphenylsilicone packings. Equation 6.34 and its consequences therefore appear to be obeyed by polymer solutions.

6.7.7 Mechanically Mixed Packings Because eq. 6.34 must on thermodynamics grounds apply to systems that exhibit complete immiscibility,[180] mechanically mixed GLC packings, that is, (support + A) + (support + S), are expected to yield partition coefficients identical to those obtained with intimately mixed packings; that is, (support + A + S). This proposition has been observed in every case in which it has been applied[35, 40, 41, 67, 68, 196] and must be regarded as further evidence of the diachoric solutions hypothesis. In addition, Freeguard and Stock[197] noted that squalane and DNP, when mixed intimately, gave solution isotherms for cyclohexane that were precisely those expected on the basis of a weighted sum of independent interactions of the solute with each pure solvent.

6.8 INTERFACE BETWEEN CONVENTIONAL AND MP SOLUTION THEORIES

The most plausible argument against the success of eq. 6.34 is that although micelle formation has been observed for liquid crystals,[198-200] soap solutions,[201-203] and other mixtures[204] it has not been shown to exist in "conventional" systems such as those listed in Table 6.14. However, ordered structure in liquids has for "simple" molecular species been said to be observed by a variety of techniques for pure substances[205-218] and mixtures[219, 220] for at least 50 years, the phenomenon originally being referred to as *cybotaxis* by Stewart[205] (synonymous in the present context with *diachoresis*). In more recent times water and alcohols have been shown to form "clusters" or "swarms" when mixed; this behavior is called the hydrophobicity effect.[221] Molecular dynamics calculations also indicate that long-term solvent clustering about a solute is to be expected for Lennard-Jones particles.[222] Convincing x-ray evidence for aggregation in *n*-alkyl halide mixtures has been presented by Brady,[223] and the thermodynamic consequences of short-range ordering phenomena in *n*-alkane mixtures have been discussed by Patterson and co-workers[191, 224] and others.[225-232] Finally, micelle formation in nonaqueous nonpolar media was, in part, the subject of a recent symposium.[233]

In contrast, Nozari and Drago[234] rejected the notion of aggregation in mixed solvents in preference to specific bimolecular interactions; for example, between pyridine and carbon tetrachloride. Similarly, Martire[167] has proposed that contact pairwise interactions account for the discrepancies noted in Table 6.3 and that eq. 6.47 is therefore inappropriate for the correlation of GLC and NMR data (although, self-evidently, eq. 6.53 could be expanded through eq. 6.54 to account for these phenomena). In addition, Parcher and Westlake[235] and Ashworth and Hooker[236] found systems for which plots of K_R versus ϕ_A were not linear; the latter study was made of mixtures of squalane and DNP with n-hexane solute at 30°C, whereas in the former binary stationary phases were generated with carriers containing condensible components.[237]

Laub, Martire, and Purnell[238] attempted to reconcile these disparate interpretations by first examining ternary n-alkane mixtures, since there are sound theoretical models (e.g., Ref. 239; Section 5.1.3) for the prediction of properties of these solutions. Figure 6.16 presents the plots of $K_R / K_{R_{(S)}}^0$ versus ϕ_A obtained by GLC by these workers; data predicted from the Orwoll-Flory[240] and Janini-Martire[239] theories are also shown. There is little to choose from these plots because *linear* least-squares treatment of any of the data sets yields correlation coefficients in excess of 0.9995. Moreover, the experimental and calculated points, when plotted in this reduced fashion, are seen to be virtually coincident, justifying at least the Janini-Martire theory. All *experimental* points, however, fall below a straight line connecting the two end points for each solute (although the *maximum* deviation from linearity amounts to only -0.5%). One may argue accordingly on statistical grounds that even though the experimental error was $\pm 1\%$ real (albeit shallow) negative curvature obtained for these systems.

Hayduk and Cheng[241] measured partition (called, in engineering terms, Ostwald) coefficients for ethane with n-hexane/n-hexadecane mixtures at 25°C; their data are shown in Fig. 6.17, where the deviation from linearity reaches a minimum of approximately -17% at $\phi_{C_6} = 0.6$. In contrast, the data of Nitta and co-workers[242] for nitrogen with iso-octane/cyclohexane mixtures at 25°C are plotted in Fig. 6.18, where the (positive) deviation is about 5% at $\phi = 0.5$.

Martire, Purnell, and co-workers[243] recently determined the partition coefficients of alkanes, alicyclics, and aromatics with squalane/DNP mixtures at 30°C in order to test the (extrapolated static) data of Ashworth and Hooker[236] as well as the supposed linearity of K_R versus ϕ_A plots. The

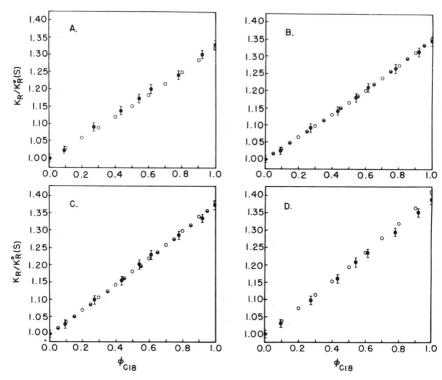

Figure 6.16 Plot of $K_R/K^0_{R(S)}$ versus ϕ_A for (A) n-pentane, (B) n-hexane, (C) n-heptane, and (D) n-octane solutes with n-octadecane/n-hexatriacontane stationary phase mixtures at 80°C. Filled circles: experimental data; open circles, Janini-Martire[239] theory; half-filled circles, Orwoll-Flory[240] theory. Data of Laub, Martire, and Purnell.[238]

results showed that K_R was not a linear function of ϕ_A (10% positive deviation from linearity) and that the curves were asymmetric toward the squalane ordinate for all but the aromatics (which were markedly asymmetric toward the DNP ordinate). *Relative* (to an arbitrary standard) partition coefficients of all solutes, however, conformed within ±2% of the diachoric solutions approximation, which was said to account for the success of the analytical applications[67, 68] of eq. 6.34. Further, partition coefficients obtained with mechanically mixed packings were within 1% of linear behavior.

Mathiasson and Jonsson[148] reported partition coefficients for 17 solutes with squalane/dodecyl laurate ester mixtures, many of which, when plotted versus ϕ_A, are linear. Table 6.15 lists the $K^0_{R(S)}$ and $K^0_{R(A)}$ data as well as the linear regression correlation coefficients (seven points) for these systems.

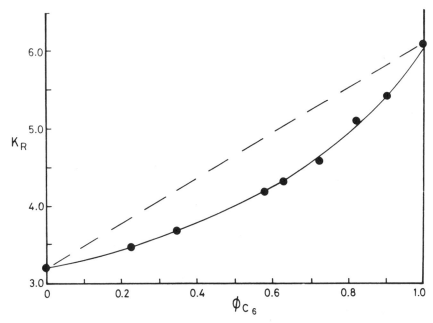

Figure 6.17 Plot of K_R versus ϕ_A for ethane with n-hexane/n-hexadecane stationary phase mixtures at 25°C. Data of Hayduk and Cheng.[241]

Table 6.15 Partition Coefficient Data[148] and Linear Regression Correlation Coefficients for Listed Solutes with Squalane and Dodecyl Laurate Ester Solvents at 60°C

Solute	$K^0_{R_{(S)}}$	$K^0_{R_{(A)}}$	r
p-Xylene	1119.	1559.	0.9996
Toluene	424.4	592.8	0.9997
Benzene	153.7	216.8	0.9994
Fluorobenzene	150.4	236.0	0.9995
Chlorobenzene	789.1	1271.	0.9998
Benzotrifluoride	174.1	300.3	0.9999
Octafluorotoluene	96.93	166.8	0.9948
Methoxybenzene	1325.	2329.	0.9992
Di-n-butyl ether	1202.	1308.	0.9688
Cyclohexane	194.7	184.3	0.9796
Methylcyclohexane	353.5	338.5	0.9652
Ethylcyclohexane	981.5	937.4	0.9634
n-Heptane	273.7	261.4	0.9649
cis-2-Heptene	282.2	293.3	0.9698
$trans$-2-Heptene	269.4	280.1	0.9702
Trichloroethene	222.5	328.3	0.9997
Tetrachloroethene	695.6	818.2	0.9978

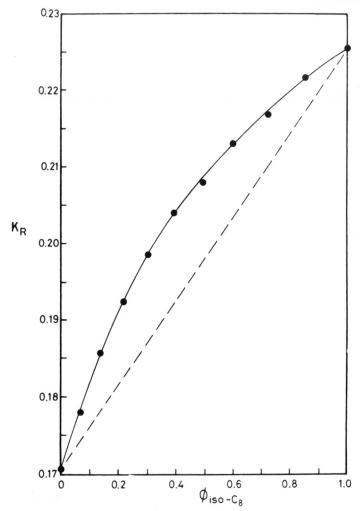

Figure 6.18 Plot of K_R versus ϕ_A for nitrogen with iso-octane/cyclohexane stationary phase mixtures at 25°C. Data of Nitta, Tatsuishi, and Katayama.[242]

In summary, several systems have to date been shown to deviate from the linearity predicted by the diachoric solutions approximation (whereas "solution" data obtained with mechanically mixed phases conform exactly). The several *hundred* ternary mixtures remaining include those reported by Martire and co-workers, Eon and Guiochon, and Purnell and co-workers already cited, for which there is no doubt that linearity obtains over the entire volume fraction range 0 to 1 and for which there has still to be developed a "conventional" alternative to the seemingly near-catholic microscopic partition model of solution.

One might, at this point, regard the entire field of complexation as a microcosm of solution studies in general, in which no one theory has proved all-encompassing. Alternatively, we are of the opinion that the gas-chromatographic investigations reviewed here have contributed substantially to (and self-evidently will continue to participate in) the advancement of both.

6.9 REFERENCES

1. H. A. Benesi and J. H. Hildebrand, *J. Am. Chem. Soc.*, **70**, 2832 (1948); **71**, 2703 (1949).

2. P. Hepp, *Justus Liebegs Ann. Chem.*, **215**, 344 (1882).

3. R. Kremann, *Chem. Ber.*, **39**, 1023 (1906).

4. A. Werner, *Chem. Ber.*, **42**, 4324 (1909).

5. P. Pfeiffer, *Justus Liebegs Ann. Chem.*, **404**, 1 (1914).

6. O. Maass and J. Russell, *J. Am. Chem. Soc.*, **40**, 1561 (1918).

7. G. C. Pimentel and A. L. McClellan, *The Hydrogen Bond*, W. H. Freeman, San Francisco, 1960.

8. G. Briegleb, *Elektronen-Donor-Acceptor Complexe*, Springer, Berlin, 1961.

9. L. J. Andrews and R. M. Keefer, *Molecular Complexes in Organic Chemistry*, Holden-Day, San Francisco, 1964.

10. J. Rose, *Molecular Complexes*, Pergamon, Oxford, England, 1967.

11. R. S. Mulliken and W. B. Person, *Molecular Complexes*, Wiley-Interscience, New York, 1969.

12. R. Foster, *Organic Charge-Transfer Complexes*, Academic, New York, 1969; *Molecular Complexes*, Vols. 1 and 2, Elek Science, London, 1974; *Molecular Association*, Vols. 1 and 2, Academic, London, 1975; *in press*.

13. T. Sinomiya, *J. Chem. Soc. Jap.*, **61**, 1221 (1940).

14. R. C. Jones and M. B. Neuworth, *J. Am. Chem. Soc.*, **66**, 1497 (1944).

15. M. Orchin and E. O. Woolfolk, *J. Am. Chem. Soc.*, **68**, 1727 (1946).

16. J. T. Ayres and C. K. Mann, *Anal. Chem.*, **36**, 2185 (1964).

17. G. Smets, V. Balogh, and Y. Castille, *J. Polym. Sci., Pt. C*, 1467 (1964).

18. D. B. Parihar, O. Prakash, I. Bajaj, R. P. Tripatti, and K. K. Verma, *J. Chromatogr.*, **59**, 457 (1971).

19. F. Mikes, V. Schurig, and E. Gil-Av, *J. Chromatogr.*, **83**, 91 (1973).

20. R. R. Heath, J. H. Tumlinson, and R. E. Doolittle, *J. Chromatogr. Sci.*, **15**, 10 (1977).

21. B. W. Bradford, D. Harvey, and D. E. Chalkley, *J. Inst. Pet.*, **41**, 375 (1955).

22. R. O. C. Norman, *Proc. Chem. Soc.*, 151 (1958).

23. S. H. Langer, C. Zahn, and G. Pantazoplos, *J. Chromatogr.*, **3**, 154 (1960).

24. A. R. Copper, C. W. P. Crowne, and P. G. Farrell, *Trans. Faraday Soc.*, **62**, 2725 (1966); **63**, 447 (1967).

25. J. P. Sheridan, M. A. Capeless, and D. E. Martire, *J. Am. Chem. Soc.*, **94**, 3298 (1972).

26. M. Orchin, *J. Org. Chem.*, **16**, 1165 (1951).

27. E. G. McRae and L. Goodman, *J. Chem. Phys.*, **29**, 334 (1958).

28. R. J. Laub, V. Ramamurthy, and R. L. Pecsok, *Anal. Chem.*, **46**, 1659 (1974).

29. G. R. Primavesi, *Nature*, **184**, 2010 (1959).

30. L. Rohrschneider, *Z. Anal. Chem.*, **170**, 256 (1959).

31. M. Singliar, A. Bobak, J. Brida, and L. Lukacovic, *Z. Anal. Chem.*, **177**, 161 (1960).

32. W. Kemula and H. Buchowski, *Bull. Acad. Pol. Sci., Ser. Sci. Chim.*, **9**, 601 (1961).

33. E. Soczewinski, *Nature*, **191**, 68 (1961).

34. E. O. A. Haahti, W. J. A. Vandenheuval, and E. C. Horning, *Anal. Biochem.*, **2**, 344 (1961).

35. T. Matsuda and H. Yatsugi, *Bunseki Kagaku*, **11**, 1116 (1962).

36. A. Waksmundzki, E. Soczewinski, and Z. Suprynowicz, *Collect. Czech. Chem. Commun.*, 2001 (1962).

37. A. A. Zhukhovitskii, M. S. Selenkina, and N. M. Turkeltaub, *Russ. J. Phys. Chem. (Engl. Trans.)*, **36**, 519 (1962).

38. H. J. Maier and O. C. Karpathy, *J. Chromatogr.*, **8**, 308 (1962).

39. R. S. Porter, R. L. Hinkins, L. Tornheim, and J. F. Jonson, *Anal. Chem.*, **36**, 260 (1964).

40. G. P. Hildebrand and C. N. Reilley, *Anal. Chem.*, **36**, 47 (1964).

41. R. A. Keller and G. H. Stewart, *Anal. Chem.*, **36**, 1186 (1964).

42. J. H. Jordan, *J. Gas Chromatogr.*, **2**, 346 (1964).

43. A. Waksmundzki, Z. Suprynowicz, and I. Miedziak, *Chem. Anal.*, **9**, 913 (1964); *J. Gas Chromatogr.*, **4**, 74 (1966); *Chem. Anal.*, **13**, 17, 635 (1968).

44. P. Chovin, *Bull. Soc. Chim. Fr.*, **104**, 726 (1964); 2124 (1965).

45. H. Pauschmann, *Z. Anal. Chem.*, **211**, 32 (1965).

46. H. Miyake, M. Mitooka, and T. Matsumoto, *Bull. Chem. Soc. Jap.*, **38**, 1062 (1965).

47. J. C. Touchstone, A. Nikolsky, and T. Murawek, *Fed. Proc.*, **24**(2), Pt. 1, 534 (1965).

48. A. Waksmundzki and Z. Suprynowicz, *J. Chromatogr.*, **18**, 232 (1965).

49. S. A. Reznikov, *Zh. Fiz. Khim.*, **42**, 1730 (1968).

50. I. M. Shevchuk, V. A. Granzhan, V. M. Sakharov, and N. V. Kovalenko, *Zh. Obshch. Khim.*, **39**, 2638 (1969).

51. M. J. Molera, J. A. G. Dominguez, and J. F. Biarge, *J. Chromatogr. Sci.*, **7**, 305 (1969).

52. E. S. Windham, *J. Assoc. Off. Anal. Chem.*, **52**, 1237 (1969).

53. R. Annino and P. F. McCrea, *Anal. Chem.*, **42**, 1486 (1970).

54. S. A. Reznikov, I. A. Aganova, and R. I. Sidorov, *Zh. Fiz. Khim.*, **44**, 1267 (1970).

55. J. C. Touchstone, T. Murawec, and A. Nikolski, *J. Chromatogr. Sci.*, **8**, 221 (1970).

56. J. C. Touchstone, C. H. Wu, A. Nikolski, and T. Murawec, *J. Chromatogr.*, **29**, 235 (1967).

57. M. Mitooka, *Nippon Kagaku Zasshi*, **84**, 923 (1963); *Bunseki Kagaku*, **20**, 1542 (1971); **21**, 189, 197, 354, 615, 717, 729, 867, 1043, 1242, 1437, 1447 (1972).

58. S. Yalkowsky, G. Flynn, and T. Slunick, *J. Pharm. Sci.*, **61**, 852 (1972).

59. G. W. Pilgrim and R. A. Keller, *J. Chromatogr. Sci.*, **11**, 206 (1973).

60. M. J. Molera, J. A. G. Dominguez, and J. F. Biarge, *J. Chromatogr. Sci.*, **11**, 538 (1973).

61. D. R. MacKenzie and R. Smol, *J. Chromatogr. Sci.*, **12**, 104 (1974).

62. A. S. Melnikov and B. V. Aivazov, *Zh. Anal. Khim.*, **29**, 365 (1974).

63. M. Suzuki, Y. Yamamoto, and T. Watanabe, *J. Assoc. Off. Anal. Chem.*, **58**, 297 (1975).

64. S. A. Reznikov and R. I. Sidorov, *Zh. Fiz. Khim.*, **49**, 376 (1975).

65. T. Czajkowska and A. Waksmundzki, *J. Chromatogr.*, **119**, 91 (1976).

66. C. L. Young, *J. Chromatogr. Sci.*, **8**, 103 (1970).

67. R. J. Laub and J. H. Purnell, *J. Chromatogr.*, **112**, 71 (1975); *Anal. Chem.*, **48**, 799, 1720 (1976).

68. R. J. Laub, J. H. Purnell, and P. S. Williams, *J. Chromatogr.*, **134**, 249 (1977); *Anal. Chim. Acta*, **95**, 135 (1977).

69. J. W. Eastman, G. Engelsma, and M. Calvin, *J. Am. Chem. Soc.*, **84**, 1339 (1962).

70. H. Kainer and W. Otting, *Chem. Ber.*, **88**, 1921 (1955).

71. R. Foster and T. J. Thomson, *Trans. Faraday Soc.*, **59**, 296 (1963).

72. Y. Matsunaga, *J. Chem. Phys.*, **41**, 1609 (1964); **42**, 1982 (1965).

73. K. Bauge and J. W. Smith, *J. Chem. Soc.*, 4244 (1964).

74. J. H. Purnell, in *Gas Chromatography 1966*, A. B. Littlewood, Ed., Institute of Petroleum, London, 1967, p. 3.

75. R. S. Juvet, Jr., and J. J. Pesek, *Anal. Chem.*, **41**, 1456 (1969).

76. R. L. Scott, *Rec. Trav. Chim. Pays-Bas*, **75**, 787 (1956).

77. P. H. Emslie, R. Foster, C. A. Fyfe, and I. Horman, *Tetrahedron*, **21**, 2843 (1965).

78. J. H. Purnell and O. P. Srivastava, *Anal. Chem.*, **45**, 1111 (1973); O. P. Srivastava, Ph.D. Dissertation, University of Wales, 1972.

79. R. J. Laub and R. L. Pecsok, *J. Chromatogr.*, **113**, 47 (1975).

80. G. L. Bertrand, D. E. Oyler, U. G. Eichelbaum, and L. G. Hepler, *Thermochim. Acta*, **7**, 87 (1973).

81. F. Van de Craats, *Anal. Chim. Acta*, **14**, 136 (1956).

82. H. M. Tenney, *Anal. Chem.*, **30**, 2 (1958).

83. M. E. Bednas and D. S. Russell, *Can. J. Chem.*, **36**, 1272 (1958).

84. C. S. G. Phillips, in *Gas Chromatography*, V. J. Coates, H. J. Noebels, and I. S. Fagerson, Eds., Academic, New York, 1958, p. 51.

85. D. W. Barber, C. S. G. Phillips, G. F. Tusa, and A. Verdin, *J. Chem. Soc.*, 18 (1959).

86. E. Gil-Av and Y. Herzberg-Minzly, *J. Am. Chem. Soc.*, **81**, 4749 (1959).

87. L. A. duPlessis and A. H. Spong, *J. Chem. Soc.*, 2027 (1959).

88. E. Gil-Av and J. Herling, *J. Phys. Chem.*, **66**, 1208 (1962).

89. M. A. Muhs and F. T. Weiss, *J. Am. Chem. Soc.*, **84**, 4697 (1962).

90. R. S. Juvet, Jr., and F. M. Wachi, *Anal. Chem.*, **32**, 290 (1960).

91. W. W. Hanneman, C. E. Spencer, and J. F. Johnson, *Anal. Chem.*, **32**, 1386 (1960).

92. G. P. Cartoni, R. S. Lowrie, C. S. G. Phillips, and L. M. Venanzi, in *Gas Chromatography 1960*, R. P. W. Scott, Ed., Butterworths, London, 1960, p. 273.

93. J. H. Bochinski, R. S. Juvet, Jr., and K. W. Gardiner, in *Ultrapurification of Semiconductor Materials*, M. S. Brook and J. K. Kennedy, Eds., Macmillan, New York, 1962, p. 239.

94. J. Herling, J. Shabtai, and E. Gil-Av, *J. Chromatogr.*, **8**, 349 (1962).

95. J. Shabtai, J. Herling, and E. Gil-Av, *J. Chromatogr.*, **11**, 32 (1963).

96. W. W. Hanneman, *J. Gas Chromatogr.*, **1**(12), 18 (1963).

97. A. Zlatkis, G. S. Chao, and H. R. Kaufman, *Anal. Chem.*, **36**, 2354 (1964).

98. R. Cvetanovic, F. J. Duncan, W. E. Falconer, and R. S. Irwin, *J. Am. Chem. Soc.*, **87**, 1827 (1965).

99. E. Gil-Av and J. Herling, in *Chromatography and Methods of Immediate Separation*, Vol. I, G. Parissakis, Ed., Association of Greek Chemists, Athens, 1966, p. 167.

100. F. M. Zado and R. S. Juvet, Jr., in *Gas Chromatography 1966*, A. B. Littlewood, Ed., Institute of Petroleum, London, 1967, p. 283.

101. R. L. Pecsok and E. Vary, *Anal. Chem.*, **39**, 289 (1967).

102. J.-M. Vergnaud, *J. Chromatogr.*, **27**, 54 (1967).

103. J. G. Atkinson, A. A. Russell, and R. S. Stuart, *Can. J. Chem.*, **45**, 1963 (1967).

104. H. Schnecko, *Anal. Chem.*, **40**, 1391 (1968).

105. D. V. Banthorpe, C. Gatford, and B. R. Hollebone, *J. Gas Chromatogr.*, **6**, 61 (1968).

106. F. Geiss, B. Versino, and H. Schlitt, *Chromatographia*, **1**, 9 (1968); *Z. Anal. Chem.*, **236**, 136 (1968).

107. C. Bighi, A. Betti, G. Saglietto, and F. Dondi, *J. Chromatogr.*, **34**, 389 (1968); **35**, 309 (1968).

108. R. S. Juvet, Jr., V. R. Shaw, and M. A. Khan, *J. Am. Chem. Soc.*, **91**, 3788 (1969).

109. B. H. Gump, *J. Chromatogr. Sci.*, **7**, 755 (1969).

110. C. Bighi, A. Betti, and F. Dondi, *J. Chromatogr.*, **39**, 125 (1969).

111. S. P. Wasik and W. Tsang, *Anal. Chem.*, **42**, 1648 (1970); *J. Phys. Chem.*, **74**, 2970 (1970).

112. E. Gil-Av and V. Shurig, *Anal. Chem.*, **43**, 2030 (1971).

113. V. Schurig, J. L. Bear, and A. Zlatkis, *Chromatographia*, **5**, 301 (1972).

114. O. Guha and J. Janak, *J. Chromatogr.*, **68**, 325 (1972).

115. C. L. deLigny, *J. Chromatogr.*, **69**, 243 (1972); *Adv. Chromatogr.*, **14**, 265 (1976).

116. C. L. deLigny, T. Van't Verlatt, and F. Karthaus, *J. Chromatogr.*, **76**, 115 (1973).

117. M. Kraitr, R. Komers, and F. Cuta, *J. Chromatogr.*, **86**, 1 (1973); *Anal. Chem.*, **46**, 974 (1974).

118. G. E. Baiulescu and V. A. Ilie, *Anal. Chem.*, **46**, 1847 (1974).

119. J. W. King, *Anal. Chem.*, **47**, 1414 (1975).

120. M. Saleem, M. A. Khan, M. Shahid, and K. Iqbal, *Chromatographia*, **8**, 699 (1975).

121. A. G. Vitenberg, B. V. Ioffe, Z. St. Dimitrova, and T. P. Strukova, *J. Chromatogr.*, **126**, 205 (1976).

122. L. A. duPlessis, *J. Gas Chromatogr.*, **1**(11), 6 (1963).

123. R. J. Cvetanovic, F. J. Duncan, and W. E. Falconer, *Can. J. Chem.*, **42**, 2410 (1964).

124. A. B. Littlewood and F. W. Willmott, *Anal. Chem.*, **38**, 1076 (1966).

125. S. H. Langer and J. H. Purnell, *J. Phys. Chem.*, **70**, 904 (1966).

126. A. V. Kogansen, G. A. Kurkchi, and O. V. Levina, in *Gas Chromatography 1966*, A. B. Littlewood, Ed., Institute of Petroleum, London, 1967 p. 35.

127. F. Ratkovics, *Magy. Kem. Foly.*, **72**, 279 (1966).

128. A. R. Cooper, C. W. P. Crowne, and P. G. Farrell, *J. Chromatogr.*, **29**, 1 (1967).

129. W. E. Falconer and R. J. Cvetanovic, *J. Chromatogr.*, **27**, 20 (1967).

130. D. F. Cadogan and J. H. Purnell, *J. Chem. Soc.*, 2133 (1968); *J. Phys. Chem.*, **73**, 3849 (1969).

131. C. W. P. Crowne, J. M. Gross, M. Harper, and P. G. Farrell, in *Fifth International Symposium on Chromatography and Electrophoresis*, Ann Arbor-Humphrey, Ann Arbor, Michigan, 1969, p. 153.

132. N. Kotsev and D. Shopov, *Dokl. Bolg. Akad. Nauk*, **21**, 889 (1968).

133. M. Taramasso and P. Fuchs, *J. Chromatogr.*, **49**, 70 (1970).

134. R. Vivilecchia and B. L. Karger, *J. Am. Chem. Soc.*, **93**, 6598 (1971).

135. C. Eon, C. Pommier, and G. Guiochon, *C. R. Acad. Sci.*, **270**, 1436 (1970); **276**, 1337 (1973); *Chromatographia*, **4**, 235, 241 (1971); *J. Phys. Chem.*, **75**, 2632 (1971).

136. D. F. Meen, F. Morris, and J. H. Purnell, *J. Chromatogr. Sci.*, **9**, 281 (1971).

137. L. Batt, G. M. Burnett, G. C. Cameron, and J. Cameron, *Chem. Commun.*, 29 (1971).

138. C. W. P. Crowne and M. F. Harper, *J. Chromatogr.*, **61**, 1 (1971).

139. C. W. P. Crowne, M. F. Harper, and P. G. Farrell, *J. Chromatogr.*, **61**, 7 (1971).

140. C. Eon and B. L. Karger, *J. Chromatogr. Sci.*, **10**, 140 (1972).

141. Y. N. Bogoslovsky, V. M. Sakharov, and I. M. Shevchuk, *J. Chromatogr.*, **69**, 17 (1972).

142. A. Bhattacharjee and A. Basu, *J. Chromatogr.*, **71**, 534 (1972).

143. B. Feibush, M. F. Richardson, R. E. Sievers, and C. S. Springer, *J. Am. Chem. Soc.*, **94**, 6717 (1972).

144. A. Janik, *Chromatographia*, **6**, 514 (1973).

145. R. A. Heacock and J. E. Forrest, *J. Chromatogr.*, **78**, 241 (1973).

146. R. J. Laub and R. L. Pecsok, *Anal. Chem.*, **46**, 1214 (1974).

147. C. Eon and G. Guiochon, *Anal. Chem.*, **46**, 1393 (1974).

148. L. Mathiasson and R. Jonsson, *J. Chromatogr.*, **101**, 339 (1974).

149. S. A. Reznikov, R. I. Sidorov, and E. R. Veber, *Zh. Fiz. Khim.*, **49**, 356 (1975).

150. S. A. Reznikov and R. I. Sidorov, *Zh. Fiz. Khim.*, **49**, 376 (1975).

151. S. A. Reznikov, Y. A. Batyrev, and R. I. Sidorov, *Zh. Fiz. Khim.*, **49**, 440 (1975).

152. S. A. Reznikov, R. I. Sidorov, E. A. Smertina, and E. R. Veber, *Zh. Fiz. Khim.*, **49**, 1851 (1975).

153. M. Lafosse and N. Thuaud-Chourrout, *Chromatographia*, **8**, 105 (1975).

154. L. Mathiasson, *J. Chromatogr.*, **114**, 39, 47 (1975).

155. C. L. deLigny, N. J. Koole, H. D. Nelson, and G. H. E. Nieuwdorp, *J. Chromatogr.*, **114**, 63 (1975).

156. M. wa. Muanda, J. B. Nagy, and O. B. Nagy, *Tetrahedron Lett.*, 3421 (1974).

157. C. A. Wellington, *Adv. Anal. Chem. Instrum.*, **11**, 237 (1973).

158. R. J. Laub and C. A. Wellington, in *Molecular Association*, Vol. 2, R. Foster, Ed., Academic, London, *in press*.

159. F. K. Nasyrova, R. S. Giniyatvillin, and M. S. Vigdergauz, in *Advances in Gas Chromatography*, Vol. IV, Pt. 1, M. S. Vigdergauz, Ed., Akademiya Nauk SSSR, Kazan, Russia, 1975, p. 147.

160. F. K. Nasyrova and M. S. Vigdergauz, in *Advances in Gas Chromatography*, Vol. IV, Pt. 1, M. S. Vigdergauz, Ed., Akademiya Nauk SSSR, Kazan, Russia, 1975. p. 157.

161. D. E. Martire and P. Riedl, *J. Phys. Chem.*, **72**, 3478 (1968).

162. J. P. Sheridan, D. E. Martire, and Y. B. Tewari, *J. Am. Chem. Soc.*, **94**, 3294 (1972).

163. J. P. Sheridan, D. E. Martire, and F. P. Banda, *J. Am. Chem. Soc.*, **95**, 4788 (1973).

164. H.-L. Liao, D. E. Martire, and J. P. Sheridan, *Anal. Chem.*, **45**, 2087 (1973).

165. S. H. Langer, B. M. Johnson, and J. R. Conder, *J. Phys. Chem.*, **72**, 4020 (1968).

166. D. F. Cadogan and D. T. Sawyer, *Anal. Chem.*, **43**, 941 (1971).

167. D. E. Martire, J. P. Sheridan, J. W. King, and S. E. O'Donnell, *J. Am. Chem. Soc.*, **98**, 3101 (1976); D. E. Martire, *Anal. Chem.*, **46**, 1712 (1974); **48**, 398 (1976).

168. C. S. Chamberlain and R. S. Drago, *J. Am. Chem. Soc.*, **98**, 6142 (1976).

169. R. S. Drago, N. O. O'Brian, and G. C. Vogel, *J. Am. Chem. Soc.*, **92**, 3926 (1970).

170. R. J. Laub, D. L. Meen, and J. H. Purnell, *unpublished work*.

171. R. C. Castells, *Chromatographia*, **6**, 57 (1973).

172. S. Carter, J. N. Murrell, and E. J. Rosch, *J. Chem. Soc.*, 2048 (1965).

173. R. E. Merrifield and W. D. Phillips, *J. Am. Chem. Soc.*, **80**, 2779 (1958).

174. M. Tamres, *J. Phys. Chem.*, **65**, 654 (1961).

175. J. M. Corkhill, R. Foster, and D. L. Hammick, *J. Chem. Soc.*, 1202 (1965).

176. K. M. C. Davis, in *Molecular Association*, Vol. 1, R. Foster, Ed., Academic, London, 1975, Ch. 3.

177. H.-L. Liao and D. E. Martire, *J. Am. Chem. Soc.*, **96**, 2058 (1974).

178. J. H. Purnell and J. M. Vargas de Andrade, *J. Am. Chem. Soc.*, **97**, 3585, 3590 (1975).

179. R. J. Laub and J. H. Purnell, *J. Am. Chem. Soc.*, **98**, 30, 35 (1976).

180. R. J. Laub, J. H. Purnell, and D. M. Summers, *J. Chem. Soc, Faraday Trans. II*, *submitted*.

181. E. V. Frisman and A. K. Dadivanian, *J. Poly. Sci., Pt. C*, **16**, 1001 (1967).

182. K. Nagai, *J. Chem. Phys.*, **49**, 4212 (1968); *J. Phys. Chem.*, **74**, 3422 (1970).

183. A. N. Gent, *Macromolecules*, **2**, 262 (1969).

184. K. Dusek and W. Prins, *Fortschr. Hochpolym. Forsch.*, **6**, 1 (1969).

185. T. Ishikawa and K. Nagai, *Polym. J.*, **1**, 116 (1970).

186. M. Fukuda, G. L. Wilkes, and R. S. Stein, *J. Polym. Sci., Pt. A-2*, **9**, 1417 (1970).

187. P. Corradine, in *Proceedings of the R. A. Welch Foundation Conference on Chemical Research*, Vol. 10, W. O. Milligan, Ed., Houston, Texas, 1967, p. 163.

188. G. S. Y. Yeh, *Pure Appl. Chem.*, **31**, 65 (1971).

189. R. G. Kirste, W. A. Kruse, and J. Schelten, *Makromol. Chem.*, **162**, 299 (1972).

190. H. Benoit, J. P. Cotton, D. Decker, B. Farnoux, J. S. Higgins, G. Jannink, R. Ober, and C. Picot, *Nature Phys. Sci.*, **245**(140), 13 (1973).

191. V. T. Lam, P. Picker, D. Patterson, and P. Tancrede, *J. Chem. Soc. Faraday Trans. II*, **70**, 1465 (1974).

192. D. D. Deshpande, D. Patterson, H. P. Schreiber, and C. S. Su, *Macromolecules*, **7**, 530 (1974).

193. R. N. Martynyuk and M. S. Vigdergauz, *Chromatographia*, **9**, 454 (1976).

194. R. J. Laub, J. H. Purnell, D. M. Summers, and P. S. Williams, *J. Chromatogr.*, *in press*.

195. J. Klein and H. Widdecke, *J. Chromatogr.*, *in press*.

196. D. F. Lynch, F. A. Palocsay, and J. J. Leary, *J. Chromatogr. Sci.*, **13**, 533 (1975).

197. G. F. Freeguard and R. Stock, in *Gas Chromatography 1962*, M. van Swaay, Ed., Butterworths, London, 1962, p. 102.

198. A. DeVries, *Mol. Cryst. Liq. Cryst.*, **10**, 219 (1970); **11**, 361 (1970).

199. G. W. Brady, C. Cohen-Addad, and E. F. X. Lyden, *J. Chem. Phys.*, **51**, 4309 (1969).

200. G. W. Brady, *J. Chem. Phys.*, **57**, 91 (1972).

201. W. Phillipoff, *Discuss. Faraday Soc.*, **11**, 96 (1951).

202. W. D. Harkins, *The Physical Chemistry of Surface Films*, Reinhold, New York, 1952.

203. H. F. Reiss and V. Luzzatti, *J. Colloid Interface Sci.*, **21**, 534 (1966).

204. A. R. Oseroff, P. W. Robbin, and M. M. Borger, *Ann. Rev. Biochem.*, **42**, 835 (1973).

205. G. W. Stewart, *Proc. Nat. Acad. Sci.*, **13**, 787 (1927); *Phys. Rev.*, **32**, 153, 558 (1928); **33**, 889 (1929); **35**, 296 (1930); **37**, 9 (1931); **39**, 176 (1932); *Proc. Iowa Acad. Sci.*, **36**, 305 (1929); **41**, 250 (1934); *Rev. Mod. Phys.*, **2**, 116 (1930); *Indian J. Phys.*, **7**, 603 (1933); *Trans. Faraday Soc.*, **29**, 982 (1933); *J. Chem. Phys.*, **2**, 147 (1934); *Kolloid-Z.*, **67**, 130 (1934).

206. G. W. Stewart and R. M. Morrow, *Proc. Nat. Acad. Sci.*, **13**, 222 (1927); *Phys. Rev.*, **30**, 232 (1927).

207. G. W. Stewart and R. L. Edwards, *Phys. Rev.*, **38**, 1575 (1931).

208. G. W. Stewart and R. D. Spangler, *Phys. Rev.*, **36**, 472 (1930).

209. G. W. Stewart and E. W. Skinner, *Phys. Rev.*, **31**, 1 (1928).

210. G. W. Stewart, R. M. Morrow, and E. W. Skinner, *Phys. Rev.*, **27**, 104 (1926).

211. G. S. Parks and C. S. Chaffee, *J. Phys. Chem.*, **31**, 439 (1927).

212. R. M. Morrow, *Phys. Rev.*, **31**, 10 (1928).

213. C. M. Sogani, *Indian J. Phys.*, **1**, 257 (1927); **2**, 97 (1927).

214. A. Müller, *Trans. Faraday Soc.*, **29**, 990 (1933).

215. F. I. G. Rawlins, *Trans. Faraday Soc.*, **29**, 993 (1933).

216. R. D. Spangler, *Phys. Rev.*, **42**, 907 (1932); **46**, 698 (1934).

217. C. A. Benz and G. W. Stewart, *Phys. Rev.*, **46**, 703 (1934).

218. C. A. Benz, *Proc. Iowa Acad. Sci.*, **41**, 249 (1934).

219. J. Rolinski, *Phys. Z.*, **29**, 658 (1928).

220. H. K. Ward, *J. Chem. Phys.*, **2**, 153 (1934).

221. F. Franks, Ed., *Physico-Chemical Processes in Mixed Aqueous Solvents*, Elsevier, New York, 1967; in *Water: A Comprehensive Treatise*, Vol. 4, Plenum, New York, 1975, pp. 1, 759.

222. K. Toukubo and K. Nakamishi, *J. Chem. Phys.*, **65**, 1937 (1976).

223. G. W. Brady, *J. Chem. Phys.*, **32**, 45 (1960); **58**, 3542 (1973); **60**, 3466 (1974); *Polym. Prepr.*, **14**, 181 (1973); *Acc. Chem. Res.*, **7**, 174 (1974).

224. M. D. Croucher and D. Patterson, *J. Chem. Soc. Faraday Trans. II*, **70**, 1479 (1974).

225. C. Clement and P. Bothorel, *J. Chim. Phys.*, 878 (1964).

226. P. Bothorel, C. Clement, and P. Maraval, *Compt. Rend.*, **264**, 658 (1967).

227. P. Bothorel, *J. Colloid Interface Sci.*, **27**, 529 (1968).

228. P. Bothorel, C. Such, and C. Clement, *J. Chim. Phys.*, **10**, 1453 (1972).

229. B. Lemaire, G. Fourche, and P. Bothorel, *C. R. Acad. Sci.*, **274**, 1481 (1972).

230. H. Quinones and P. Bothorel, *C. R. Acad. Sci.*, **277**, 133 (1973).

231. P. Bothorel and G. Fourche, *J. Chem. Soc. Faraday Trans. II*, **69**, 441 (1973).

232. G. D. Patterson and P. J. Flory, *J. Chem. Soc. Faraday Trans. II*, **68**, 1098 (1972).

233. K. L. Mittal, Ed., *International Symposium on Micellizations, Solubilization, and Micro-emulsions*, Vols. 1 and 2, Plenum, New York, *in press*.

234. M. S. Nozari and R. S. Drago, *J. Am. Chem. Soc.*, **94**, 6877 (1972).

235. J. F. Parcher and T. N. Westlake, *J. Phys. Chem.*, **81**, 307 (1977).

236. A. J. Ashworth and D. M. Hooker, *J. Chromatogr.*, **131**, 399 (1977).

237. J. F. Parcher and T. N. Westlake, *J. Chromatogr. Sci., in press.*

238. R. J. Laub, D. E. Martire, and J. H. Purnell, *J. Chem. Soc. Faraday Trans. I*, **73**, 1686 (1977); *J. Chem. Soc., Faraday Trans. II*, **74**, 213 (1978).

239. G. M. Janini and D. E. Martire, *J. Chem. Soc. Faraday Trans. II*, **70**, 837 (1974).

240. R. A. Orwoll and P. J. Flory, *J. Am. Chem. Soc.*, **89**, 6814, 6822 (1967).

241. W. Hayduk and S. C. Cheng, *Can. J. Chem. Eng.*, **48**, 93 (1970).

242. T. Nitta, A. Tatsuishi, and T. Katayama, *J. Chem. Eng. Jap.*, **6**, 475 (1973).

243. M. W. P. Harbison, D. E. Martire, R. J. Laub, J. H. Purnell, and P. S. Williams, *J. Chem. Soc. Faraday Trans. I, in preparation.*

CHAPTER 7

Thermodynamics
of Adsorption

In addition to the well-known comprehensive treatments of the phenomenon of adsorption,[1, 2] gas-solid chromatography (GSC) has been the subject of a number of reviews[3-15] to which the reader is referred for details regarding, for example, the kinetics of sorption-desorption equilibria[16-30] and the relevance of site distribution phenomena.[31-34] The physical and chemical properties[35-40] of silica,[41-42] zeolites,[43-46] graphitized carbon,[47, 48] and synthetic[49, 50] and surface-modified[51-53] adsorbents have in addition been summarized.[54] Here, alternatively, are surveyed the various *applications* of GSC that are relevant to the thermodynamics of adsorption. Broadly speaking, they fall into two interrelated areas, namely, the determination of adsorption isotherms and heats of adsorption. The measurement of specific surface areas, moreover, has become an important application of GSC and for convenience of presentation, is described as well.

7.1 ADSORPTION ISOTHERMS

Adsorption isotherms are determined by GSC most readily via the FA or FACP techniques detailed in Section 2.1.3. Elution by characteristic point[55, 56] (ECP) has also been employed, although, in general, it is not applicable over extended ranges of solute mole fraction. It does, however, find use in investigations at low-solute-surface coverage; for example, in polymer studies.[57, 58]

Table 7.1 presents a few representative examples of systems for which the GSC method has been used; they are important, *inter alia*, in the elucidation of adsorbent structure, pigments and catalysis, and various fundamental studies of chemi- and physisorption (e.g., Ref. 1) in addition to providing data from which heats of adsorption and surface areas are

213

Table 7.1 Examples of the Determination of Adsorption Isotherms by GSC

Solutes	Adsorbents	Temperature Range (°C)	Ref.
Cyclohexane, benzene	Charcoal	77, 100	59
Carbon dioxide	Silica, charcoal		60
Nitrogen	13X Molecular sieve, silica, alumina, calcined alumina, 0.6% Pt on alumina	−196	61
n-Hexane, benzene (isobars)	10% MoO_3 on alumina, 13% alumina on silica, calcined alumina	177 to 427	62
Sulfur, uranium hexafluoride	Unspecified	60	63
Nitrogen, argon, oxygen, carbon monoxide, hexafluoroethane	Carbon black, bone mineral	−200 to 5	64
Methane, propane	Silica	−25 to 35	65, 66
Nitrogen	Etched glass	0	67
Xenon	LiX, NaX		68
Carbon dioxide	Zeolites		69
Water	Synthetic fibers	Ambient	70
Methane	Fused silica	−92 to 32	71
Various	Porous polymer (Porapak Q)		72
Nitrogen	Porous polymers (Porapak N, P, Q, R, S, and T)	−196	73
Oxygen, nitrogen	5A Molecular sieve	10 to 50	74
Amines	Cobalt, alumina	60 to 300	75
Methane	Fused silica	−80 to −20	76
Cyclohexane, cyclohexene	Silica	126	77
n-Pentane	Silica	20 to 150	78
Various	Porous polymer (Tenax)		79
Benzene	Fused silica		80
n-Hexane	Graphitized carbon		81
n-Decane	poly(Methyl methacrylate)	25 to 60	82
Various	Polymers		83
n-Alkanes, dioxane, alcohols	Cotton, ramie, cellulose fibers	25 to 75	84, 85
Various	poly(vinyl chloride), poly(vinyl fluoride), poly(vinylidene chloride)		86–88
H_2, D_2^*	4A, 5A, 13X Molecular sieves, silica, charcoals	−200 to −180	89

*(static apparatus employing thermal conductivity detector)

214

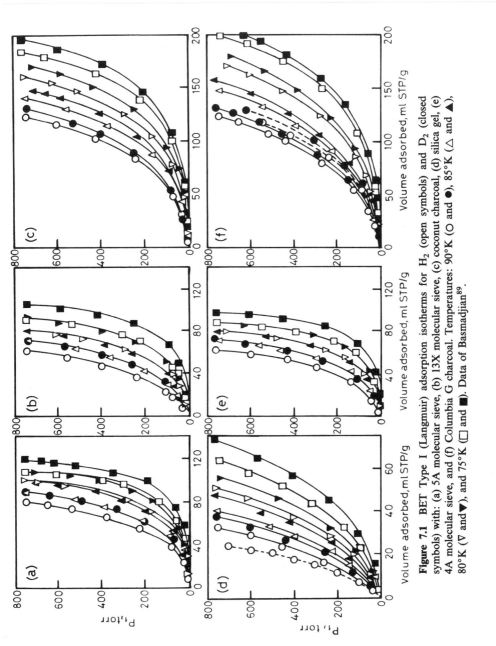

Figure 7.1 BET Type I (Langmuir) adsorption isotherms for H_2 (open symbols) and D_2 (closed symbols) with: (a) 5A molecular sieve, (b) 13X molecular sieve, (c) coconut charcoal, (d) silica gel, (e) 4A molecular sieve, and (f) Columbia G charcoal. Temperatures: 90°K (○ and ●), 85°K (△ and ▲), 80°K (▽ and ▼), and 75°K (□ and ■). Data of Basmadjian[89].

215

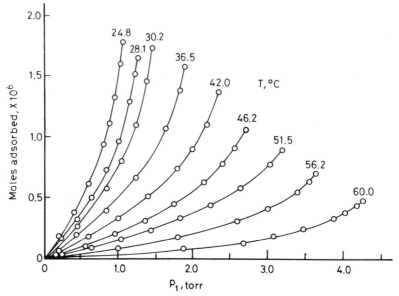

Figure 7.2 BET Type III adsorption isotherms at indicated temperatures for *n*-decane with *poly*(methyl methacrylate) beads. Data of Gray and Guillet[82].

derived. Figures 7.1 and 7.2 show typical plots of isotherm data, the former being H_2 and D_2 with various adsorbents[89] and the latter[82] for *n*-decane with poly(methyl methacrylate).

7.2   HEATS OF ADSORPTION

At low surface coverage, that is, in the Henry's law region (in which symmetric and sample-size-independent peaks obtain) the thermodynamic properties of adsorption are determined in GSC via the relation analogous to that for GLC:

$$\ln V_g^0 = \frac{\Delta H_a^0}{RT} + C \tag{7.1}$$

where ΔH_a^0 refers to the solute heat of adsorption at zero surface coverage. Plots of $\ln V_g^0$ versus $1/T$ thus yield ΔH_a^0 from the slope. At finite solute surface concentration (nonlinear isotherms)

$$\ln p_1 = \frac{\Delta H_a^s}{RT} + C' \tag{7.2}$$

where p_1 is the solute partial pressure and ΔH_a^s is the *isosteric* heat of adsorption (plots of $\ln p_1$ versus $1/T$ being called *isosteres*). Table 7.2 lists several examples in which heats of adsorption have been determined by GSC and Table 7.3 presents a typical set of gas-chromatographic data.

Table 7.2   Examples of the Determination of Heats of Adsorption by GSC

Solutes	Adsorbents	Temperature Range (°C)	Ref.
Argon, oxygen, nitrogen, carbon monoxide, methane	Charcoal, silica		95
Alkanes, nitrogen	Charcoals	97, 254	96
Nitrogen, benzene	13X Molecular sieve, silica, alumina, calcined alumina, 0.6% Pt on alumina	− 196	61
n-Hexane, benzene	Molybdina/alumina, alumina/silica	177 to 427	62
Alcohols	Graphitized carbon	72 to 200	97
Nitrogen, oxygen, argon, methane	Bone minerals, graphitized carbons	0 to 130	98
Hydrogen, oxygen, nitrogen, methane, carbon monoxide, krypton, xenon	5A Molecular sieve	40 to 100	99
Nitrogen, oxygen, argon, carbon monoxide, hexafluoroethane	Bone minerals, graphitized carbon	− 100 to 5	64
Water, methanol, ammonia, methylamine	NaX zeolites	23	100
n-Alkanes, alcohols, aromatics	Siliceous fillers		101
Methane, ethane	Silica	25	66
Various	Silica/copper salts	Unspec.	41
n-Pentane, methanol, ethanol, *n*-propanol	Graphitized carbon/PEG 1500	50 to 100	103
Propylene, acetic acid	Graphitized carbon	75 to 102	104
Amines	Cobalt, alumina	60 to 300	75
n-Alkanes	Nickel	200 to 300	105
Various	Graphitized carbon	110 to 395	106
Propylene and oxidation products	Molybdate catalysts		107
Benzene	Fused silica		80
n-Decane, dioxane, alcohols	Cotton, ramie fibers	25 to 75	85
Methane, *n*-pentane, benzene	Graphitized carbon	60 to 160	52
Aliphatics, aromatics	Exchanged zeolites	65 to 120	108

Table 7.3 Heats of Adsorption ΔH_a^0 for Listed Solutes with Cobalt and Manganese Chlorides[109]

Solute	Cobalt Chloride		Manganese Chloride	
	T (°C)	$-\Delta H_a^0$ (kcal/mole)	T (°C)	$-\Delta H_a^0$ (kcal/mole)
n-Pentane	50–80	8.29	70–95	5.61
n-Hexane	50–80	8.21	70–95	8.09
n-Heptane	50–80	9.56	70–95	9.14
n-Octane	50–80	10.3	70–95	11.2
n-Nonane	50–80	11.5	...	...
1-Chloropentane	50–80	11.3	...	...
2-Chloropentane	50–80	11.6	70–95	11.7
3-Chloropentane	50–80	8.59	70–95	11.3
1-Chlorobutane	50–80	9.36	70–95	11.8
2-Chlorobutane	50–80	9.40	70–95	7.85
1-Chloro-1-butene	50–80	10.3	70–95	11.3
2-Chloro-1-butene	50–80	10.4	70–95	11.9
4-Chloro-1-butene	50–80	10.9	70–95	13.8
1-Pentene	50–80	5.58	70–95	11.0
1-Pentyne	50–80	10.2	...	...
1-Hexene	50–80	8.34	70–95	12.2
1-Hexyne	50–80	13.6	...	...
Benzene	50–80	10.7	70–95	13.2
Chlorobenzene	50–80	10.7	70–95	13.6
Toluene	50–80	11.8	109–128	13.2
o-Xylene	50–80	12.6	140–160	14.1
m-Xylene	50–80	12.3	140–160	14.2
p-Xylene	50–80	11.6	140–160	14.3

   Liquid- and salt-coated adsorbents have lately received renewed attention primarily because of the analytical separations possible with these materials[47, 53, 110] (although considerable care is required in their preparation[110]). Table 7.4 shows, for example, heats of adsorption for a variety of solutes with several phthalocyanine-coated graphitized carbon packings.[111] In a similar series of studies Sawyer and co-workers[112–119] found in accordance with the suggestion of Kiselev[5] that adsorption phenomena could be divided into "specific" and "nonspecific" interactions, values of the former being assigned to functional groups. These were determined by constructing plots of $\ln V_N$ versus $1/T$ for a wide range of solutes with various salt-coated aluminas and silicas, straight lines drawn through homologous series (e.g., n-alkanes, 1-olefins, etc.) for each packing, and the vertical distances D (in terms of $\ln V_N$ "units") between the alkane lines and the

Table 7.4 Heats of Adsorption ΔH_a^0 for Named Solutes with Phthalocyanine-Coated Graphitized Carbon[111] at 100–150°C

	$-\Delta H_a^0$ (kcal/mole)					
	None	Ph	CuPh	CoPh	ZnPh	NiPh
Benzene	9.8	8.3	10.1	11.2	11.1	—
Acetone	8.3	7.1	7.7	10.8	13.5	13.1
Diethyl ether	8.8	6.8	7.5	10.0	13.1	10.8
n-Pentane	8.9	6.9	7.5	7.5	7.7	7.8
Methyl propionate	—	8.6	9.3	11.6	13.5	15.9
n-Hexane	10.4	8.2	8.8	8.7	9.2	9.1
Di-n-propyl ether	—	9.0	10.1	12.3	14.9	13.8
n-Heptane	12.5	9.5	9.9	9.7	10.3	10.1
Toluene	11.6	9.2	10.6	11.4	11.5	—
Ethylbenzene	12.7	10.0	11.1	11.6	11.7	—
iso-Propylbenzene	13.2	10.3	11.3	11.7	11.7	—
m-Xylene	—	10.7	11.8	12.9	13.3	—
p-Xylene	—	11.3	12.0	12.9	13.3	—
o-Xylene	—	11.3	12.3	13.6	13.6	—
n-Propylbenzene	14.3	11.3	11.8	12.6	12.3	—
Pyridine	10.0	8.3	10.1	16.8	—	—
Aniline	13.1	9.1	11.3	15.0	15.6	21.1
N-Methylaniline	—	10.1	13.1	16.7	16.6	16.6
N,N-Dimethylaniline	14.0	11.3	14.1	15.7	16.1	13.7

solutes of interest then measured:

$$\Delta G_{\text{spec}} = -RT\ln\left(\frac{V_N}{V_N^{\text{alkane}}}\right) = -RTD \qquad (7.3)$$

Table 7.5 illustrates the variation of substituent ΔH_{spec} and ΔS_{spec} data with salt-coated adsorbents.

Heats of adsorption, measured by GSC, find general agreement with those measured by static (e.g., calorimetric) techniques.[61, 95, 97, 120–122] Kiselev and co-workers,[97] however, noted that chromatographic heats, for example, for alcohols with graphitized carbon at zero surface coverage, were some 5 kcal/mole lower than isosteric heats at low surface coverage that were measured calorimetrically. The former, in contrast, agreed impressively with calculated values. (n-Alkane solute data measured separately by the two techniques were in good agreement.) These results were said to reflect the association of alcohols at the lowest concentrations

Table 7.5 ΔH_{spec} **(kcal/mole) and** ΔS_{spec} **(eu) Values for Named Columns and Substituents at 150–250°C**[114]

Substituent	10% Na$_2$SO$_4$/naw Al$_2$O$_3$		10% Na$_2$SO$_4$/aw Al$_2$O$_3$		10% NaCl/naw Al$_2$O$_3$		10% Na$_2$SO$_4$/ Porasil C	
	ΔH_{spec}	ΔS_{spec}	ΔH_{spec}	ΔS_{spec}	ΔH_{spec}	ΔS_{spec}	ΔH_{spec}	ΔS_{spec}
(H$_2$)	−0.23	14.98	−0.30	15.61	0.29	15.99	1.33	16.69
—CH$_2$—	1.34	1.46	1.33	1.44	1.27	1.35	0.86	0.97
π-terminal	1.10	1.43	1.32	1.71	0.83	1.04	1.02	1.52
π-trans	1.06	1.61	1.31	1.95	0.76	1.14	0.92	1.01
π-cis	1.34	1.86	1.85	2.66	1.05	1.45	0.92	1.01
π-conjugated	1.61	2.34	2.38	3.38	1.52	2.19	1.26	1.72
π-aromatic	1.35	2.00	1.31	1.57	0.74	0.87	1.11	1.69
ϕ-CH$_3$	1.40	1.57	1.70	2.04	1.38	1.60	1.83	2.59
ϕ-C$_2$H$_5$	2.63	2.85	3.28	4.11	2.44	2.58	2.77	3.69
ϕ-i-C$_3$H$_7$	2.90	2.64	3.73	4.52	3.10	3.24	3.44	4.46
ϕ-t-C$_4$H$_9$	4.46	4.83	4.56	5.38	3.94	4.05	4.20	5.35
ϕ-CF$_3$	1.85	2.06	...	...	...	...	−0.81	−2.17
ϕ-F	1.00	1.25	0.60	0.94	0.28	0.11	−0.90	−1.70
ϕ-Cl	2.17	2.47	2.17	2.81	0.90	0.14	−0.49	−1.74
ϕ-Br	3.00	3.26	3.09	3.66	2.45	2.33	0.22	−0.87
ϕ-I	4.12	4.28	4.00	4.22	3.66	3.51	1.01	0.08
ϕ-OCH$_3$	...	...	...	...	...	...	4.82	5.30

reached in the static experiments but at levels several orders of magnitude lower, that is, in the range in which gas-chromatographic detectors are used, the solutes were sufficiently dilute to be completely dissociated. In this regard ΔH_a^0 data obtained via GSC may well be superior to those obtained from nonchromatographic techniques.

7.3   SURFACE AREAS

The well-known "isotherm" equation of Brunauer, Emmett, and Teller[123] in the form

$$\frac{p_1}{V(p_1^0 - p_1)} = \frac{1}{V_M c} + \frac{(c-1)}{V_M c} \frac{p_1}{p_1^0} \tag{7.4}$$

has been used for many years for the determination of surface areas by static as well as what is commonly regarded as chromatographic methods (V is the total volume of adsorbed solute at vapor pressure p_1, i.e., an isotherm point, V_M is the volume required to form a monolayer, and c is the "BET constant"). Plots of the lhs versus p_1/p_1^0 yield straight lines of slope $(c-1)/V_M c$ and intercept $1/V_M c$ from which V_M is obtained. The surface area A_S of the adsorbent is then found by multiplying the number of molecules in V_M by the cross-sectional area σ of the adsorbate, where the number of molecules n in a milliliter of gas at STP is taken to be Avogadro's number divided by the volume of one mole of gas: $n = 2.687 \times 10^{19}$ molecules/ml. The surface area is therefore

$$A_S = 2.687 \times 10^{19} \sigma V_M \tag{7.5}$$

and the specific surface area S, the surface area per gram adsorbent, is

$$S = \frac{A_S}{w_S} \tag{7.6}$$

where w_S is the weight of adsorbent used; σ data reported by Ettre[124] are given in Tables 7.6 and 7.7.

The gas-chromatographic method of the determination of V_M was originally developed by Nelsen and Eggertsen[125] in 1958 and employs a binary carrier consisting of nitrogen and helium. The column containing the adsorbent has only to be maintained at room temperature. At a given moment the column is immersed in liquid nitrogen which causes the carrier

Table 7.6 Required Data for Nitrogen for Surface Area Measurements[124]

T (°K)	p_1^0 (torr)	σ [(Å)^2/molecule]	Surface/Volume Ratio [m^2/ml (STP)]
77.36	760	16.268	4.3721
77.47	770	16.275	4.3739
77.59	780	16.281	4.3756
77.69	790	16.287	4.3773
77.80	800	16.294	4.3790
77.91	810	16.301	4.3808
78.02	820	16.308	4.3826
78.12	830	16.315	4.3844
78.23	840	16.321	4.3862
78.33	850	16.327	4.3880
78.43	860	16.334	4.3898
78.53	870	16.340	4.3915
78.63	880	16.346	4.3932
78.73	890	16.353	4.3949
78.83	900	16.359	4.3966

Table 7.7 Cross-Sectional Areas for Named Gases[124]

Gas	T (°C)	σ [(Å)^2/molecule]
Krypton	−195	18.5
Argon	−195	14.6
	−183	14.4
Oxygen	−183	14.3
Carbon monoxide	−183	16.6
Carbon dioxide	−78	19.5
	−56.6	17.0
Ethane	−183	22.5
n-Butane	0	38.4
1-Butene	0	40.6
Water	25	10.8
Nitrous oxide	−78	20.4

nitrogen to condense out of the mobile phase onto the adsorbent and results in a negative "adsorption" peak in the recorder trace. At a later time the liquid nitrogen bath is removed and the carrier gas returns to its original composition, resulting in a positive "desorption" peak. The area of either peak is directly proportional to V. (In practice the desorption peak is generally used because it is almost always symmetric, whereas the adsorption peak is usually asymmetric.) By varying the nitrogen/helium ratio of the carrier stream several V values can be determined, thus providing sufficient data for calculation of the surface area from the BET equation. The apparatus of Nelsen and Eggertsen[125] is shown in Fig. 7.3 and a typical adsorption-desorption chromatogram, in Fig. 7.4. A cold trap is used immediately before the sample tube to remove water and other contaminants in the carrier gas. The bypass line is also helpful when high flow rates are employed, for the flow rate (hence the baseline) will be unstable for a short period after the coolant flask is removed (and desorption begins); in other words, the bypass line is used until the baseline stabilizes. The line may not be necessary at low flow rates because temperature and pressure equilibration will be fast. A critical feature of the apparatus is the flow control of both gases because the BET plotting procedure requires accurate p_1 data. Thus precision regulators are necessary in both the helium and nitrogen streams. Equally important is the measurement of the desorption peak area and calibration of the TC detector. Nevertheless, these requirements are easily handled, as illustrated in Table 7.8 where static and GSC surface area data are compared.

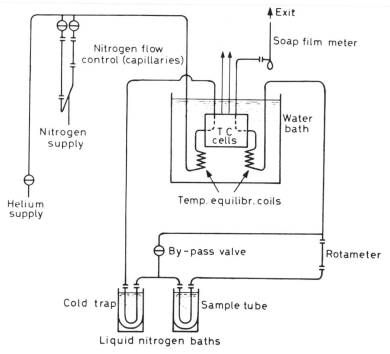

Figure 7.3 Schematic of the original system of Nelsen and Eggertsen[125]

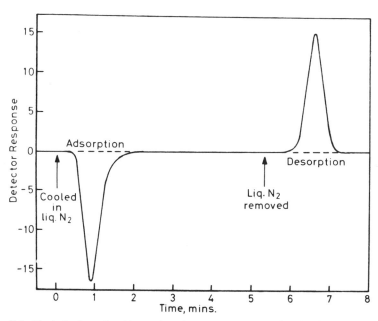

Figure 7.4 Typical adsorption/desorption chromatogram; note that the desorption peak is symmetric whereas the adsorption peak is not.

Table 7.8   Comparison of Surface Area Measurements
by Static and GSC BET Methods[125]

	Surface Area (m^2/g)	
Adsorbent	Static	GSC
Firebrick	3.1	3.4
Furnace black	24	25.7
Used silica/alumina catalyst	103	101
Fresh silica/alumina catalyst	438	455
Alumina	237	231

The Nelsen–Eggertsen method has been considerably refined over the years: first, the use of two separate cylinders for N_2 and He is obviated if one cylinder composed of a known N_2/He ratio is employed. Such mixtures are now available from supply houses and the two carrier streams therefore may be replaced by the number of tanks corresponding to the number of BET points desired. Roth and Ellwood[126] found no difference in results obtained with separate tanks and a manifold or single cylinders containing commercial mixtures. They also noted that if the flow rate through the sample tube is no greater than 25 ml/min the bypass line is unnecessary. Daeschner and Stross[127] critically analyzed sources of error in the GSC technique in 1962, these being primarily the linearity of the detector response and flow rate changes during desorption. The use of a gas injection valve with several loops (each of known volume) was employed to calibrate their detector accurately; flow-rate fluctuations were kept to a minimum by partial venting of the carrier pressure just before desorption was allowed to occur. (As a result of their studies a commercial instrument, the Perkin–Elmer Model 212 "sorptiometer," was developed and is still widely used.) Cahen and Marechal[128] found that water vapor in the carrier streams caused inaccurate results but could be eliminated by using only glass and/or metal (as opposed to plastic) gas transfer lines and drying tubes. They also found that a stream splitter placed just before the detector improved the accuracy of measurement of desorption peak areas; that is, injection of a large volume of nitrogen as a calibration standard was not required. Derby and LaMont[129] and Smolkova and co-workers[130] showed that the method was applicable to samples of surface area of as little as 0.003 m^2/g. Several workers have reported GSC-dedicated computer algorithms[131–133] and Tremaine and Gray described the calculation of areas from ECP data[134] after Stock.[135] Extensive[134–162] use has indeed been made of variants of GSC for the determination of BET (nitrogen adsorbate) surface areas.

The surface area of adsorbents available to organic compounds (larger molecular dimensions than N_2; cf. Table 7.7) is often of predominant interest. Perrett and Purnell[163] and others[84, 85, 108, 164] have shown that when this is the case nitrogen is an inappropriate choice of adsorbate, for because of its size it may have access to the area within pores that may not be available to the (larger) compounds of concern. GSC surface areas determined with organic adsorbates are, in fact invariably found to be less than those measured with nitrogen, as indicated in Table 7.9.

Table 7.9 Examples of Surface Area Measurements with Organic Adsorbates

Adsorbent	Adsorbate	T (°C)	σ (Mean) $(\mathring{A})^2$	S (Mean) (m^2/g)
A. Data of Gray and co-workers[84, 85]				
Cotton fiber	n-Octane	25.8	45.8	1.63
	n-Decane	25.8, 35.5, 45.0, 56.2	(52.2)	(1.60 ± 0.03)
	n-Dodecane	56.1	58.4	1.59
	1,4-Dioxane	25.7, 40.2	(29.9)	(1.68 ± 0.04)
	1-Butanol	25.7, 49.8	(31.4)	(1.62 ± 0.01)
	Nitrogen	−196.	16.2	1.9
B. Data of Dyer and co-workers[108] (unspecified temperature)				
Zirconium phosphate	n-Hexane			10.8
	Cyclohexane			10.2
	1-Hexene			10.4
	Benzene			10.0
	Nitrogen			32.5
Lithium zirconium phosphate	n-Hexane			23.8
	Cyclohexane			26.3
	1-Hexene			23.0
	Nitrogen			42.2
Sodium zirconium phosphate	n-Hexane			13.9
	Cyclohexane			14.1
	Benzene			14.3
	1-Hexene			13.3
	Cyclohexene			14.9
	Nitrogen			23.6
Potassium zirconium phosphate	Benzene			10.4
	Cyclohexane			10.0
	n-Hexane			10.2
	Nitrogen			20.1

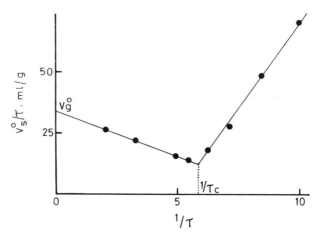

Figure 7.5 Plot of V_S^0/τ *vs.* $1/\tau$ for *n*-pentane with 1-octadecanol on silica gel at the melting point of the stationary phase (58°C). Data of Serpinet[168].

Serpinet and co-workers[165–168] (cf., in addition, Refs. 169, 170) have recently described a GC method, substantially different from those presented above, that makes use of thin-film formation of stationary phases on hydrophilic (adsorbent) supports. Figure 7.5 shows a plot of V_S^0/τ versus $1/\tau$ for *n*-pentane with 1-octadecanol on silica gel, where V_S^0 is the solute net retention volume (corrected to 0°C) per gram *support* and τ is the weight of stationary phase per gram support. The shape of the curve, reminiscent of plots of V_N/V_L versus $1/V_L$ described in Section 2.2.4 (cf. Fig. 3.A.iii in Ref. 171), is determined by the intersection of the two straight lines. A perpendicular dropped from the point of intersection to the abscissa gives a value for what is described as the critical (monolayer) weight of liquid phase per gram support from which the surface area may be calculated:

$$S = \frac{\tau_c N \sigma}{MW_L} \tag{7.7}$$

where σ is the molecular area of the liquid phase at the column temperature as before [21.0 (Å)^2 for 1-octadecanol at 58°C]. The critical load may also be determined with a single column containing a stationary phase loading slightly in excess of τ_c as illustrated in Fig. 7.6:

$$\tau = \frac{\Delta V_S^0}{V_g^0} \tag{7.8}$$

$$\tau_c = \tau - \tau_b \tag{7.9}$$

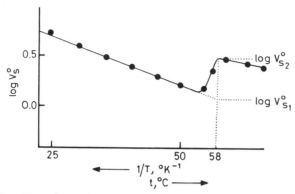

Figure 7.6 Plot of log V_S^0 vs. $1/T$ for *n*-octane with 2.70% w/w 1-octadecanol on Spherosil XOC 005 fused silica. Data of Serpinet[168].

Table 7.10   Comparison of Surface Area Measurements by BET and Thin-Film Techniques[168]

Adsorbent	S (m²/g) BET	Thin-Film
A. 1-Octadecanol stationary phase		
Chromosorb P	3.6	3.60
Spherosil XOC 005	11.1	11.08
Spherosil XOB 015	26.2	25.5
"Exal" α-alumina (I)	6.5	7.04
"Exal" α-alumina (II)	14.5	14.4
"Exal" α-alumina (III)	28	27.6
Chromosorb G	...	0.41
B. Dibutyl sulfone stationary phase		
Chromosorb P	3.6	3.58
Spherosil XOC 005	11.1	11.08
Spherosil XOA 200	162.5	162.8
"Exal" α-alumina (III)	28	26.2

where V_g^0 (the bulk-liquid specific retention volume) must be measured in a separate experiment (e.g., Fig. 7.5). Once a particular adsorbent has been characterized with the known area of the "standard" stationary phase the molecular area of other liquids may be back-calculated by repeating the experiment with the latter solvent. The area of dibutyl sulfone was found by this method, for example, to be 42.8 $(\mathring{A})^2$.

Table 7.10 compares surface areas measured by the BET (nitrogen adsorbate) and Serpinet (thin-film) techniques.

The mean pore diameter of the materials in Table 7.10 ranges from 150 to 10,000 $\mathring{A}$ and indicates that the thin-film technique may well be applicable to microporous adsorbents, provided that adsorbates of suitably small dimensions are used. Indeed, given the ease and accuracy of this technique and the fact that an elution gas chromatograph is the sole instrumentation required, the thin-film method of Serpinet shows much promise of supplanting altogether the BET alternatives described above.

7.4 REFERENCES

1. D. M. Young and A. D. Crowell, *Physical Adsorption of Gases*, Butterworths, London, 1962; S. Ross and J. P. Olivier, *On Physical Adsorption*, Interscience, New York, 1964; J. H. deBoer, *The Dynamical Character of Adsorption*, 2nd ed., Oxford University Press, Oxford, England, 1968.

2. L. R. Snyder, *Principles of Adsorption Chromatography*, Marcel Dekker, New York, 1968; in E. Heftmann, Ed., *Chromatography*, 2nd ed., Reinhold, New York, 1967, Chapter 4.

3. R. S. Petrova, E. V. Khrapova, and K. D. Shcherbakova, in *Gas Chromatography 1962*, M. van Swaay, Ed., Butterworths, London, 1962, p. 18.

4. D. H. Everett and F. S. Stone, Eds., *Structure and Properties of Porous Materials*, Butterworths, London, 1958; D. H. Everett, in *Gas Chromatography 1964*, A. Goldup, Ed., Institute of Petroleum, London, 1965, p. 219; D. H. Everett, Ed., *Colloid Science*, Vol. 1, The Chemical Society, London, 1973; D. H. Everett and R. H. Ottewill, *Proceedings of the International Symposium on Surface Area Determination*, Butterworths, London, 1970.

5. A. V. Kiselev, in *Gas Chromatography 1964*, A. Goldup, Ed., Institute of Petroleum, London, 1965, p. 238; in *Aspects in Gas Chromatography*, H. G. Struppe, Ed., Akademie Verlag, Berlin, 1971, p. 48; *Adv. Chromatogr.*, **4**, 113 (1967); *Usp. Khromatogr.*, 33 (1972).

6. C. G. Scott and C. S. G. Phillips, in *Gas Chromatography 1964*, A. Goldup, Ed., Institute of Petroleum, London, 1965, p. 266.

7. A. V. Kiselev and Y. I. Yashin, *Gazo-Adsorbtsionnaya Khromatografiya*, Nauka, Moscow, 1967; *Gas-Adsorption Chromatography*, Plenum, New York, 1969.

8. H. W. Habgood, in *The Solid-Gas Interface*, Vol. 2, E. A. Flood, Ed., Marcel Dekker, New York, 1967, p. 611.

9. S. G. Ash, D. H. Everett, and G. H. Findenegg, *Trans. Faraday Soc.*, **66**, 708 (1970).

10. E. Boucher and D. H. Everett, *Trans. Faraday Soc.*, **67**, 2720 (1971).

11. S. J. Gregg and K. S. W. Sing, *Adsorption, Surface Area, and Porosity*, Academic, London, 1967; S. J. Gregg, *The Surface Chemistry of Solids*, 2nd ed., Chapman and Hall, London, 1961.

12. A. J. Tench, D. Giles, and J. F. J. Kibblewhite, *Trans. Faraday Soc.*, **67**, 854 (1971).

13. V. L. Anokhin, *Zh. Fiz. Khim.*, **46**, 2630, 2853, 2857 (1972).

14. V. G. Berezkin, *Usp. Khromatogr.*, 215 (1972).

15. N. N. Avgul, A. V. Kiselev, and P. D. Poshkus, *Adsorption of Gases and Vapors on Homogeneous Surfaces*, Khimia, Moscow, 1975.

16. D. P. Timofeev, *Dokl. Akad. Nauk SSSR*, **131**, 1390 (1960).

17. V. V. Rachinskii, *The General Theory of Sorption Dynamics and Chromatography*, Nauka, Moscow, 1964.

18. P. E. Eberly, Jr., *J. Appl. Chem.*, **14**, 330 (1964); *Ind. Eng. Chem. Fundam.*, **8**, 25 (1969).

19. O. Grubner, M. Ralek, and A. Zikanov, *Collect. Czech. Chem. Commun.*, **31**, 2629 (1966).

20. C. Vidal-Madjar and G. Guiochon, *J. Phys. Chem.*, **71**, 4031 (1967); in *Aspects in Gas Chromatography*, H. G. Struppe, Ed., Akademie Verlag, Berlin, 1971, p. 191.

21. G. Padberg and J. M. Smith, *J. Catal.*, **12**, 172 (1968).

22. J. Kwok, E. R. Fett, and G. A. Mickelson, *J. Gas Chromatogr.*, **6**, 491 (1968).

23. L. R. Snyder, *J. Chromatogr.*, **36**, 455 (1968).

24. M. Kocirik, P. Seidl, and J. Dubsky, *Chromatographia*, **3**, 78 (1970).

25. A. K. Moreland and L. B. Rogers, *Sep. Sci.*, **6**, 1 (1971).

26. N. R. Rakshieva, J. Novak, S. Wicar, and J. Janak, *J. Chromatogr.*, **91**, 51 (1974).

27. B. M. Zhitomirskii, A. V. Aganov, A. D. Berman, and M. I. Yanovskii, *J. Chromatogr.*, **94**, 1 (1974).

28. V. R. Choudhary, *J. Chromatogr.*, **98**, 491 (1974).

29. V. R. Choudhary and P. G. Menon, *J. Chromatogr.*, **116**, 431 (1976).

30. C. L. deLigny, *J. Chromatogr.*, **36**, 50 (1968).

31. A. Waksmundzki, W. Rudzinski, and Z. Suprynowicz, *J. Gas Chromatogr.*, **4**, 93 (1966).

32. R. Rowan, Jr., and J. B. Sorrell, *Anal. Chem.*, **42**, 1716 (1970).

33. W. Rudzinski, A. Waksmundzki, R. Leboda, and M. Jaroniec, *Chromatographia*, **7**, 663 (1974).

34. A. Waksmundzki, M. Jaroniec, and Z. Suprynowicz, *J. Chromatogr.*, **110**, 381 (1975).

35. N. C. Saha and D. S. Mathur, *J. Chromatogr.*, **81**, 207 (1973).

36. V. R. Choudhary and L. K. Doraiswamy, *Ind. Eng. Chem. Prod. Res. Dev.*, **10**, 218 (1971).

37. A. Waksmundzki, R. Leboda, W. Rudzinski, A. Ksiezycki, B. Koczorowicz, J. Wojciechowska-Zwolska, and B. Woszek, *Przem. Chem.*, **54**, 403 (1975).

38. A. B. Littlewood, *Gas Chromatography*, 2nd ed., Academic, New York, 1970, Ch. 4.

39. P. G. Jeffrey and P. J. Kipping, *Gas Analysis by Gas Chromatography*, 2nd ed., Pergamon, London, 1972.

40. R. L. Grob, in *Modern Practice of Gas Chromatography*, R. L. Grob, Ed., Wiley-Interscience, New York, 1977, pp. 103–110; R. L. Grob, in *Progress in Analytical Chemistry*, Vol. 8, I. V. Simmons and G. W. Ewing, Eds., Plenum, New York, 1976, pp. 151–194.

41. A. G. Datar and P. S. Ramanathan, *J. Chromatogr.*, **114**, 29 (1975).

42. M. Popl, V. Dolansky, and J. Mostecky, *J. Chromatogr.*, **117**, 117 (1976).

43. R. M. Barber, *Ber. Bunsenges. Phys. Chem.*, **69**, 786 (1965).

44. G. Alberti and U. Costantino, *J. Chromatogr.*, **102**, 5 (1974).

45. D. W. Breck, *Zeolite Molecular Sieves*, Wiley-Interscience, New York, 1974.

46. V. Patzelova, *Chem. Zvesti*, **29**, 331 (1975).

47. C. Vidal-Madjar, M.-F. Gonnord, and G. Guiochon, *J. Colloid Interface Sci.*, **52**, 102 (1975); *Adv. Chromatogr.*, **13**, 177 (1975).

48. S. A. Rang, O. G. Eisen, A. V. Kiselev, A. E. Meister, and K. D. Shcherbakova, *Chromatographia*, **8**, 327 (1975).

49. W. R. Supina, *The Packed Column in Gas Chromatography*, Supelco, Bellafonte, Pennsylvania, 1974.

50. S. B. Dave, *J. Chromatogr. Sci.*, **7**, 389 (1969); *Ind. Eng. Chem. Prod. Des. Dev.*, **14**, 85 (1975).

51. E. V. Kalashnikova, A. V. Kiselev, A. M. Makogon, and K. D. Shcherbakova, *Chromatographia*, **8**, 399 (1975).

52. C. Vidal-Madjar, M.-F. Gonnord, M. Goedert, and G. Guiochon, *J. Phys. Chem.*, **79**, 732 (1975).

53. A. DiCorcia and A. Liberti, *Adv. Chromatogr.*, **14**, 305 (1976).

54. B. G. Linsen, Ed., *Physical and Chemical Aspects of Adsorbents and Catalysts*, Academic, London, 1970.

55. E. Cremer and H. F. Huber, *Angew. Chem.*, **73**, 461 (1961); in *Gas Chromatography*, N. Brenner, J. E. Callen, and M. D. Weiss, Eds., Academic, New York, 1962, p. 169.

56. J. F. K. Huber and A. I. M. Keulemans, in *Gas Chromatography 1962*, M. van Swaay, Ed., Butterworths, London, 1962, p. 26.

57. J. E. Guillet, *Adv. Anal. Chem. Instrum.*, **11**, 187 (1973).

58. J.-M. Braun and J. E. Guillet, *Adv. Polym. Sci.*, **21**, 108 (1976).

59. D. H. James and C. S. G. Phillips, *J. Chem. Soc.*, 1066 (1954).

60. S. Toth and L. Graf, *Acta Chim. Acad. Sci. Hung.*, **22**, 231 (1960).

61. P. E. Eberly, Jr., *J. Phys. Chem.*, **65**, 68 (1961).

62. P. E. Eberly, Jr., and C. N. Kimberlin, Jr., *Trans. Faraday Soc.*, **57**, 1169 (1961).

63. D. R. Owens, A. G. Hamlin, and T. R. Phillips, *Nature*, **201**, 901 (1964).

64. R. A. Beebe, P. L. Evans, T. C. W. Kleinsteuber, and L. W. Richards, *J. Phys. Chem.*, **70**, 1009 (1966).

65. J. J. Haydel and R. Kobayashi, *Ind. Eng. Chem. Fundam.*, **6**, 546 (1967).

66. S. Masukawa and R. Kobayashi, *J. Chem. Eng. Data*, **13**, 197 (1968); *J. Gas Chromatogr.*, **6**, 461 (1968); *AIChE J.*, **14**, 740 (1968); **15**, 190 (1969).

67. G. Alberini, F. Bruner, and G. Devitofrancesco, *Anal. Chem.*, **41**, 1940 (1969).

68. B. G. Aristov, V. Bosacek, and A. V. Kiselev, *Zh. Fiz. Khim.*, **43**, 292 (1969).

69. L. D. Belyakova, V. L. Keybal, and A. V. Kiselev, *Zh. Fiz. Khim.*, **44**, 2345 (1970).

70. B. Chabert, J. Chaucard, and G. Edel, *C. R. Acad. Sci.*, **271**, 38 (1970).

71. Y. Hori and R. Kobayashi, *J. Chem. Phys.*, **54**, 1226 (1971).

72. M. Gassiot-Matas and B. Monrabal-Bas, *Chromatographia*, **3**, 547 (1971).

73. M. F. Burke and D. G. Ackerman, Jr., *Anal. Chem.*, **43**, 573 (1971).

74. E. Van Der Vlist and J. Van Der Meijden, *J. Chromatogr.*, **79**, 1 (1973).

75. J. Volf, J. Koubek, and J. Pasek, *J. Chromatogr.*, **81**, 9 (1973).

76. Y. Hori and R. Kobayashi, *Ind. Eng. Chem. Fundam.*, **12**, 26 (1973).

77. A. Waksmundzki, W. Rudzinski, Z. Suprynowicz, R. Leboda, and M. Lason, *J. Chromatogr.*, **92**, 9 (1974).

78. S. M. Yanovskii, I. A. Silaeva, and O. N. Alksnis, *J. Chromatogr.*, **93**, 464 (1974).

79. M. Novotny, M.-L. Lee, and K. D. Bartle, *Chromatographia*, **7**, 333 (1974).

80. G. Deininger, J. Asshauer, and I. Halasz, *Chromatographia*, **8**, 143 (1975).

81. P. Valentin and G. Guiochon, *J. Chromatogr. Sci.*, **14**, 132 (1976).

82. D. G. Gray and J. E. Guillet, *Macromolecules*, **5**, 316 (1972).

83. G. Edel and B. Chabert, *C. R. Acad. Sci.*, **267**, 54 (1968); **268**, 226 (1969).

84. U. B. Mohlin and D. G. Gray, *J. Colloid Interface Sci.*, **47**, 747 (1974).

85. P. R. Tremaine and D. G. Gray, *J. Chem. Soc. Faraday Trans. I*, **71**, 2170 (1975).

86. H. F. Stoeckli, *Helv. Chim. Acta*, **55**, 101 (1972).

87. C. Janneret and H. F. Stoeckli, *Helv. Chim. Acta*, **56**, 2509 (1973).

88. J. P. Houriet, P. Ghiste, and H. F. Stoeckli, *Helv. Chim. Acta*, **57**, 851 (1974).

89. D. Basmadjian, *Can. J. Chem.*, **38**, 141, 149 (1960).

90. L. D. Belyakova, A. V. Kiselev, and N. V. Kovaleva, *Bull. Soc. Chim. Fr.*, 285 (1967).

91. S. P. Dzhavadov, A. V. Kiselev, and Y. S. Nikitin, *Zh. Fiz. Khim.*, **41**, 1131 (1967).

92. G. A. Gaziev, O. V. Isayev, L. Y. Derlyukova, and L. Y. Margolis, *Neftekhim.*, **9**, 457 (1969).

93. J. Czubryt and H. D. Gesser, *J. Chromatogr.*, **59**, 1 (1971).

94. J. R. Conder, *Chromatographia*, **7**, 387 (1974).

95. S. A. Greene and H. Pust, *J. Phys. Chem.*, **62**, 55 (1958).

96. H. W. Habgood and J. F. Hanlan, *Can. J. Chem.*, **37**, 843 (1959).

97. L. D. Belyakova, A. V. Kiselev, and N. N. Kovaleva, *Anal. Chem.*, **36**, 1517 (1964).

98. R. L. Gale and R. A. Beebe, *J. Phys. Chem.*, **68**, 555 (1964).

99. R. Aubeau, J. Leroy, and L. Champeix, *J. Chromatogr.*, **19**, 249 (1965).

100. N. N. Avgul, A. V. Kiselev, L. Y. Kurdyukova, and M. V. Serdobov, *Zh. Fiz. Khim.*, **42**, 188 (1968).

101. B. I. Tulbovich and E. I. Priimak, *Zh. Fiz. Khim.*, **43**, 362 (1969).

102. Y. Moro-oka, *Trans. Faraday Soc.*, **67**, 3381 (1971).

103. A. DiCorcia, A. Liberti, and R. Samperi, *Anal. Chem.*, **45**, 1228 (1973).

104. A. DiCorcia and R. Samperi, *J. Chromatogr.*, **77**, 277 (1973).

105. D. S. Mathur, U. D. Chaubey, and A. Sinha, *J. Chromatogr.*, **99**, 281 (1974).

106. A. V. Kiselev, K. D. Shcherbakova, and D. P. Poshkus, *J. Chromatogr. Sci.*, **12**, 788 (1974).

107. J. Forys and B. Grzybowska, *Bull. Acad. Pol. Sci., Ser. Sci. Chim.*, **23**, 269 (1975).

108. A. Dyer, D. Leigh, and W. E. Sharples, *J. Chromatogr.*, **118**, 319 (1976).

109. E. J. McGonigle and R. L. Grob, *J. Chromatogr.*, **101**, 39 (1974).

110. W. Al-Thamir, R. J. Laub, and J. H. Purnell, paper presented at Twelfth International Symposium on Advances in Chromatography, Amsterdam, 1977; *J. Chromatogr.*, **142**, 3 (1977).

111. C. Vidal-Madjar and G. Guiochon, *J. Chromatogr. Sci.*, **9**, 664 (1971).

112. D. J. Brookman and D. T. Sawyer, *Anal. Chem.*, **40**, 106, 1368, 2013 (1968).

113. G. L. Hargrove and D. T. Sawyer, *Anal. Chem.*, **40**, 409 (1968).

114. D. T. Sawyer and D. J. Brookman, *Anal. Chem.*, **40**, 1847 (1968).

115. A. F. Isbell, Jr., and D. T. Sawyer, *Anal. Chem.*, **41**, 1381 (1969).

116. D. F. Cadogan and D. T. Sawyer, *Anal. Chem.*, **42**, 190 (1970); **43**, 941 (1971).

117. R. L. McCreery and D. T. Sawyer, *J. Chromatogr. Sci.*, **8**, 122 (1970).

118. J. Okamura and D. T. Sawyer, *Anal. Chem.*, **43**, 1730 (1971).

119. J. N. Gerber and D. T. Sawyer, *Anal. Chem.*, **44**, 1199 (1972).

120. R. A. Beebe and P. H. Emmett, *J. Phys. Chem.*, **65**, 184 (1961).

121. K. Kochloefl, P. Schneider, R. Komers, and F. Jost, *Collect. Czech. Chem. Commun.*, **32**, 2456 (1967).

122. R. Komers and K. Kochloefl, *Collect. Czech. Chem. Commun.*, **32**, 3679 (1967).

123. S. Brunauer, P. H. Emmett, and E. Teller, *J. Am. Chem. Soc.*, **60**, 309 (1938).

124. L. S. Ettre, Application Bulletin No. SO-AP-002, Perkin-Elmer Corp., Norwalk, Connecticut, 1966.

125. F. M. Nelsen and F. T. Eggertsen, *Anal. Chem.*, **30**, 1387 (1958).

126. J. F. Roth and R. J. Ellwood, *Anal. Chem.*, **31**, 1738 (1959).

127. H. W. Daeschner and F. H. Stross, *Anal. Chem.*, **34**, 1150 (1962).

128. R. M. Cahen and J. Marechal, *Anal. Chem.*, **35**, 259 (1963).

129. J. V. Derby and B. D. LaMont, paper presented at Eleventh Pittsburgh Conference on Analytical Chemistry and Applied Spectroscopy, February 1960.

130. E. Smolkova, O. Grubner, and L. Feltl, in *Gas Chromatography, 1965*, East German Academy of Science, Berlin, 1965, p. 509.

131. R. W. Freedman, *Lab. Pract.*, **17**, 710 (1968).

132. M. F. Burke and D. G. Ackerman, *Anal. Chem.*, **43**, 573 (1971).

133. R. N. Nikolov, L. D. Petkova, and M. D. Shopova, *Chromatographia*, **7**, 376 (1974).

134. P. R. Tremaine and D. G. Gray, *Anal. Chem.*, **48**, 380 (1976).

135. R. Stock, *Anal. Chem.*, **33**, 966 (1961).

136. E. Cremer, *Z. Anal. Chem.*, **170**, 219 (1959); *Angew, Chem.*, **71**, 512 (1959).

137. F. Wolf and H. Bayer, *Chem. Technol.*, **11**, 142 (1959).

138. L. S. Ettre, *J. Chromatogr.*, **4**, 166 (1960); in *Pigment Handbook*, Vol. III, T. C. Patton, Ed., Wiley, New York, 1973, p. 139.

139. G. A. Gaziev, M. I. Yanovskii, and V. V. Brazhnikov, *Kinet. Katal.*, **1**, 548 (1960).

140. L. S. Ettre, N. Brenner, and E. W. Cieplinski, *Z. Phys. Chem.*, **219**, 17 (1962).

141. R. S. Hansen, J. A. Murphy, and T. C. McGee, *Trans. Faraday Soc.*, **60**, 597 (1964).

142. A. V. Kiselev, R. S. Petrova, and K. D. Shcherbakova, *Kinet. Katal.*, **5**, 526 (1964).

143. P. J. Kipping, P. G. Jeffery, and C. A. Savage, *Res. Dev. Ind.*, (39), 18 (1965).

144. C. S. Brooks, *Soil Sci.*, **99**, 182 (1965).

145. T. B. Gavrilova and A. V. Kiselev, *Zh. Fiz. Khim.*, **39**, 2582 (1965).

146. Y. Kuge and Y. Yoshikawa, *Bull. Chem. Soc. Japan*, **38**, 948 (1965).

147. A. V. Kiselev and Y. I. Yashin, *Zh. Fiz. Khim.*, **40**, 944 (1966).

148. G. N. Fadeyev and T. B. Gavrilova, *Zh. Fiz. Khim.*, **42**, 3075 (1968).

149. N. K. Bebris, Y. Y. Gienko, G. Y. Yaytseva, A. V. Kiselev, G. L. Kustova, B. A. Lipkind, Y. S. Nikitvin, and Y. I. Yashin, *Neftekhim.*, **10**, 733 (1970).

150. J. R. Wallace, P. J. Kozak, and F. Noel, *SPE J.*, **26**, 43 (1970).

151. M. Krejci and D. Kourilova, *Chromatographia*, **4**, 48 (1971).

152. C. L. Guillemin, M. LePage, and A. J. deVries, *J. Chromatogr. Sci.*, **9**, 470 (1971).

153. C. L. Guillemin, M. Deleuil, S. Cirendini, and J. Vermont, *Anal. Chem.*, **43**, 2015 (1971).

154. M. G. Farey and B. G. Tucker, *Anal. Chem.*, **43**, 1307 (1971).

155. N. K. Bebris, A. V. Kiselev, B. Y. Mokeev, Y. S. Nikitvin, Y. I. Yashin, and G. E. Zaizeva, *Chromatographia*, **4**, 93 (1971).

156. I. L. Maryasin, S. L. Pishchulina, I. S. Rafalkes, T. D. Tiekunova, and L. M. Boradina, *Zavod Lab.*, **37**, 41 (1971).

157. S. K. Ghosh, H. S. Sarkar, and N. C. Saha, *Technology*, **9**, 330 (1972).

158. W. Rudzinski, A. Waksmundzki, Z. Suprynowicz, and J. Rayss, *J. Chromatogr.*, **72**, 221 (1972).

159. E. Sund, E. Haanaes, O. Smidsrod, and J. Ugelstadt, *J. Appl. Polym. Sci.*, **16**, 1869 (1972).

160. S. Ghosh, H. Sarkar, and N. C. Saha, *J. Chromatogr.*, **74**, 171 (1972).

161. R. Salovey, R. Cortellucci, and A. Roaldi, *Polym. Eng. Sci.*, **14**, 120 (1974).

162. R. L. Grob, M. A. Kaiser, and M. J. O'Brien, *Am. Lab*, **7**(6), 13 (1975); **7**(8), 33 (1975).

163. R. H. Perrett and J. H. Purnell, *J. Chromatogr.*, **7**, 455 (1962).

164. S. Thiabault, J. Amouroux, and J. Talbot, *Bull. Soc. Chim. Fr.*, 1319 (1968).

165. J. Serpinet and J. Robin, *C. R. Acad. Sci.*, **272**, 1765 (1971).

166. G. Untz and J. Serpinet, *C. R. Acad. Sci.*, **273**, 392 (1971); *Bull. Soc. Chim. Fr.*, 1591, 1595 (1973).

167. J. Serpinet, G. Untz, C. Gachet, L. de Mourgues, and M. Perrin, *J. Chim. Phys.*, **71**, 949 (1974).

168. J. Serpinet, *J. Chromatogr.*, **68**, 9 (1972); **77**, 289 (1973); **119**, 483 (1976); *C. R. Acad. Sci.*, **275**, 985 (1972); *J. Chromatogr. Sci.*, **12**, 832 (1974); *Chromatographia*, **8**, 18 (1975); *Anal. Chem.*, **48**, 2264 (1976).

169. M. Krejci, *Collect. Czech. Chem. Commun.*, **32**, 1152 (1967).

170. V. G. Berezkin, D. Kourilova, M. Krejci, and V. M. Fateeva, *J. Chromatogr.*, **78**, 261 (1973).

171. J. R. Conder, D. C. Locke, and J. H. Purnell, *J. Phys. Chem.*, **73**, 700 (1969).

PART THREE
Kinetics

Rate Constants and On-Column Reactions

8.1 RATES OF REACTION IN DYNAMIC SYSTEMS

Gas chromatography is widely used today to study reaction rates, energies, and mechanisms. Most frequently the GC is used merely as an analytical tool to follow the appearance of products (and/or intermediates) and the disappearance of reactants in a reaction vessel or a precolumn microreactor (e.g., see Refs. 1 and 2). Even in gas kinetics (analytical) GC is used almost exclusively (rather than a mass spectrometer) because of cost and simplicity of operation (e.g., Ref. 3). There are other instances, however, in which a solute reacts or interacts *on the column* in a rate-controlled manner; kinetics data therefore become available directly from elution behavior.

Consider an empty tube through which a carrier gas is flowing. Suppose that the gas contains a reactant, A, that reacts irreversibly in the tube via a first-order mechanism. The mass-balance equation for the reactant in an infinitesimal segment, dV, of the tube may be written in terms of the concentration of A within the segment C_A, the difference in concentration between that leaving the segment and that entering, dC_A, and the change in the number of moles, dn_A, with time:

$$\frac{dn_A}{dt} = -kC_A\,dV - f\,dC_A \tag{8.1}$$

where k is the first-order rate constant and f is the flow rate of carrier through the segment. If A is fed continually into the tube, a steady-state condition is reached such that $dn_A/dt = 0$, at which point

$$-kC_A\,dV = f\,dC_A \tag{8.2}$$

or

$$-\frac{k}{f}dV = \frac{dC_A}{C_A} \tag{8.3}$$

Integration of eq. 8.3 over the length of the reactor tube gives

$$-\frac{2.3kV_t}{F} = \log\frac{C_A^{\text{out}}}{C_A^{\text{in}}} \tag{8.4}$$

where V_t is the total reactor volume and C_A^{out} and C_A^{in} are the concentrations of A at the outlet and inlet of the tube, respectively. The quotient V_t/F can be replaced by the time t that a molecule requires to pass through the reactor; t is then called the *contact time* of the reactant. Log($C_A^{\text{out}}/C_A^{\text{in}}$) is plotted versus $2.303t$ to obtain k as the slope of the straight line.

Equation 8.4 must be modified when used in GC to include the effects of a packing (which may be an adsorbent or a liquid film on a support) and the fact that the reactant is introduced as an injected "pulse" rather than continuously. Several approaches have been developed but the most direct is that reported in 1969 by Langer and co-workers[4] who reconsidered the mass-balance relation

$$v_M\left(\frac{\partial C_A^M}{\partial t}\right) + v_L\left(\frac{\partial C_A^L}{\partial t}\right) = -v_M\bar{u}_S\left(\frac{\partial C_A^M}{\partial t}\right) - v_M r_M - v_L r_L \tag{8.5}$$

where v_M and v_L are the volumes of the mobile and stationary phases per unit column volume, $(\partial C_A^i/\partial t)$ and $(\partial C_A^i/\partial x)$ represent the changes in C_A^i with time t and distance x, along the column in the mobile M and stationary L phases, $\bar{u}_S$ is the average linear velocity of the solute, and r_M and r_L are the rates of reaction per unit volume of mobile and stationary phases, respectively. The quotient $(\partial C_A^L/\partial t)/(\partial C_A^M/\partial t)$ is just the partition coefficient K_R, so that eq. 8.5 may be written as

$$(v_M + K_R v_L)\left(\frac{\partial C_A^M}{\partial t}\right) = -v_M\bar{u}_S\left(\frac{\partial C_A^M}{\partial x}\right) - v_M r_M - v_L r_L \tag{8.6}$$

Equations 8.5 or 8.6 indicate that in a thin cross section of a GC column the amount of reactant A changes because of flow as well as because of the reactions that occur in both the mobile and stationary phases. These equations are the analogues of the steady-state tube, that is eq. 8.1, except that the presence of two phases in the GC column has now been taken into account.

Because, for first-order irreversible reactions, $r_i = k_i C_A^i$, where k_i is the rate constant in the ith phase, eq. 8.6 can be rewritten as

$$\alpha\left(\frac{\partial C_A^M}{\partial t}\right) + \left(\frac{\partial C_A^M}{\partial x}\right) + \beta = 0 \tag{8.7}$$

where

$$\alpha = \frac{v_M + v_L K_R}{v_M \bar{u}_S} \tag{8.8}$$

$$\beta = \frac{v_M k_M + v_L k_L K_R}{v_M \bar{u}_S} \tag{8.9}$$

Recalling that

$$t_M = \frac{L}{\bar{u}_S} \tag{8.10}$$

$$\frac{t_L}{t_M} = \frac{v_L}{v_M} K_R \tag{8.11}$$

where t_L is the solute residence time in the liquid phase ($\equiv t_R'$) and t_M is the residence time in the mobile phase

$$\alpha = \frac{t_M + t_L}{L} = \frac{t_R}{L} \tag{8.12}$$

$$\beta = \frac{k_M t_M + k_L t_L}{L} \tag{8.13}$$

Combining the above with eq. 8.7 provides[4]

$$C_A^M(x,t) = \phi(t - \alpha x)\exp(-\beta x) \tag{8.14}$$

where the mobile-phase concentration of A at any time t and distance x along the column is given by the product of the input function $\phi(t)$ (which describes the waveform of the injection pulse; cf. Section 3.2.4) and the term $\exp(-\beta x)$. At $x=0$ and at $x=L$ (the two ends of the column)

$$C_A^M(0,t) = \phi(t) \tag{8.15}$$

$$C_A^M(L,t) = \phi(t - \alpha L)\exp(-\beta L) \tag{8.16}$$

The total number of moles of reactant entering the column n_{in} is given by

$$n_{in} = jF_c\int_0^\infty C_A^M(0,t)\,dt = jF_c\int_0^\infty \phi(t)\,dt \tag{8.17}$$

(where jF_c is the average carrier flow rate which has been fully corrected for temperature and pressure) and the number leaving n_{out} by

$$n_{out} = jF_c \int_0^\infty C_A^M(L, t)\, dt = jF_c \int_0^\infty \phi(t - \alpha L)\, dt \qquad (8.18)$$

The input function ϕ is bounded by the limits $\phi(t) = 0$ when $t < 0$ and $\int_0^\infty \phi(t)\, dt = \int_0^\infty \phi(t - \alpha L)\, dt$; eqs. 8.17 and 8.18 can therefore be combined to yield

$$\ln\left(\frac{n_{in}}{n_{out}}\right) = k_M t_M + k_L t_L \qquad (8.19)$$

If $t_M \ll t_L$ (usually the case in GC), then

$$\ln\left(\frac{n_{in}}{n_{out}}\right) \cong k_L t_L = k_L t'_R \qquad (8.20)$$

and k_L may be found from the slope of plots of $\ln(n_{in}/n_{out})$ versus t'_R.

Kallen and Heilbronner[5] and Nakagaki and Nishino[6] have presented a model of the elution process in the case of reacting solutes in terms of the column plate number N:

$$k_L t_L + k_M t_M = N\left[1 - \left(\frac{n_{out}}{n_{in}}\right)^{1/N}\right] \qquad (8.21)$$

where k_i is now expressed in terms of the number of theoretical plates of the reactant peak. When k_M is small, plots of $N[1 - (n_{out}/n_{in})^{1/N}]$ versus t'_R yield straight lines of slope k_L. Langer and Patton[7] have shown, however, that eqs. 8.20 and 8.21 are identical: as $N \to \infty$,

$$\lim_{N \to \infty}(k_L t_L + k_M t_M) = \lim_{N \to \infty} \frac{\left[1 - (n_{out}/n_{in})^{1/N}\right]}{N^{-1}}$$

$$= \lim_{N \to \infty} \frac{(n_{out}/n_{in})^{1/N} \ln(n_{out}/n_{in}) N^{-2}}{-N^{-2}}$$

$$= -\ln\left(\frac{n_{out}}{n_{in}}\right) \lim_{N \to \infty} \left(\frac{n_{out}}{n_{in}}\right)^{1/N}$$

$$= \ln\left(\frac{n_{out}}{n_{in}}\right) \qquad (8.22)$$

Thus, if the reactant solute yields 1000 or more theoretical plates on the column, agreement between the two methods will be within 0.1%.

Bassett and Habgood[8] have presented an approach that differs slightly from the above. They considered the case of gas-solid chromatography in which a reaction was said to occur only when the reactant was adsorbed onto the surface of the adsorbent. Consider a small section of column that contains a weight, dw_S, of adsorbent (analogous to dV_L ml of liquid in the GLC case). The n number of moles passing through the section partition between the stationary and mobile phases as usual. Letting v_M represent the volume of mobile phase per gram adsorbent, the mobile-phase volume of the segment is given by $v_M dw_S$ and the number of moles of solute in the segment, by

$$n_1 = \frac{p_1 v_M dw_S}{RT} + K_S p_1 dw_S \tag{8.23}$$

where p_1 is the solute vapor pressure at the column temperature and K_S is the solute partition coefficient (moles/gram atmosphere) defined for the present purposes as

$$K_S = \frac{v_S}{RT} \tag{8.24}$$

where v_S is the solute retention volume (corrected for column dead space) per gram adsorbent. Substituting eq. 8.24 into 8.23 gives

$$n_1 = \frac{p_1(v_M + v_S)dw_S}{RT} \tag{8.25}$$

The change of solute mole number with time within the segment is

$$-\frac{dn_1}{dt} = kK_S p_1 dw_S = \frac{kK_S n_1 RT}{v_M + v_S} \tag{8.26}$$

where k is the rate constant. Integrating over all segments,

$$\ln\left(\frac{1}{1-x}\right) = \frac{kK_S RTt}{v_M + v_S} \tag{8.27}$$

where x is the fraction of solute that has reacted after a time, t. The adjusted solute retention volume V_R' is given by $(v_M + v_S)w_S$ (where w_S is the total weight of adsorbent in the column), so that

$$\ln\left(\frac{1}{1-x}\right) = \frac{kV_R'}{F_c} \tag{8.28}$$

Note that the sum $(v_M + v_S)$ expresses the GSC mobile and stationary phase contributions to the total adjusted retention volume analogous to the GLC case cited above in which the sum $(k_M t_M + k_L t_L)$ was used; the two sums express the same quantities when $k_M = 0$:

$$\frac{kV'_R}{F_c} = kt'_R \qquad (8.29)$$

but t'_R is proportional to t_S, that is, $[(t_M + t_S) - t_M]$. Thus

$$\frac{kV'_R}{F_c} \propto kt_S \qquad (8.30)$$

Inclusion of the pressure correction factor j provides

$$\ln\left(\frac{1}{1-x}\right) = \frac{jRTw_S kK_S}{F_c} = \frac{kV_N}{F_c} \qquad (8.31)$$

In an actual experiment the solute retention time, thence K_S, is first measured. The term $\ln(1/1-x)$ is then found by comparing the reactant peak area to a predetermined calibration chart. Knowledge of the amount injected and the amount of solute that emerges thereby allows calculation of x. A plot of $\ln(1/1-x)$ versus V_N/F_c is constructed next to determine the rate constant k from the slope of the straight line.

A few examples of the gas-chromatographic determination of rate constants are outlined below; other methods have been described by several groups over the years[9-35] (including a variety of techniques pioneered by Phillips and co-workers[36-44]) reviewed by Berezkin,[2] van Swaay,[45] Phillips,[46] Suzuki and Smith,[47] and Langer and Patton.[7]

8.1.1 Decomposition of Cyclopropane to Propylene

The initial work of Bassett and Habgood[8] concerned the use of various catalysts to promote the reaction cyclopropane→propylene. Figure 8.1 shows two chromatograms—the upper for cyclopropane and the lower for propylene—both on a column containing nickel-exchanged 13X molecular sieve. The propylene peak in the first chromatogram is relatively low and diffuse, as would be expected, for it is produced continually along the length of the column. In contrast, the propylene peak in the lower chromatogram is relatively sharp. These chromatograms were used as follows: first, x was measured at various flow rates; then $\ln(1/1-x)$ was plotted versus $1/F_c$, as shown in Fig. 8.2. The product kK_S for the system shown (370°C) was 0.2×10^{-5} mole/g atm sec from the slope of the line; $\ln K_S$ varied in the usual manner with reciprocal temperature and was found to be 9.32×10^{-5}

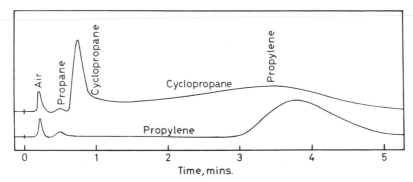

Figure 8.1 Chromatograms illustrating the reaction of cyclopropane to produce propylene: upper trace, cyclopropane injected; lower trace, propylene injected. Column: nickel-exchanged 13X molecular sieve at 225°C; corrected flow rate: 79 ml/min. Data of Bassett and Habgood.[8]

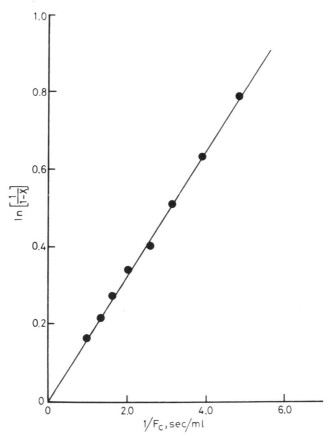

Figure 8.2 Plot of $\ln(1/1-x)$ versus $1/F_c$ for the conversion of cyclopropane to propylene at 370°C by nickel-exchanged 13X molecular sieve. Data of Bassett and Habgood.[8]

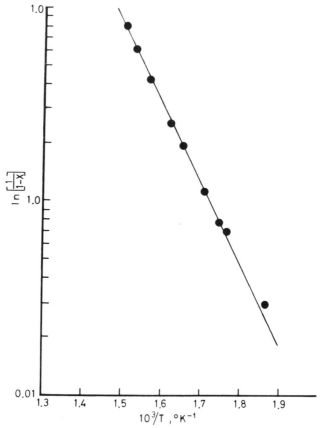

Figure 8.3 Plot of $\ln[\ln(1/1-x)]$ versus $10^3/T$ for the cyclopropane/propylene conversion described in Figs. 8.1 and 8.2. Data of Bassett and Habgood.[8]

mole/g atm at this temperature. Thus k was given by

$$k = \frac{2.0 \times 10^{-6} \text{ mole/g atm sec}}{9.32 \times 10^{-5} \text{ mole/g atm}} = 2.15 \times 10^{-2} \text{ sec}^{-1}$$

Alternatively, $\ln[\ln(1/1-x)]$ was plotted versus $10^3/T$ as shown in Fig. 8.3. The value of the product kK_S was found by this method to be 7.7×10^{-5} mole/g atm at 370°C, yielding a rate constant of 8.25×10^{-1}/sec. The discrepancy between the two rate constants was not discussed but may be due to the use of a K_S value that was extrapolated from the range 102–163°C.

8.1.2 Diels-Alder Reactions In 1961 Gil-Av and Herzberg-Minzly[48] studied the Diels-Alder reaction of dienes with a chloromaleic anhydride stationary phase. In practice, rather than measure $\ln(1/1-x)$ these workers used an internal standard (butane or heptane) or the zero-reaction peak area and measured the relative disappearance of diene. Plots of $\ln$ (zero-reaction peak area/reaction peak area) versus the retention time of the reactant were constructed over a range of flow rates (4-110 ml/min); the results are shown in Fig. 8.4 from which the following rate constants were found: *trans,trans*-2,4-hexadiene, 2.12×10^{-3}/sec; isoprene, 1.70×10^{-3}/sec; and 1,3-butadiene, 1.43×10^{-3}/sec. This study is of particular interest because Diels-Alder reactions are known to be bimolecular. In this

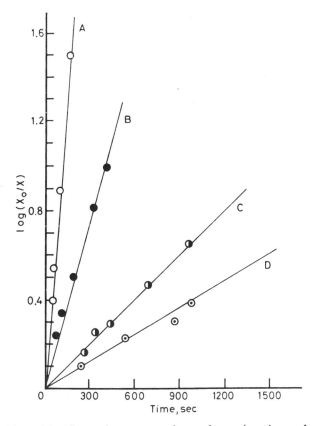

Figure 8.4 Plots of log(diene solute zero-reaction peak area/reaction peak area) versus solute retention time obtained by varying the flow rate from 4 to 110 ml/min at 40°C. Stationary phase: chloromaleic anhydride. Solutes: *trans,trans*-2,4-hexadiene (O), *trans*-1,3-pentadiene (●), isoprene (◐), and 1,3-butadiene (◉). Data of Gil-Av and Herzberg-Minzly.[48]

instance, however, use of the dienophile as the stationary phase rendered the process pseudo-first order; that is, because one reactant is present in excess its concentration changes very little when a few microliters of diene are injected onto the column. Berezkin and co-workers[23] have reported similar investigations with maleic anhydride.

8.1.3 Unimolecular Dissociation of Dicyclopentadiene

Pratt and Langer[50] investigated the dissociation of dicyclopentadiene (to form cyclo-pentadiene) in 1969. This reaction was chosen because it had previously been examined by pressure[51] and colorimetric and volumetric[52] techniques. For the study ln(reactant area/standard area) was plotted versus retention time over a range of flow rates using the stationary phases Apiezon L, silicone oil DC 550, and polyphenyl ether (six rings). A typical reaction chromatogram is shown in Fig. 8.5. The ratio k_L/k_M was found to be close to unity for the GLC solvents and somewhat less for the static paraffin data,[52] as shown in Table 8.1; $t_R \gg t_A$ for these solvents, and so the overall GLC rate constant should be approximately the same as the static gas phase result. That this was the case is shown by the Arrhenius plot of the rate constant data in Fig. 8.6, in which the points were obtained by GLC and the solid line by the static pressure[51] technique. Langer and co-

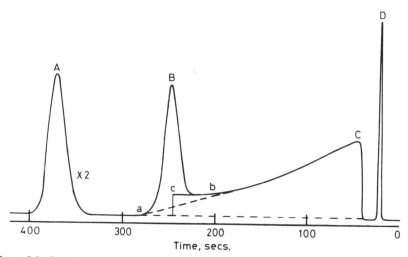

Figure 8.5 Reaction chromatogram for dicyclopentadiene (B) with Apiezon L solvent at 209.6°C. (A), internal standard (1-bromo-3-chlorobenzene); (C), cyclopentadiene; (D), air. The product area bounded by $\overline{abC}$ is approximated by the area bounded by $\overline{cbC}$; the reactant peak area is then corrected with this approximation for the small amount of peak overlap. Data of Pratt and Langer.[50]

Table 8.1 Rate Constant Ratios for the Dissociation of Dicyclopentadiene[50]

	GLC			Static
	k_L/k_M			
T (°K)	Silicone DC 550	Polyphenyl Ether	Apiezon L	Paraffin
453.0	1.095	1.242	0.818	0.574
462.9	0.878	1.030	0.817	0.581
472.8	0.955	0.974	0.812	0.587
482.8	0.972	0.965	0.804	0.594
492.9	0.964	1.074	0.828	0.600
502.8	1.007	1.185	...	0.606

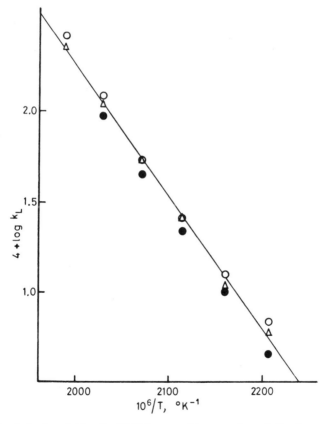

Figure 8.6 Arrhenius plot for the GLC decomposition rate data of dicyclopentadiene: (●), Apiezon L; (△), DC 550 silicone oil; (○), polyphenyl ether (six rings); line, data of Kistiakowsky and co-workers[51] obtained via static gas-phase techniques. Data of Pratt and Langer.[50]

workers[1] have also shown that reaction chromatography can be used to obtain pure samples of cyclopentadiene, formaldehyde, and other unstable laboratory reagents under the proper conditions.

8.1.4 Isotrope Interconversion LeRoy and co-workers[53, 54] have studied the kinetics of ortho-para hydrogen interconversion on alumina by GSC. Assuming that the conversion is a simple first-order process, the integrated rate equation (at the column temperature, here 77° K), is given by

$$P_e - P_t = (P_e - P_i)\exp(-k_c t) \tag{8.32}$$

where P_e, P_i, and P_t are the equilibrium and inlet amounts of the para form and the amount leaving the column after a residence time t; k_c is the sum of the forward and reverse rate constants

$$p\text{-}H_2 \underset{k_0}{\overset{k_p}{\rightleftarrows}} o\text{-}H_2 \tag{8.33}$$

$$k_c = k_p + k_0$$

Rate constants were, as above, determined from peak areas and retention

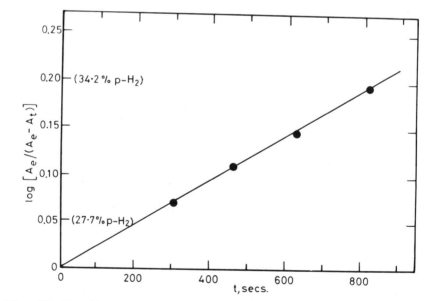

Figure 8.7 Plot of $\log[A_e/(A_e - A_t)]$ versus t for parahydrogen with alumina at 77°K. Data of Quickert and LeRoy.[53]

times; here

$$\ln \frac{A_e}{A_e - A_t} = k_c t \qquad (8.34)$$

where t is found by dividing the total column dead volume plus the volume of adsorbed hydrogen (determined by warming the column to room temperature) by the flow rate. Figure 8.7 shows a plot of $\log A_e/(A_e - A_t)$] versus t from which k_c was found to be $5.53 \times 10^{-4}/\text{sec}$.

8.1.5 Higher Order Irreversible Reactions and Reversible Reactions
Langer and Patton[7] reviewed the literature concerning higher order irreversible and reversible reactions. To date, no simple method has been devised to cope with these situations; treatments such as those by Keller and Giddings[55] and Klinkenberg[56] have resulted in intractable equations. The GC technique is therefore limited at present to the determination of first-order (or pseudo-first-order) rate constants for irreversible reactions.

8.2 GC COLUMNS AS ISOTOPE EXCHANGE VESSELS

Isotope-labeled organic and inorganic compounds are used in many chemical and biological fields but the preparation of small quantities of several such species by conventional techniques is difficult.[57] Tadmor[58] demonstrated, however, that a GC column could be used simultaneously as an isotope exchange vessel and as an analytical separations method, as had earlier been suggested by Schmidt-Bleek and co-workers[59] and Tadmor.[60] Thus a mixture of unlabeled compounds is injected onto a column containing the exchanging nucleus (e.g., ^{36}Cl) and separated (and labeled) compounds, collected at the column exit.

In initial studies Tadmor[58] used a Sil-O-Cel firebrick support and $H^{36}Cl$ as the exchanger; the solutes were $GeCl_4$, $SnCl_4$, $AsCl_3$, PCl_3, and $FeCl_3$. $H^{36}Cl$ was deposited on the support by injecting it directly onto the column or by slurrying the support with aqueous $H^{36}Cl$, followed by drying and packing the "coated" Sil-O-Cel into a column. In either case the nonsorbed $H^{36}Cl$ was removed by conditioning the column at 100–300°C (the detector used throughout was a liquid beta counter). It was found that exchange occurred even when the treated support was coated with a liquid phase, as shown in Fig. 8.8, in which the effects of support particle size and liquid loading are also illustrated.

Tadmor[58] also devised a method of determining the exchange constants of solutes as a function of time for a given column, flow rate, and

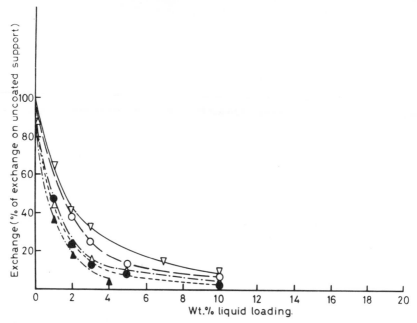

Figure 8.8   Effect of particle size on percent exchange. Liquid phase, n-decane; supports: ▲, 100/140-mesh Alundum; △, 100/140-mesh Sil-O-Cel; ●, 70/100-mesh Sil-O-Cel; ○, 30/50-mesh Sil-O-Cel; ▽, 10/12-mesh Sil-O-Cel. Data of Tadmor.[58]

temperature. Letting K represent the amount of isotope exchange that occurs in a given period, it follows that

$$\frac{KV_t}{F} = \log\left(\frac{bx'}{bx' - nax}\right) \tag{8.35}$$

where V_t is the total column volume, F is the flow rate, b is the number of moles of labeled solvent adsorbed on the support, x' is the percentage of the theoretical (100%) amount of exchange that the solvent undergoes when a solute interacts with it, n is the number of exchangeable atoms in the solute (the solvent is assumed to have only one), a is the number of moles of solute injected, and x is the percentage of exchange that the solute undergoes on passage through the column. Tadmor plotted V_t/F versus $\log[bx'/(bx' - nax)]$ for several different columns and flow rates and found that eq. 8.35 was indeed obeyed for flow rates of 5 to 40 ml/min. Equally straight lines were also found when $\log[bx'/(bx' - nax)]$ was plotted versus N, which indicated that the exchange efficiency was a function of the number of theoretical plates.

Tadmor's isotope exchange method is not limited solely to inorganic solutes; for example, Burlingame and co-workers[61,62] have applied the technique to the deuteration and ^{18}O-labeling of a variety of compounds, in which an average of 96% ^2H exchange was obtained for enolizable ketonic hydrogen. These workers also noted that liquid-column chromatographic exchange was, in general, less efficient than GLC, which may be due to differences in the number of theoretical plates inherent in the two techniques. "High-performance" liquid chromatography may prove useful in this regard.

8.3 REFERENCES

1. R. S. Juvet, Jr., and J. Chiu, *J. Am. Chem. Soc.*, **83**, 1560 (1961); S. Dal Nogare and R. S. Juvet, Jr., *Gas-Liquid Chromatography*, Interscience, New York, 1962, pp. 394–399.

2. P. Steingaszner, in *Ancillary Techniques of Gas Chromatography*, L. S. Ettre and W. H. McFadden, Eds., Wiley, New York, 1969, p. 13; C. C. Cassil, R. P. Stanovick, and R. F. Cook, in *Residue Reviews*, Vol. 26, F. A. Gunther, Ed., Springer Verlag, New York, 1969, p. 63; V. G. Berezkin, *Analiticheskaya Reaktsionnaya Gazovaya*, Nauka, Moscow, 1966; *Analytical Reaction Gas Chromatography*, Plenum, New York 1968; S. Z. Roginskii, M. I. Yanovskii, and A. D. Berman, in *Aspects in Gas Chromatography*, H. G. Struppe, Ed., Akademie Verlag, Berlin, 1971, p. 96; *The Fundamentals of Chromatographic Applications in Catalysis*, Nauka, Moscow, 1972.

3. P. D. Pacey and J. H. Purnell, *J. Chem. Soc. Faraday Trans. I*, **68**, 1462 (1972); *Ind. Eng. Chem. Fundam.*, **11**, 233 (1972); *Int. J. Chem. Kinet.*, **4**, 657 (1972).

4. S. H. Langer, J. Y. Yurchak, and J. E. Patton, *Ind. Eng. Chem.*, **61**, 10 (1969).

5. J. Kallen and E. Heilbronner, *Helv. Chim. Acta*, **43**, 489 (1960).

6. M. Nakagaki and M. Nishino, *Yakugaki Zasshi*, **85**, 305 (1965).

7. S. H. Langer and J. E. Patton, *Adv. Anal. Chem. Instrum.*, **11**, 293 (1973).

8. D. W. Bassett and H. W. Habgood, *J. Phys. Chem.*, **64**, 769 (1960).

9. R. J. Kokes, H. Tobin, and P. H. Emmett, *J. Am. Chem. Soc.*, **77**, 5860 (1955).

10. N. H. Ray, *Analyst*, **80**, 853, 957 (1955).

11. A. E. Martin and J. Smart, *Nature*, **175**, 422 (1955).

12. G. E. Green, *Nature*, **180**, 295 (1957).

13. S. Z. Roginskii, M. I. Yanovskii, and G. A. Gaziev, *Dokl. Akad. Nauk SSSR*, **140**, 1125 (1961).

14. T. R. Phillips and D. R. Owens, in *Gas Chromatography 1960*, R. P. W. Scott, Ed., Butterworths, London, 1960, p. 308.

15. S. Z. Roginskii, M. I. Yanovskii, and G. A. Gaziev, *Kinet. Katal.*, **3**, 529 (1962).

16. E. M. Magee, *Ind. Eng. Chem. Fundam.*, **2**, 32 (1963).

17. F. E. Gore, *Ind. Eng. Chem. Proc. Des. Dev.*, **6**, 10 (1967).

18. S. Z. Roginskii, E. I. Semenenko, and M. I. Yanovskii, *Dokl. Akad. Nauk SSSR*, **153**, 383 (1963).

19. E. I. Semenenko, S. Z. Roginskii, and M. I. Yanovskii, *Kinet. Katal.*, **5**, 490 (1964); **6**, 320 (1965).

20. B. D. Blaustein and G. M. Feldman, *Anal. Chem.*, **36**, 65 (1964).

21. S. Z. Roginskii and A. L. Rozental, *Dokl. Akad. Nauk SSSR*, **146**, 152 (1962); *Kinet. Katal.*, **5**, 104 (1964).

22. G. A. Gaziev, V. Y. Filinovskii, and M. I. Yanovskii, *Kinet. Katal.*, **4**, 688 (1963).

23. V. G. Berezkin, V. S. Kruglikova, and N. A. Belikova, *Dokl. Akad. Nauk SSSR*, **158**, 182 (1964).

24. V. G. Berezkin, V. S. Kruglikova, and V. E. Shiryaeva, *Neftekhim.*, **6**, 630 (1966).

25. V. R. Choudhary and L. K. Doraiswamy, *Ind. Eng. Chem. Prod. Res. Devel.*, **10**, 218 (1971).

26. N. C. Saha and D. S. Mathur, *J. Chromatogr.*, **81**, 207 (1973).

27. D. G. Anderson, K. E. Isakson, J. T. Vandenberg, M. Y. T. Jao, D. J. Tessari, and L. C. Afremow, *Anal. Chem.*, **47**, 1008 (1975).

28. P. Schulz, *Anal. Chem.*, **47**, 1979 (1975).

29. A. N. Genkin and N. A. Petrova, *J. Chromatogr.*, **105**, 25 (1975).

30. E. Cremer and R. Kramer, *J. Chromatogr.*, **107**, 253 (1975).

31. R. Kramer, *J. Chromatogr.*, **107**, 241 (1975).

32. S. H. Langer, J. E. Patton, and J. Coca, *Chem. Ind.*, 1346 (1970).

33. J. E. Patton, H. Kung, and S. H. Langer, *J. Chromatogr.*, **104**, 73 (1975).

34. K. Yamaoka and T. Nakagawa, *J. Chromatogr.*, **117**, 1 (1976).

35. S. H. Langer, H. R. Melton, T. D. Griffith, and J. Coca, *J. Chromatogr.*, **122**, 487 (1976).

36. D. W. Barber, C. S. G. Phillips, G. F. Tusa, and A. Verdin, *J. Chem. Soc.*, 18 (1959).

37. G. P. Cartoni, R. S. Lowrie, C. S. G. Phillips, and L. M. Venanzi, in *Gas Chromatography 1960*, R. P. W. Scott, Ed., Butterworths, London, 1960, p. 273.

38. C. G. Scott and C. S. G. Phillips, in *Gas Chromatography 1964*, A. Goldup, Ed., Institute of Petroleum, London, 1965, p. 266.

39. A. O. S. Maczek and C. S. G. Phillips, *J. Chromatogr.*, **29**, 7 (1967).

40. C. S. G. Phillips, A. J. Hart-Davis, R. G. L. Saul, and J. Wormald, *J. Gas Chromatogr.*, **5**, 424 (1967).

41. C. S. G. Phillips, M. J. Walker, C. R. McIlwrick, and P. A. Rosser, *J. Chromatogr. Sci.*, **8**, 401 (1970).

42. R. Lane, B. Lane, and C. S. G. Phillips, *J. Catal.*, **18**, 281 (1970).

43. C. S. G. Phillips and C. R. McIlwrick, *Anal. Chem.*, **45**, 782 (1973).

44. K. F. Scott and C. S. G. Phillips, *J. Chromatogr.*, **112**, 61 (1975).

45. M. van Swaay, *Adv. Chromatogr.*, **8**, 363 (1969).

46. C. S. G. Phillips, in *Gas Chromatography 1970*, R. Stock, Ed., Institute of Petroleum, London, 1971, p. 1.

47. M. Suzuki and J. M. Smith, *Adv. Chromatogr.*, **13**, 213 (1975).

48. E. Gil-Av and Y. Herzberg-Minzly, *Proc. Chem. Soc.*, 316 (1961).

49. E. Gil-Av and Y. Herzberg-Minzly, *J. Chromatogr.*, **13**, 1 (1964).

50. G. L. Pratt and S. H. Langer, *J. Phys. Chem.*, **73**, 2095 (1969).

51. J. B. Harkness, G. B. Kistiakowsky, and W. H. Mears, *J. Chem. Phys.*, **5**, 682 (1937).

52. B. S. Khambata and A. Wasserman, *J. Chem. Soc.*, 375 (1939).

53. K. A. Quickert and D. J. LeRoy, *Can. J. Chem.*, **48**, 2532 (1970).

54. W. R. Schulz and D. J. LeRoy, *Can. J. Chem.*, **42**, 2480 (1964).

55. R. A. Keller and J. C. Giddings, *J. Chromatogr.*, **3**, 205 (1960).

56. A. Klinkenberg, *Chem. Eng. Sci.*, **15**, 255 (1961).

57. H. Budzikiewicz, C. Djerassi, and D. H. Williams, *Structure Elucidation of Natural Products by Mass Spectrometry*, Vol. 1, Holden-Day, San Francisco, 1964, Ch. 2.

58. J. Tadmor, *Anal. Chem.*, **36**, 1565 (1964).

59. F. Schmidt-Bleek, G. Stocklin, and W. Herr, *Angew. Chem.*, **72**, 778 (1960).

60. J .Tadmor, *J. Inorg. Nucl. Chem.*, **23**, 158 (1961); *Chromatogr. Rev.*, **5**, 223 (1963); *Anal. Chem.*, **38**, 1624 (1966).

61. W. J. Richter, M. Senn, and A. L. Burlingame, *Tetrahedron Lett.*, 1235 (1965).

62. M. Senn, W. J. Richter, and A. L. Burlingame, *J. Am. Chem. Soc.*, **87**, 680 (1965).

CHAPTER 9

Diffusion Phenomena

9.1 DIFFUSION COEFFICIENTS

The dynamics of chromatography as expressed in various rate theories has been reviewed frequently.[1-17] However, treatments relating experimentally observed band broadening in packed columns to a summation of gas-and liquid-phase diffusion and mass transfer nonequilibria remain, to date, somewhat equivocal, due, mainly, to the complicated nature of flow around (and/or through) porous support media (e.g., Ref. 13) and the lack of coherent data regarding the modes of dispersion of liquid phases on such materials (Section 2.2.5). Diffusion data derived from van Deemter-type plots must therefore be interpreted with some caution. In contrast, several techniques have been developed for the unequivocal measurement of diffusivity by elution through empty tubes; these are considered here because they represent by far the simplest and most accurate chromatographic techniques available.

9.1.1 Gas-Phase Diffusion Coefficients

Maynard and Grushka[18] and Marrero and Mason[19] have presented comprehensive reviews of the measurement of gas-phase diffusion coefficients D_{12}, the former including a list of GC data said to be complete through 1972. Many were determined by what has come to be known as the chromatographic peak-broadening technique developed simultaneously by Giddings and co-workers[20-22] and Bohemen and Purnell[23] (which followed from the work by Taylor[24] and Aris[25]):

$$2D_{\text{eff}} = 2D_{12} + \frac{\bar{u}^2 r^2}{24D_{12}} \tag{9.1}$$

where D_{eff} is the effective (Taylor) axial diffusion coefficient that takes into account the radial velocity dispersion and radial diffusional spreading, $\bar{u}$ is the solute linear velocity, and r is the tube radius. In terms of the peak variance (cf. Fig. 2.1) in units of length,

$$\sigma^2 = \frac{2D_{12}L}{\bar{u}} + \frac{r^2\bar{u}L}{24D_{12}} \tag{9.2}$$

Because, in GC, $H = \sigma^2/L$, eq. 9.2 becomes

$$H = \frac{2D_{12}}{\bar{u}} + \frac{r^2\bar{u}}{24D_{12}} \tag{9.3}$$

Equation 9.3 is precisely the result that is obtained from the van Deemter (eq. 2.39), Golay (2.41), or Giddings (2.44) relations in the absence of a packing ($\gamma = 1$) and sorbent phase ($k' = 0$) in the limit of negligible pressure drop across an empty tube. [Because, under these conditions, the mean carrier velocity is identical to that at the outlet (L/t_A), the term u_o may be replaced by $\bar{u}$ and the superscript o, dropped from the diffusion coefficient symbol.]

Bohemen and Purnell[23] found that the second term on the rhs of eq. 9.3 contributed at most 3% to the plate height; consequently it was ignored and plots of H versus $1/\bar{u}$ (found to be linear) for the elution of solutes through empty tubes were used to determine diffusion coefficients. Alternatively, Giddings and co-workers[20-22] solved eq. 9.3 for D_{12}:

$$D_{12} = \frac{\bar{u}}{4}\left[H \pm \left(H^2 - \frac{r^3}{3}\right)^{\frac{1}{2}}\right] \tag{9.4}$$

Diffusion coefficients were measured in this instance by passing solutes through an empty tube, measuring H for each, and calculating D_{12} from eq. 9.4, the positive root of which is the significant one at low carrier velocities ($< \bar{u}_{opt}$), whereas the negative root is meaningful at high $\bar{u}$ ($> \bar{u}_{opt}$). In practice[22,26] the choice of root is facilitated by determining u_{opt} separately, where

$$\bar{u}_{opt} = \frac{4\sqrt{3}\, D_{12}}{r} \tag{9.5}$$

Use of eq. 9.4 requires careful measurement of $\bar{u}$, H, r, and the column length. Accuracy of the technique also depends on minimization of the system dead volume. Initially, Giddings and Seager[20] sought to account for these effects with the use of a 30-m×1/4-in. O.D. tube in combination with a 1-m tube of the same material in a commercial instrument. Subtraction of the peak variance of the latter from the former was said to give diffusion coefficient data (for a 29-m tube) free of end effects. Later[22] a 50-m by 1/4-in. OD tube was employed in conjunction with an on-column injection valve. The volume of the diffusion tube was 343.7 ml, the injection volume, 0.5 ml, and the FID detector volume, estimated at 0.38 ml. Thus the system volume was maximized with respect to the extra-system dead volume. Other sources of error were considered negligible in comparison with the (manual) measurement of H and σ^2, the error on the latter being estimated at ±2%.

Knox and McLaren[27] in 1964 proposed a stop-flow technique that has recently been the subject of an extensive error analysis (which does not differ significantly from the form given in Section 3.1) by Cloete and co-workers.[28] According to eq. 9.2, the total variance σ_t^2 *per unit time* is

$$\frac{\sigma_t^2}{t} = 2D_{eff} \qquad (9.5)$$

Now suppose that midway during elution (begun at time $t_{initial}$) the flow is stopped at some time t_1 and later restarted at a time t_2. A variance, $\sigma_D^2 = 2D_{12}(t_2 - t_1)$ ($\bar{u} = 0$), will have occurred in addition to that normally found for uninterrupted flow. Further, the *uninterrupted* portion of the variance is given by $\sigma_C^2 = 2D_{eff}(t_1 + t_3)$, where $t_3 = t_{final} - t_2$. Considering the total variance to be the sum of the individual variances,

$$\sigma_D^2 \geqslant \sigma_t^2 = \sigma_C^2 + \Sigma\sigma_i^2 \qquad (9.6)$$

where $\Sigma\sigma_i^2$ represents all other instrumental sources of band broadening. Each of the terms in eq. 9.6 is a constant except for σ_D^2. The quantity $\sigma_C^2 + \Sigma\sigma_i^2$, is therefore first determined by a normal-flow experiment, σ_D^2, from the difference between stopped-flow and normal-flow runs, and D_{12}, found from the slope ($2D_{12}$) of plots of σ_D^2 versus ($t_2 - t_1$). Alternatively, σ_t^2 may be determined from the intercept of such plots (at which point $\sigma_D^2 = \sigma_t^2$).

Table 9.1 presents recently reported examples of the use of both the stop-flow and uninterrupted-flow GC techniques. Comparison with literature values in the first of these indicates that gas chromatographic data are likely to be accurate to ±2% with manual injection and data reduction that could be improved to better than 1% if these features were automated.[29]

Table 9.1 Examples of the Measurement of Binary Vapor-Phase Diffusion Coefficients by Gas Chromatography

Solute/Carrier Pair	T (°K)	GC		Literature	
		D_{12} (cm²/sec)	Precision (%)	D_{12} (cm²/sec)	Precision (%)
A. Data of Cloete, Smuts, and De Clerk[28] (Stop-Flow Technique)					
Methane/helium	294.7	0.662	0.4	0.656	3.0
	296.6	0.670	0.5	0.663	3.0
	372.0	0.999	0.6	0.986	3.5
Ethane/helium	294.5	0.482	0.1	0.501	…
	298.1	0.516	0.9	0.511	…
Propane/helium	294.4	0.392	0.2	…	…
	296.6	0.411	0.4	…	…
n-Butane/helium	292.9	0.330	0.1	0.354	0.3
	296.6	0.347	0.3	0.361	0.3
2,2-Dimethylpropane/helium	292.8	0.297	0.3	…	…
	296.8	0.317	0.7	…	…
Sulfur hexafluoride/helium	294.6	0.377	0.7	0.403	3.0
Helium/argon	298.1	0.761	0.5	0.747	1.0
	298.5	0.788	0.8	0.749	1.0
Neon/argon	297.0	0.311	1.9	0.319	1.0
Nitrogen/argon	296.2	0.195	0.5	0.193	2.0
	297.6	0.190	…	0.195	2.0
	483.2	0.446	0.3	0.456	3.0
Methane/argon	298.5	0.219	0.2	0.205	3.0

Table 9.1 (*Continued*)

Solute/Carrier Pair	GC			Literature	
	T (°K)	D_{12} (cm²/sec)	Precision (%)	D_{12} (cm²/sec)	Precision (%)
Ethane/argon	302.5	0.149	0.8	0.114	⋯
	483.2	0.357	0.4	⋯	⋯
Propane/argon	301.8	0.247	1.2	⋯	⋯
	483.2	0.096	0.5	0.104	0.4
n-Butane/argon	297.1	0.096	0.9	0.082	⋯
	298.5	0.096	0.5	0.104	0.4
	476.7	0.276	0.6	⋯	⋯
2,2-Dimethylpropane/argon	297.9	0.077	0.1	⋯	⋯
Sulfur hexafluoride/argon	298.7	0.078	0.6	0.081	3.0
Helium/nitrogen	294.5	0.699	0.7	0.703	2.0
Methane/nitrogen	294.4	0.226	0.6	0.209	⋯
Ethane/nitrogen	301.0	0.167	0.3	0.150	⋯
Propane/nitrogen	292.8	0.113	0.5	0.151	⋯
	297.5	0.122	0.8	0.154	⋯
n-Butane/nitrogen	293.8	0.097	0.4	0.094	⋯
				0.090	⋯
				0.110	⋯
	371.9	0.141	1.4	⋯	⋯
	479.1	0.238	0.9	⋯	⋯
	481.2	0.237	0.6	⋯	⋯

2,2-Dimethylpropane/nitrogen	293.6	0.078	0.4	...	...
Sulfur hexafluoride/nitrogen	297.4	0.089	0.6	...	...
	294.2	0.092	0.4	0.093	3.0

B. Data of Grushka and Schnipelsky[29] (Helium Carrier; 100°C)

Solute	D_{12} $\left(\dfrac{cm^2}{sec}\right)$	$\left(\dfrac{Precision}{\%}\right)$
n-Heptane	0.3544	0.62
3-Methylhexane	0.3586	0.36
2-Methylhexane	0.3626	0.54
2,4-Dimethylpentane	0.3635	0.27
2,2-Dimethylpentane	0.3720	0.11
2,3-Dimethylpentane	0.3758	0.61
3,3-Dimethylpentane	0.3843	0.59
2,2,3-Trimethylpentane	0.3858	0.32

9.1.2 Liquid-Phase Diffusion Coefficients Grushka and Kitka,[30] Pratt and Wakeham,[31,32] and others[33,36] have recently devised what amounts to a liquid-chromatographic method of determining liquid-phase diffusion coefficients, D_{13}. Because diffusion in liquids is roughly 10^{-5} times that in gases, the first term on the rhs of eq. 9.3 becomes negligibly small at low mobile-phase velocities:

$$H = \frac{r^2 \bar{u}^2}{24 D_{13}} \tag{9.7}$$

and D_{13} may be calculated directly from the solute plate height obtained from its elution with a (liquid) carrier through an empty tube. Table 9.2 lists D_{13} data determined by this technique.

Table 9.2   Examples of the Determination of D_{13} Data by Liquid Chromatography

A. Benzine in n-Heptane[30]

T ($°C$)	D_{13} $(cm^2/sec, \times 10^5)$			
	LC	Ref. 37	Ref. 38	Ref. 39
40	4.45	4.39	4.74	4.30
50	5.05	4.90	5.33	4.95
60	5.51	5.53	5.90	5.61
70	6.31	6.26	6.49	6.25
80	7.00	7.15	7.07	6.90

B. Phenones in n-Heptane[30]

Solute	D_{13} $(cm^2/sec, \times 10^5)$				
	40°C	50°C	60°C	70°C	80°C
Acetophenone	2.26	2.60	2.97	3.36	3.94
n-Propiophenone	2.16	2.37	2.79	3.13	3.66
n-Butyrophenone	2.08	2.14	2.58	2.70	3.29
iso-Butyrophenone	2.02	2.09	2.39	2.65	3.13
Valerophenone	1.94	2.04	2.41	2.63	3.17
iso-Valerophenone	1.86	1.95	2.26	2.49	2.91
n-Hexaphenone	1.86	1.98	2.30	2.50	3.01
n-Heptaphenone	1.81	1.94	2.25	2.43	2.87
n-Octaphenone	1.77	1.89	2.18	2.36	2.74
n-Nonaphenone	1.71	1.81	2.11	2.29	2.61
n-Decaphenone	1.69	1.77	2.06	2.12	2.17
Myristophenone	1.64	1.69	1.98	2.06	2.47

The measurement of D_{13} by gas chromatography is severely hampered (regardless of the form of van Deemter equation used) because the stationary-phase film thickness on diatomaceous-earth supports normally employed is indeterminant. This difficulty likely extends to glass beads as well, for they are not wetted by low molecular-weight solvents most often of interest, *inter alia*, hydrocarbons. Polymer stationary phases, on the other hand, appear not only to wet glass beads but to give coatings that are even and reproducible films. So far, Guillet and co-workers[40,41] appear to be the only workers to have taken advantage of these phenomena in order to measure D_{13} values for (probe) solutes with polymeric liquids. Referring back to the abbreviated form of the van Deemter relation, eq. 2.40, at sufficiently high flow velocity $B/\bar{u} \rightarrow 0$, A and $C_M \bar{u}$ remain suitably small, and C_L can be determined from the slope of the linear region of H versus $\bar{u}$ plots: $H \cong C_L \bar{u}$. D_{13} is then calculated from the relation

$$D_{13} = \frac{8}{\pi^2} \frac{d_L^2}{C_L} \left[\frac{(k')^2}{(1+k')^2} \right] \tag{9.8}$$

where d_L is calculated from the density of the polymer and the specific surface area of the beads (which requires a carefully sized support). Table 9.3 lists the diffusion coefficients obtained by Gray and Guillet[40] for three solutes with a low-density polyethylene stationary phase.

Table 9.3 D_{13} **Data for Named Solutes with Polyethylene Stationary Phase**[40]

Solute	T (°C)	C_L (sec, $\times 10^3$)	D_{13} (cm^2/sec, $\times 10^8$)
n-Decane	30	30.0 ± 1.8	0.35
	50	24.6 ± 0.7	1.03
	60	31.8 ± 0.7	1.00
	65	29.9 ± 1.5	1.28
	80	38.2 ± 1.0	1.34
n-Tetradecane	125	20 ± 7	0.85
	140	13.0 ± 0.06	2.2
	150	10.3 ± 0.4	3.7
	160	12.5 ± 0.6	4.1
	170	8.6 ± 0.5	7.4
Benzene	25	114 ± 6	0.82
	25	$\cdots$	1.18[a]

[a]Static method of McCall and Slichter[42]

9.2   LENNARD-JONES POTENTIAL CONSTANTS

The Lennard-Jones equation is[43]

$$\phi(r_{12}) = 4\varepsilon_{12}\left[\left(\frac{\sigma_{12}}{r_{12}}\right)^{12} - \left(\frac{\sigma_{12}}{r_{12}}\right)^{6}\right] \qquad (9.9)$$

where $\phi(r)$ is the potential energy of two spherical nonpolar molecules, r_{12} is the intermolecular center-to-center distance, σ_{12} is the collision cross section, and ε_{12} is a molecular interaction term. The potential energy of the entire system is found by summing over all pairwise interactions, that is, $\Sigma\phi(r_{ij})$. The quantities σ_{12} and ε_{12} are called the Lennard-Jones potential constants and are related to intermolecular diffusion coefficients by[43]:

$$D_{12} = 2.628 \times 10^{-3}\left(\frac{MW_1 + MW_2}{2\,MW_1 MW_2}\right)^{\frac{1}{2}}\left(\frac{T^{\frac{3}{2}}}{\sigma_{12}^2 P \Omega_{12}^{(1,1)*}}\right) \qquad (9.10)$$

where T is in $°K$, P is the system pressure, and $\Omega_{12}^{(1,1)*}$ is the first-order collision integral which takes into account initial relative speeds, reduced masses, and deflection angles of collision impacts.[44] Fortunately, Hirschfelder, Curtiss, and Byrd[43] have tabulated values of $\Omega_{12}^{(l,s)*}$ as a function of T^*, where

$$T^* = \frac{Tk}{\varepsilon} \qquad (9.11)$$

The combining rules[43]

$$\sigma_{12} = \frac{\sigma_{11} + \sigma_{12}}{2} \qquad (9.12)$$

$$\varepsilon_{12} = (\varepsilon_{11}\varepsilon_{22})^{\frac{1}{2}} \qquad (9.13)$$

are often used to find T^* (thence $\Omega_{ij}^{(l,s)*}$) and σ_{12} in order to calculate D_{12}. Here, however, chromatographic diffusion coefficients are used to determine σ_{12} and ε_{12} (or ε_{12}/k, the quantity most often tabulated). Ultimately, a method of measuring σ_{ii} and ε_{ii} is desired because eqs. 9.11–9.13 could then be used to calculate D_{12} values directly. The probable success of this approach is illustrated in Table 9.4, in which calculated and experimental diffusion data are compared. The calculated values were obtained from the combining rules (eqs. 9.12 and 9.13) and eqs. 9.10 and 9.11. The experimental values were determined from viscosity measurements.[44,45]

So far Schnipelsky and Grushka[46] and Cloete and co-workers[47] appear to be the only workers to have attempted to determine Lennard-Jones poten-

Table 9.4 Comparison of Calculated[43] and Experimental[44,45] Interdiffusion Coefficients

Interdiffusing Pair	T (°K)	Calculated σ_{12} (Å)	Calculated ε_{12}/k (°K)	Calculated D_{12} (cm²/sec)	Experimental D_{12} (cm²/sec)
Ar/He	273.2	3.059	21.3	0.653	0.641
N_2/H_2	273.2	3.325	55.2	0.656	0.674
N_2/O_2	273.2	3.557	102	0.175	0.181
N_2/CO	273.2	3.636	100	0.174	0.192
N_2/CO_2	273.2	3.839	132	0.130	0.144
H_2/O_2	273.2	3.201	61.4	0.689	0.697
H_2/CO	273.2	3.279	60.6	0.661	0.651
H_2/CO_2	273.2	3.482	79.5	0.544	0.550
CO/O_2	273.2	3.512	112	0.175	0.185
CO_2/O_2	273.2	3.715	147	0.128	0.139
CO_2/CO	273.2	3.793	145	0.128	0.137
CO_2/N_2O	273.2	3.938	204	0.092	0.096
CO_2/CH_4	298.2	3.939	161	0.138	0.153
N_2/C_2H_4	298.2	3.957	137	0.156	0.163
N_2/C_2H_6	298.2	4.050	145	0.144	0.148
$N_2/n\text{-}C_4H_{10}$	298.2	4.339	194	0.0986	0.0960
$N_2/i\text{-}C_4H_{10}$	298.2	4.511	169	0.0970	0.0908
$N_2/cis\text{-}2\text{-}C_4H_8$	298.2	4.467	188	0.0947	0.0950

tial constants from purely GC data. The former group determined diffusion coefficients at two temperatures, T_1 and T_2; they next divided $D_{12}(T_2)$ by $D_{12}(T_1)$:

$$\left[\frac{D_{12}(T_2)}{D_{12}(T_1)}\right]\left(\frac{T_1}{T_2}\right)^{\frac{3}{2}} = \frac{\Omega_{12}^{(1,1)*}(T_1)}{\Omega_{12}^{(1,1)*}(T_2)} \tag{9.14}$$

which yields the ratio of collision integrals. The T^* versus $\Omega_{i,j}^{(l,s)*}$ tabulation in Ref. 43 (Chapter 4) was then used to find the two values of $\Omega_{12}^{(1,1)*}(T)$ which satisfied eq. 9.14 [hence $T^*(T)$] from which ε_{12}/k and σ_{12} were calculated. To determine σ_{ii} and ε_{ii} diffusion coefficients were measured in two different carrier gases; D_{12} for one carrier in the other was also obtained. The three measurements were then repeated at a second temperature. Solving the combining rules for σ_{ii} and ε_{ii} for the three gases,

$$\sigma_{ii} = \sigma_{ij} + \sigma_{im} - \sigma_{jm} \tag{9.15}$$

$$\varepsilon_{ii} = \frac{\varepsilon_{ij}\varepsilon_{im}}{\varepsilon_{jm}} \tag{9.16}$$

Table 9.5   Lennard-Jones Potential Constants for Named Binary Mixtures[46]

Mixture	ε_{12}/k (°K)	σ_{12} (Å)
n-Hexane/helium	52	4.24
n-Hexane/argon	174	4.78
n-Heptane/helium	55	4.60
n-Heptane/argon	185	4.97
n-Octane/helium	58	5.05
n-Octane/argon	195	5.49
Helium/argon	38	2.96

Table 9.6   Lennard-Jones Potential Constants for Pure Compounds[46]

Compound	ε_{11}/k (°K)	Literature Value[43]	σ_{11} (Å)	Literature Value[43]
n-Hexane	238	423	6.06	5.92
n-Heptane	268	...	6.61	...
n-Octane	298	320	7.58	7.45
Helium	11.3	10.2	2.51	2.56
Argon	128	120	3.41	3.41

which gave six equations in six unknowns from which σ_{11} and ε_{11} were calculated. The (experimental) σ_{12} and ε_{12}/k data determined by Schnipelsky and Grushka[46] are listed in Table 9.5 and the (calculated) σ_{11} and ε_{11}/k values in Table 9.6. With the exception of the ε_{11}/k value for n-hexane, the agreement with literature values is respectable.

9.3   REFERENCES

1. J. H. Beynon, S. Clough, D. A. Crooks, and G. R. Lester, *Trans. Faraday Soc.*, **54**, 705 (1958).

2. J. H. Purnell, *Gas Chromatography*, Wiley, New York, 1962, Chapters 8 and 9.

3. A. B. Littlewood, in *Gas Chromatography 1964*, A. Goldup, Ed., Institute of Petroleum, London, 1965, p. 77.

4. J. C. Giddings, *Dynamics of Chromatography*, Marcel Dekker, New York, 1965.

5. T. W. Smuts and V. Pretorius, in *Gas Chromatography 1966*, A. B. Littlewood, Ed., Institute of Petroleum, London, 1967, p. 75.

6. J. C. Giddings, *J. Chem. Educ.*, **44**, 704 (1967).

7. G. Guiochon, *Chromatogr. Rev.*, **8**, 1 (1965).

8. J. H. Purnell, *Ann. Rev. Phys. Chem.*, **18**, 81 (1967).

9. C. Landault and G. Guiochon, *Chromatographia*, **1**, 277 (1968).

10. J. C. Giddings, M. N. Myers, and J. W. King, *J. Chromatogr. Sci.*, **7**, 276 (1969).

11. E. A. Walker and J. P. Palframan, *Analyst*, **94**, 609 (1969).

12. A. B. Littlewood, *Gas Chromatography*, Academic, New York, 1970, Chapter 6.

13. *Flow Through Porous Media*, A.C.S. Publications, Washington, D.C., 1970.

14. R. M. Bethea and P. C. Bensten, *J. Chromatogr. Sci.*, **10**, 575 (1972).

15. O. K. Guha, J. Novak, and J. Janak, *J. Chromatogr.*, **84**, 7 (1973).

16. E. Grushka, L. R. Snyder, and J. H. Knox, *J. Chromatogr. Sci.*, **13**, 25 (1975).

17. R. L. Grob, in *Modern Practice of Gas Chromatography*, R. L. Grob, Ed., Wiley-Interscience, New York, 1977, pp. 65–81.

18. V. R. Maynard and E. Grushka, *Adv. Chromatogr.*, **12**, 99 (1975).

19. T. R. Marrero and E. A. Mason, *J. Phys. Chem. Ref. Data*, **1**, 3 (1972).

20. J. C. Giddings and S. L. Seager, *J. Chem. Phys.*, **35**, 2242 (1958); *Ind. Eng. Chem. Fundam.*, **1**, 277 (1962).

21. S. L. Seager, L. R. Geertson, and J. C. Giddings, *J. Chem. Eng. Data*, **8,** 168 (1963).

22. E. N. Fuller, K. Ensley, and J. C. Giddings, *J. Phys. Chem.*, **73**, 3679 (1969).

23. J. Bohemen and J. H. Purnell, *J. Chem. Soc.*, 360, 2630 (1961).

24. G. Taylor, *Proc. Roy. Soc. Ser. A*, **219**, 186 (1953); **223**, 446 (1954); **225**, 473 (1954).

25. R. Aris, *Proc. Roy. Soc. Ser. A*, **235**, 67 (1956).

26. E. Grushka and V. R. Maynard, *J. Chem. Educ.*, **49**, 565 (1972); *Chem. Technol.*, **4**, 560 (1974).

27. J. H. Knox and L. McLaren, *Anal. Chem.*, **36**, 1477 (1964).

28. C. E. Cloete, T. W. Smuts, and K. De Clerk, *J. Chromatogr.*, **120**, 1, 17 (1976).

29. E. Grushka and P. Shnipelsky, *J. Phys. Chem.*, **78**, 1428 (1974).

30. E. Grushka and E. J. Kitka, Jr., *J. Phys. Chem.*, **78**, 2297 (1974); **79**, 2199 (1975); *J. Am. Chem. Soc.*, **98**, 643 (1976).

31. K. C. Pratt, O. H. Slater, and W. A. Wakeham, *Chem. Eng. Sci.*, **28**, 1901 (1973).

32. K. C. Pratt and W. A. Wakeham, *Proc. Roy. Soc. Ser. A*, **336**, 393 (1974); **342**, 401 (1975); *J. Phys. Chem.*, **79**, 2198 (1975).

33. A. C. Ouano, *Ind. Eng. Chem. Fundam.*, **11**, 268 (1972).

34. A. C. Ouano and J. A. Carothers, *J. Phys. Chem.*, **79**, 1314 (1975).

35. H. Komiyama and J. M. Smith, *J. Chem. Eng. Data*, **19**, 384 (1974).

36. M. K. Tham and K. E. Gubbins, *J. Chem. Soc. Faraday Trans. I*, **68**, 1339 (1972).

37. L. R. Wilke and P. Chang, *AIChE J.*, **1**, 264 (1955).

38. S. A. Sanni, C. J. D. Fell, and P. Hutchinson, *J. Chem. Eng. Data*, **16**, 424 (1971).

39. W. F. Calus and M. T. Tyn, *J. Chem. Eng. Data*, **18**, 377 (1973).

40. D. G. Gray and J. E. Guillet, *Macromolecules*, **6**, 223 (1973).

41. J.-M. Braun, S. Poos, and J. E. Guillet, *J. Polym. Sci., Polym. Lett.*, in press.

42. D. W. McCall and W. P. Slichter, *J. Am. Chem. Soc.*, **80**, 1861 (1958).

43. J. O. Hirschfelder, C. F. Curtiss, and R. B. Byrd, *Molecular Theory of Gases and Liquids*, Wiley, New York, 1954, Chapters 4 and 8.

44. S. Chapman and T. G. Cowling, *The Mathematical Theory of Non-Uniform Gases*, Cambridge University Press, Cambridge, England, 1939.

45. C. A. Boyd, N. Stein, V. Steingrimson, and W. F. Rumpel, *J. Chem. Phys.*, **19**, 548 (1951).

46. P. N. Schnipelsky and E. Grushka, *J. Phys. Chem.*, in press.

47. C. E. Cloete, T. W. Smuts, and K. De Clerk, *J. Chromatogr.*, **120**, 29 (1976).

PART FOUR
Properties of Pure Substances

CHAPTER 10   MOLECULAR PROPERTIES

Molecular Properties

10.1 VAPOR PRESSURE

Gas chromatographs are often indispensible features of apparatus designed to measure instrinsic molecular properties. One example is the determination of vapor pressure described by Friedrich and Stammbach.[1] Figure 10.1 shows their instrument: the thermostated column, 6, is packed with a mechanical mixture of the solid (whose vapor pressure is to be determined) and an inert support such as Celite (liquid materials are coated directly onto the support in approximately 20–40% w/w). Carrier gas is allowed to flow through the saturation column into a GC column, 7, which is cooled by a dry ice-acetone bath. A small heater and pressure regulator are connected at the outlet of the saturator. The total gas flow through the system is measured by water displacement, 8. At a given time the gas flow is stopped and the GC column is transferred to a gas chromatograph in which the adsorbed materials are eluted in the usual manner. An internal standard is also injected. The weight of each material, w_1, transferred from the saturator to the GC column can then be determined from peak areas. Because the volume occupied by an ideal gas at STP is 22.415 l/mole,

$$p_1^0 = \frac{w_1(22.415)\,T_s p_s}{MW_1 V_1(273.15)} \tag{10.1}$$

The volume occupied by the solute at the saturator temperature T_s is just the volume of carrier gas required to elute it from the saturator V_s. This volume is measured at the outlet of the GC column and must be corrected to the volume corresponding to the outlet of the saturator column by multiplying by the ratio p_a/p_s, where p_a is the outlet (usually atmospheric)

269

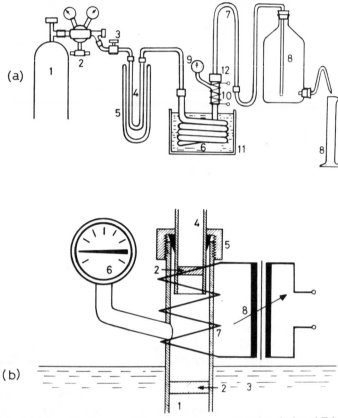

Figure 10.1 (a) Schematic diagram of the vapor pressure saturation device of Friedrich and Stammbach.[1] 1, nitrogen cylinder; 2, pressure regulator; 3, precision pressure regulator; 4, precolumn containing 5A molecular sieve; 5, Dewar flask; 6, saturator column; 7, GC column; 8, carrier gas volume measurement by water displacement; 9, pressure gauge; 10, heater; 11, constant-temperature bath; 12, dry-ice cooling sleeve. (b) Expanded diagram of the saturator-GC column interconnections. 1, saturator column; 2, glass wool; 3, constant-temperature bath; 4, GC column; 5, Swagelok fittings; 6, pressure gauge; 7, heating wire; 8, Variac.

pressure and p_s is the saturator outlet pressure:

$$p_1^0 = \frac{w_1(22.415)T_s p_s^2}{MW_1(273.15)p_a} \tag{10.2}$$

Friedrich and Stammbach[1] used eq. 10.2 to determine the vapor pressures of substituted triazines; Table 10.1 gives the Antoine constants and vapor pressures of several of the herbicides at 20°C. This method is obviously of

Table 10.1 Vapor Pressures and Antoine Constants for Named Triazines[1] $\left(\log p_1^0 = A - \dfrac{B}{T}\right)$

Triazine	A	B	p_1^0 (torr, at 20°C)
Atrazine	13.766	5945	3.0×10^{-7}
Propazine	14.754	6533	2.9×10^{-8}
Simazine	15.107	6833	6.1×10^{-9}
Atratone	11.303	4933	2.9×10^{-6}
Prometone	10.794	4817	2.3×10^{-6}
Simetone	11.894	5130	2.4×10^{-6}
Ametryne	11.911	5270	8.4×10^{-7}
Prometryne	11.841	5222	1.0×10^{-6}
Simetryne	11.914	5293	7.1×10^{-7}
GS 34360	12.101	5302	1.0×10^{-6}

considerable value because other techniques such as headspace analysis[2] would be difficult to apply to compounds with vapor pressures as low as those shown.

Eggertsen and co-workers,[3] Groszek,[4] Mackle and co-workers,[5] and others[6-9] have used variants of the Friedrich-Stammbach procedure for a variety of solutes. Griffiths and Phillips,[10] the first to determine vapor pressures by GC, employed displacement analysis for esters, ketones, ethers, and aniline. Voykevich and co-workers[11] have measured the vapor pressures of perfumes. Rose and Schrodt[12] pointed out that because plots of $\log V_g^0$ versus $\log p_1^0$ are linear for homologous series the vapor pressure of a higher member may be found by extrapolation from three or more lower order compounds with known vapor pressures. Most recently Dimov and co-workers[13] have reported the determination of vapor pressures from retention indices.

10.2 HEAT OF VAPORIZATION

Heats of vaporization are determined from GC data from the variation of $\log V_g^0$ with $\log p_1^0$ (slope of $-\Delta H_v / RT$). The technique has not been widely applied, however, because ΔH_v is readily calculated from the Clausius-Clapeyron equation

$$\ln p_1^0 = \frac{\Delta H_v}{RT} + C \tag{10.3}$$

It is useful, nonetheless, when vapor pressure data are unavailable, as in the case of volatile inorganic solutes.[14, 15] Alternatively, Mackle and co-workers[5] used vapor pressures obtained from GC in conjunction with eq. 10.3 to determine heats of vaporization.

Agreement with data from nonchromatographic techniques has, when reported,[16, 17] been good.

10.3 BOILING POINT

Homologues of known boiling point may be used to construct $\log K_R$ versus bp plots from which unknowns are determined graphically. Linear behavior is expected for homologous solutes because the heats of vaporization are linearly related. Further, any other parameter that reflects K_R may be employed; for example, t'_R, α, or retention indices. A comprehensive study of 324 alkanes and olefins was conducted by Matukuma[18] who also reported several discrepancies between A.P.I.-listed[19] boiling points and those he found from retention data. One of the worst was 2-methyl-3-iso-propylhexane which had a listed bp of 166.7°C but which gave a bp of 155.22°C by GC. Subsequently, the A.P.I. listings were revised in accordance with Matukuma's findings. Walraven and Ladon[20] later extended Matukuma's work and, in addition to finding several more boiling point errors, they also questioned the cis-trans assignments of several A.P.I. standards (geometric configuration is considered below). Baumann and co-workers[21] have determined the boiling points of phenyl-substituted alkanes by GC. Martin and co-workers,[22] following Grant and Vaughn,[23] used Trouton's approximation,[24]

$$\frac{\Delta H_v}{T_b} \cong 21 \tag{10.4}$$

to express retention data as a function of boiling point T_b:

$$\ln t'_R \cong -\frac{21 T_b}{RT} + C \tag{10.5}$$

where C is a column constant for a given homologous series. Boiling points for several phenothiazine derivatives determined in this manner agreed well with those calculated from reduced-pressure data.

Baumann and co-workers[21] compared the boiling points found from $\log t'_R$ versus T_b plots for two stationary phases, Apiezon L and SF 96. A portion of their data, given in Table 10.2, shows good internal agreement.

Table 10.2 Boiling Point Data by Gas Chromatography[21]

	T_b (°C)				
	GLC		Literature		
Solute	Apiezon L	SF 96	Ref. 25	Ref. 26	Ref. 27
5-Phenyldecane	273.8	274.7			275
4-Phenyldecane	275.7	276.3			
3-Phenyldecane	279.3	280.0			
2-Phenyldecane	286.3	286.2	307		299
1-Phenyldecane	297.9	297.9			
6-Phenylundecane	288.7	290.2			285
5-Phenylundecane	289.9	290.2			
4-Phenylundecane	292.0	291.8			
3-Phenylundecane	295.0	295.6			
2-Phenylundecane	301.9	301.6	314		298
1-Phenylundecane	313.2	313.2			
6-Phenyldodecane	304.0	304.5	302	304	209
5-Phenyldodecane	304.8	305.1	307	304	316
4-Phenyldodecane	307.3	307.2	295	314	343
3-Phenyldodecane	310.9	310.7	310	311	318
2-Phenyldodecane	317.2	316.5	313	312	310
1-Phenyldodecane	327.6	327.6			

The listed literature data, on the other hand, disagree by as much as 100°C. Willis[28] extended Baumann's work to all positional isomers of phenylundecane through phenylhexadecane by temperature-programming capillary columns. Sojak and co-workers[29–32] have determined the boiling points of alkanes, olefins, and aromatics by gas chromatography, and in 1971 Gesheva and co-workers[33] fixed the boiling points of alkyl astatides by GC. Most recently, Sultanov and Arustamova[34] measured T_b for 68 positional isomers of decane by correlating known boiling points with retention indices. Agreement of these data with those reported by Matukuma[18] was within ±0.2°C.

10.4 MOLECULAR WEIGHT

The gas density balance detector was historically the first to be used in gas chromatography.[35, 36] It was also employed to measure molecular weights in one of the first GC physicochemical investigations. (Although in the present context it might well be regarded as a device ancillary to gas chromatography, continuing interest[37-56] in apparatus of this kind for molecular weight measurements justifies a brief discussion of its principles of operation.)

In 1946 Claesson[57] used mercury manometers to measure the density difference between two gas streams. According to the ideal gas law

$$\rho = \frac{p_1 MW_1}{RT} \tag{10.6}$$

where p is the gas stream pressure, ρ, its density, and MW, its molecular weight. The apparatus was bulky, however, and therefore not convenient for GC purposes. The innovation of Martin and James was to miniaturize the device by replacing the mercury manometers with thermocouples. Figure 10.2 shows their gas density balance detector in schematic form. Effluent from the column enters at A, splits into two streams and recombines to exit at E; N and N' are push rods used to balance the split ratio. The reference side admits pure carrier gas at F which passes by push rods, P and P', N and N', and also exits at E. The short path between L and L'(M) contains two thermocouples, one near L and the other near L'. Contained in M is a small Nichrome wire heater. When the gas in M is heated and there is zero gas flow across L, L', the two thermocouples will

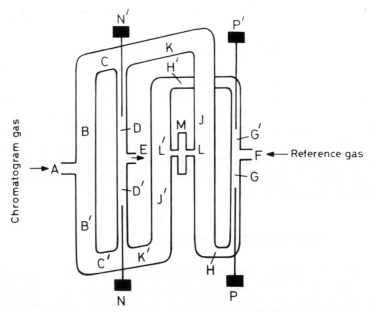

Figure 10.2 Schematic diagram of the Martin and James[35, 36] gas density balance GC detector.

be heated to the same extent and will show no net temperature gradient. When a solute heavier than the carrier gas enters at A, the gas streams in B and B' become unbalanced. Because the pressures at D and D' tend to equalize, the gas streams from K and K' also become unbalanced. This is reflected by a small flow through the junction L, L', from right to left and results in the thermocouple at L' being heated more than that at L. Thus the recorded temperature difference is a direct function of the molecular weight, first realized by Liberti and co-workers[58] in 1956. Suppose that the chromatogram of a mixture containing an unknown, X, of molecular weight, MW_x, and an internal standard, k, of molecular weight, MW_k, appears as the solid trace in Fig. 10.3. If a different carrier gas is used at the same flow rate, the chromatogram will appear as the dashed line in Fig. 10.3; that is, the retention times will be approximately identical with both carriers but the areas will differ. The densities of the reference and column effluent gases are given by

$$\text{first reference carrier: } \rho_1 = \frac{pMW_1}{RT} \tag{10.7}$$

$$\text{second reference carrier: } \rho_2 = \frac{pMW_2}{RT} \tag{10.8}$$

column effluent with carrier No. 1:

$$\rho_{X_1} = \frac{p(X_1MW_1 + X_XMW_X)}{RT} \tag{10.9}$$

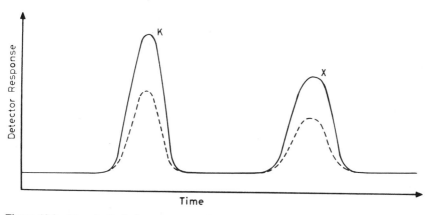

Figure 10.3 Hypothetical chromatogram of an unknown, X, and an internal standard, k. Solid line: carrier 1; dotted line: carrier 2. Detector: gas density balance.

column effluent with carrier No. 2:

$$\rho_{X_2} = \frac{p(X_2 MW_2 + X_X MW_X)}{RT} \tag{10.10}$$

Thus

$$\rho_{X_1} - \rho_1 = \Delta\rho_1 = \frac{pX_X(MW_X - MW_1)}{RT} \tag{10.11}$$

and

$$\rho_{X_2} - \rho_2 = \Delta\rho_2 = \frac{pX_X(MW_X - MW_2)}{RT} \tag{10.12}$$

where p is the total pressure and X_i represents the mole fraction of the indicated species. The total amount of X that passes through the column with carrier 1 is given by

$$Q_{X_1} = \int_0^\infty \frac{X_X MW_X p}{RT} dv = \left(\frac{MW_X}{MW_X - MW_1}\right) \int_0^\infty \Delta\rho_{X_1} dv \tag{10.13}$$

Similarly for the internal standard k

$$Q_{k_1} = \left(\frac{MW_k}{MW_k - MW_1}\right) \int_0^\infty \Delta\rho_{k_1} dv \tag{10.14}$$

Because the integral of the solute gas-phase density over the entire time of the run corresponds to a constant times the total solute peak area

$$\int_0^\infty \Delta\rho\, dv = \text{const} \int_0^\infty \Delta(\text{detector response})\, dv = (\text{const})(\text{peak area}) \tag{10.15}$$

Thus equating the ratios of Q_{X_1}, Q_{X_2}, Q_{k_1}, and Q_{k_2} yields

$$MW_X = \frac{A_{k_2} MW_2 Z - A_{k_1} MW_1}{A_{k_2} Z - A_{k_1}} \tag{10.16}$$

where

$$Z = \frac{A_{X_1}(MW_k - MW_1)}{A_{X_2}(MW_k - MW_2)} \tag{10.17}$$

and where $A_{X_{1,2}}$ and $A_{k_{1,2}}$ are the peak areas of X and k, respectively, with carrier 1 or 2. Liberti and co-workers[58] initially used dodecane ($MW = 170.3$) as the unknown and found a molecular weight of 163 by the peak area method; the accuracy was thus claimed to be $\pm 5\%$.

Using a slightly different approach, Phillips and Timms[59] measured the pressure and volume of an amount of unknown which was then passed through the gas density balance. In this case eq. 10.13 reduces to:

$$pV = \frac{KA_X}{MW_X - MW_1} \tag{10.18}$$

where K is a new constant determined with a standard solute. The relation can be simplified to

$$p_X V_X = \frac{g_X RT}{MW_X} = \frac{\text{const}}{MW_X} \tag{10.19}$$

where the gas density balance is used merely to measure the weight of solute g_X. The extra step of measuring pV extends the time of molecular weight determinations to about an hour, but the accuracy of the procedure is better than $\pm 1\%$, a fivefold improvement over that of Liberti. (Karasek and Laub[60] reported a similar method that eliminates the need for a GC column altogether. They determined g_X by weighing a syringe before and after injection of a solute into an evacuated chamber by which procedure eq. 10.19 is reduced to

$$p_X = \frac{g_X}{MW_X}(\text{const})$$

Thus a plot of p_X versus g_X should be linear of slope, $1/MW_X$, and zero intercept, which was verified for several different solutes.)

In the GC techniques described the moleculer weights of both carrier gases were less than that of the unknown. Parsons[61] has proposed that one of the carriers should have a molecular weight that is larger than that of the unknown; that is, the two gases should bracket the solute. The relative detector response to a given unknown is given by

$$Y = \frac{A_{X_c}}{A_{k_c}[MW_k/(MW_k - MW_c)]} = \frac{A_{X_c}}{A_{k_c}} - \frac{A_{X_c}MW_c}{A_{k_c}MW_k} \tag{10.20}$$

where a subscript, c, indicates the carrier. Now, if Y is plotted versus MW_c, a straight line should be obtained that crosses the MW_c axis; the point of intersection on the abscissa is MW_X. Parsons used this method to determine the molecular weights of pyrolysis products of polyethylene glycol adipate and polypropylene glycol adipate; Fig. 10.4 shows plots of Y versus MW_c for two such products, acetaldehyde and propionaldehyde (identified by IR subsequent to the molecular weight determinations). He also noted that carrier gases of molecular weight up to 200 (octafluorocyclobutane) are available so that the bracketing technique will accommodate most GC solutes. Even if this were not the case, plots of Y versus MW_c could still be extrapolated to the abscissa. In any event, carriers of higher molecular weight give greater accuracy. Table 10.3 lists the data for several "unknown" solutes and carriers determined in this manner.

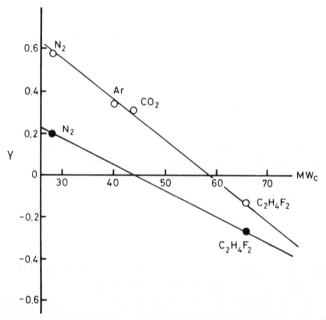

Figure 10.4 Plot of Y versus MW_c for two pyrolysis products with various carrier gases. Data of Parsons.[61]

Prominent contributing factors to renewed interest in the gas density balance detector have been the recent commercial availability of such devices and the inability of mass spectrometry to yield accurate molecular weights (because of molecular-ion instability) in at least several cases.[64]

Martire and Purnell[65] developed a technique for measuring MW_L, the molecular weight of the liquid phase. It can be shown that MW_L is given

Table 10.3 Molecular Weights of Named "Unknowns" Determined by the Bracketing Technique of Parsons[61]

Unknown	MW (g/mole)	Internal Standard	Bracketing Carriers	MW Found	Error
Carbon tetrachloride	153.8	Benzene	CCl_2F_2, C_4F_8	155.2	$+1.4$
Chloroform	119.4	Acetone	CCl_2F_2, C_4F_8	120.0	$+0.6$
Toluene	92.2	Benzene	CCl_2F_2, CF_4	93.5	$+1.3$
Acetone	58.1	Cyclohexane	N_2, Ar, CO_2, $C_2H_4F_2$	59.0	$+0.9$

by

$$\frac{1}{MW_L} = \frac{\ln V_g^0 + \ln p_1^0 + \ln \overline{V}_1 - \ln(273R) + \chi_H + 1}{\overline{V}_1 \rho_L} \tag{10.21}$$

For two closely related solutes X and Y in solvent L eq. 10.21 can be written as

$$\ln\left[\frac{(p_1^0)_Y}{(p_1^0)_X}\right] + \ln\left[\frac{(V_g^0)_{Y,L}}{(V_g^0)_{X,L}}\right] = \ln\left[\frac{(\overline{V}_1)_X}{(\overline{V}_1)_Y}\right]$$

$$+ \frac{\rho_L}{MW_L}\left[(\overline{V}_1)_Y - (\overline{V}_1)_X\right] + (\chi_{H,X,L} - \chi_{H,Y,L}) \tag{10.22}$$

For a different solvent L' eq. 10.22 becomes

$$\ln\left[\frac{(p_1^0)_Y}{(p_1^0)_X}\right] + \ln\left[\frac{(V_g^0)_{Y,L'}}{(V_g^0)_{X,L'}}\right] = \ln\left[\frac{(\overline{V}_1)_X}{(\overline{V}_1)_Y}\right]$$

$$+ \frac{\rho_{L'}}{MW_{L'}}\left[(\overline{V}_1)_Y - (\overline{V}_1)_X\right] + (\chi_{H,X,L'} - \chi_{H,Y,L'}) \tag{10.23}$$

If L and L' are similar types of solvents (e.g., hydrocarbons, polyethylene glycols, and methylsilicone oils), the following approximations are valid:

$$\chi_{H,X,L} \cong \chi_{H,X,L'}$$

$$\chi_{H,Y,L} \cong \chi_{H,Y,L'}$$

$$(\chi_{H,X,L} - \chi_{H,Y,L}) \cong (\chi_{H,X,L'} - \chi_{H,Y,L'})$$

Dividing eq. 10.22 by 10.23 and rearranging therefore gives

$$\ln\left\{\frac{[(V_g^0)_Y/(V_g^0)_X]_L}{[(V_g^0)_Y/(V_g^0)_X]_{L'}}\right\} = \left(\frac{\rho_L}{MW_L} - \frac{\rho_{L'}}{MW_{L'}}\right)\left[(\bar{V}_1)_Y - (\bar{V}_1)_X\right] \quad (10.24)$$

Upon substitution of R_L and $R_{L'}$ for $[(V_g^0)_Y/(V_g^0)_X]_L$ and $[(V_g^0)_Y/(V_g^0)_X]_{L'}$

$$\frac{1}{MW_{L'}} = \frac{\rho_L}{\rho_{L'}MW_L} - \frac{\ln(R_L/R_{L'})}{\rho_{L'}\Delta\bar{V}_1} \quad (10.25)$$

where $\Delta\bar{V}_1$ is the difference between the solute molar volumes $[(\bar{V}_1)_Y - (\bar{V}_1)_X]$. Because the ratio of specific retention volumes is employed, only relative adjusted retention data (α values) need be determined. This considerably simplifies the required measurements.

Martire and Purnell[65] tested eq. 10.25 by determining the molecular weights of squalane and polypropylene glycol (PPG) 1200; n-eicosane and polypropylene glycol 400 were used as the respective standard solvents and n-hexane/n-heptane and n-heptane/n-octane as the solutes. The data are presented in Table 10.4 from which Martire and Purnell found, respectively, 422.0 and 1220 g/mole for squalane and PPG 1200, which were in good agreement with the actual values 422.8 and 1260 g/mole. The accuracy of this method clearly depends on the accuracy with which the ratio $R_L/R_{L'}$ is determined. With precision apparatus it should be possible to measure molecular weights in the range $< 10^4$ g/mole.

Martire[66] has since reconsidered this technique from the standpoint of the use of molal-based activity coefficients (the simplest of his three proposed methods makes use of retention ratios as before). Statistical

Table 10.4   Data of Martire and Purnell[65] Used to Test Eq. 10.25

Solute Pair (X, Y)	Solvent	T (°C)	R	$(\bar{V}_1)_X$, (ml/mole)	$(\bar{V}_1)_Y$, (ml/mole)	MW_L, (g/mole)	ρ_L, (g/ml)
n-Hexane/ n-heptane	n-Eicosane	53.2	2.715	137.2	152.9	282.54	0.7667
n-Hexane/ n-heptane	Squalane	53.2	2.676	137.2	152.9	?	0.7886
n-Heptane/ n-octane	PPG 400	61.8	2.326	154.9	171.0	409.0	0.9779
n-Heptane/ n-octane	PPG 1200	61.8	2.267	154.9	171.0	?	0.9727

analysis indicated, however, that a standard deviation of only 0.007 on the retention data will produce a relative error of 10% as $MW_L \rightarrow 5000$, which again emphasizes the need for precision instrumentation when gas chromatography is applied to measurements of this kind.

10.5 MOLECULAR GEOMETRY

The establishment of molecular geometry is not as straightforward as one might suppose and some confusion in the literature has resulted because of misassignments. Gas chromatography provides a relatively straightforward method of identifying cis-trans isomers, shown by Hively,[67] one of the first to make use of GC for these species, who employed several stationary phases to determine the relative retention times of a variety of olefins. In cases in which the stereochemistry had been established unambiguously it was found that for a squalane column the trans form of *disubstituted* olefins eluted before the cis form. Conversely, for *trisubstituted* olefins cis compounds eluted before their trans conjugates. Table 10.5 lists the cis/trans α values for selected olefins* with squalane at 27 and 49°C.

Matukuma,[18] Walraven and Ladon,[20] and others[70-75] later extended Hively's findings and noted that several A.P.I. standards had been mislabeled. Cornforth and co-workers[76] have independently verified the *E-Z*

Table 10.5 α Values (cis/trans) for Olefins with Squalane[67] at Two Temperatures

Olefin	$\alpha_{cis/trans}$	
	27°C	49°C
2-Butene	1.12	1.12
2-Pentene	1.04	1.05
2-Hexene	1.07	1.07
3-Hexene	0.99	1.00
2-Heptene	1.04	1.05
3-Heptene	1.02	1.03
3-Methyl-2-pentene	0.89	0.90
3,4-Dimethyl-2-pentene	0.91	0.93
3-Methyl-2-hexene	0.92	0.93
3-Methyl-3-hexene	0.93	0.93

*There is no difficulty in the nomenclature of these olefins but for higher trisubstituted compounds the terminology is less clear; the naming of tri- and tetrasubstituted olefins is currently undergoing revision; the *E*- and *Z*-rules (*entgegen*, opposed, and *zusammen*, together) are favored at present.[68, 69]

configurations of a variety of olefins, confirming the GC results. The importance of these investigations can be gauged by the fact that J. W. Cornforth was awarded the 1975 Nobel Prize in Chemistry for his substantial efforts in this field.

Recently, Pelter and co-workers[77, 78] have established the configuration of several E-Z pairs by GLC and NMR and by stereoselective synthesis. The GLC method was particularly rewarding: several E-Z pairs of 5-substituted-5-decenes and 7-substituted-7-tetradecenes were chromatographed on Apiezon M and N and the α values for each isomer taken relative to n-dodecane. Log α versus carbon number plots were then constructed for the following:

$$n\text{-}C_4H_9\diagdown \qquad n\text{-}C_4H_9$$
$$C{=}C$$
$$H\diagup \qquad \diagdown R$$

where

$$R = CH_3{-}, n\text{-}C_2H_5{-}, \text{ and } n\text{-}C_4H_9{-}$$

$$n\text{-}C_6H_{13}\diagdown \qquad n\text{-}C_6H_{13}$$
$$C{=}C$$
$$H\diagup \qquad \diagdown R'$$

where

$$R' = CH_3{-}, n\text{-}C_2H_5{-}, \text{ and } n\text{-}C_3H_7{-}$$

The results are shown in Figs. 10.5 and 10.6. For the $n\text{-}C_4H_9$ compounds the two lines should converge at $R = n\text{-}C_4H_9{-}$ because each E and each Z set forms an independent homologous series that becomes identical at carbon number = 4. Similarly, for the $n\text{-}C_6H_{13}$ compounds convergence was expected (and found) at $R' = n\text{-}C_6H_{13}{-}$ (carbon number = 6). The GLC method was used in this case to substantiate the stereoselectivity of certain organic reactions but the results also confirm that, for trialkyl-substituted olefins, the Z-isomer boils at a lower temperature (hence elutes earlier from "nonpolar" columns) than the E-isomer.

Diastereoisomerism is of considerable importance in biological chemistry, and it is therefore not surprising that gas chromatographers have

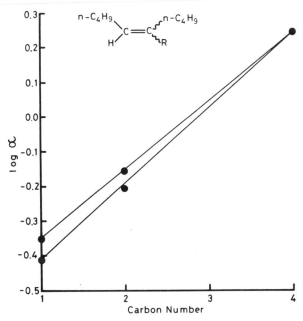

Figure 10.5 Plots of log α versus carbon number for *E*- (upper line) and *Z*-geometric isomers of 5-substituted-5-decenes. Data of Pelter and co-workers.[78]

investigated the resolution of these species. Two different approaches have been used: the compounds to be separated are first derivatized diastereo-isomerically[79-82] or optically active stationary phases are employed to resolve the (underivatized) mixture.[83-98] The latter is a straightforward analytical application (albeit the stationary phase is unusual). König[99] recently combined the two methods; that is, he used an optically active stationary phase to separate diastereoisomeric derivatives. The derivatization method has recently been summarized by Halpern;[100] the work of Westley and Halpern[101] is given here as an example. These workers found that *N*-trifluoroacetyl-(TFA)-*S*-prolyl chloride* was a particularly useful reagent because it is available in optically pure form, does not racemize during derivatization, and gives rapid and quantitative coupling reactions. The general reaction scheme, shown in Fig. 10.7, is illustrated with amino acids (hydroxyamino acids must first be converted to trimethylsilyl ethers before coupling). The resultant product of the scheme is a mixture of two optical isomers of each amino acid denoted by *SR* and *SS* (the first *S* refers to the *N*-TFA-*S*-prolyl group and the second letter to the amino

*The *R*-*S* nomenclature[68, 69, 102] for optical isomers is employed in these examples.

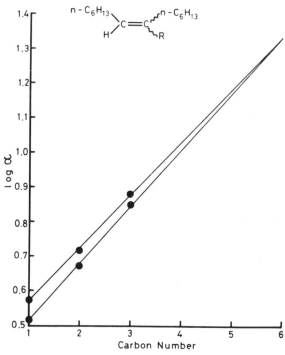

Figure 10.6 Plots of $\log \alpha$ versus carbon number for E- (upper line) and Z-geometric isomers of 7-substituted-7-tetradecenes. Data of Pelter and co-workers.[78]

acid which can be R or S). Once prepared, separation of an SR/SS mixture is straightforward; examples of the resolution of optically active amino acids are given in Table 10.6. Westley and Halpern[101] were also able to apply the technique to N-chloroalkanoyl valine methyl esters, racemic amines, α- and β-phenylethylamines, cyclic amines, and various amides and esters. Finally, they proposed a general scheme for the absolute assignment of stereochemistry by GLC using optically active derivatizing agents: it was found in all cases that the α values for SS/RR and SR/RS pairs were unity but that SR/SS and SR/RR pairs were easily resolved. The determination of stereochemistry thus becomes a matter of deduction; for example, suppose that alanine has been produced by what is thought to be a new stereoselective synthetic method but the stereochemistry of the product is not known. The first step in its identification is to derivatize the sample with N-TFA-S-prolyl chloride, as in Fig. 10.7, followed by chromatographing the product with SE-30. The procedure is then repeated with N-TFA-R-prolyl chloride. Two retention times are thus found; the possi-

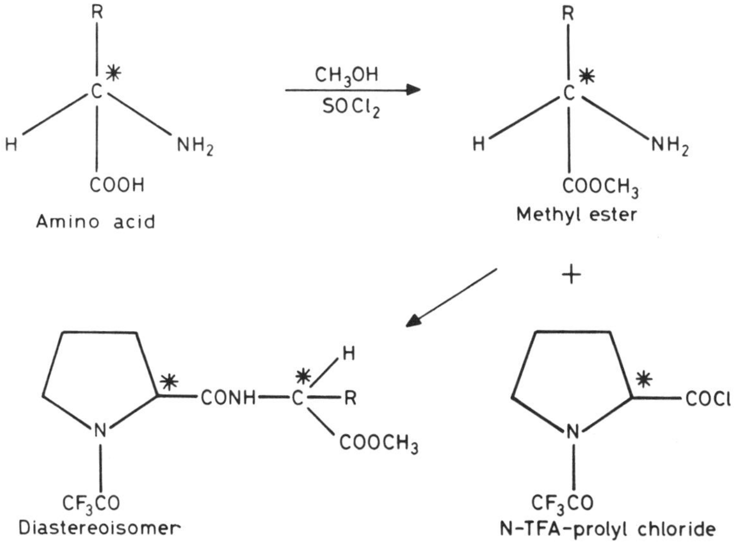

Figure 10.7 Method of preparation of N-TFA-S-prolyl amino acid derivatives. After Westley and Halpern.[101]

bilities as to their identity are tabulated below:

Unknown = S-Alanine

Run	Resolving	GLC Results	
No.	Agent	First Peak	Second Peak
1	S	...	SS
2	R	RS	...

Unknown = R-Alanine

Run	Resolving	GLC Results	
No.	Agent	First Peak	Second Peak
1	S	SR	...
2	R	...	RR

If the product were S-alanine, run No. 1 would produce a peak with a longer retention time than the peak in run No. 2. Conversely if the sample were R-alanine, the peak of run No. 1 would elute before the peak in run No. 2. Only two chromatograms are therefore needed to establish the

Table 10.6   α Values of SS/SR Pairs of N-Trifluoroacetyl-S-Prolyl Peptide Methyl Esters[101]

Amino Acid	T (°C)	Liquid Phase	α (SS/SR)
Alanine	176	SE-30	1.14
Valine	176	SE-30	1.15
Leucine	176	SE-30	1.09
Proline	176	SE-30	1.11
Serine-TMS	185	PEGA	1.22
Threonine-TMS	185	PEGA	1.23
γ-Hydroxyproline-TMS	185	PEGA	1.21
Aspartic acid	185	PEGA	1.08
Glutamic acid	185	PEGA	1.15
Methionine	185	PEGA	1.14
Phenylglycine	220	PEGA	1.12
Phenylalanine	220	PEGA	1.04

stereochemistry of the unknown. If the sample had been a completely new chemical compound, however, the SR/SS relative retention behavior may be reversed; that is, SS may elute earlier than SR. Homologues of known stereochemistry must be run in these cases to establish the order of elution of the SS derivative with respect to the SR isomer. Once the order is known R and S derivatives of the unknown are chromatographed and its stereochemistry established deductively as above.

Because the analysis of natural products rarely involves more than a few milligrams of sample and a gas chromatograph with an FID will detect microgram to subnanogram amounts, far greater than the sensitivity of a polarimeter, the elucidation scheme of Westley and Halpern appears to be an attractive technique for the determination of stereochemistry.

10.6   BOND ANGLE DEFORMATION

Laub, Ramamurthy, and Pecsok[103] have attempted to measure bond angle deformations by examining charge transfer K' values (cf. eq. 6.22) of substituted aromatic amines and β-ionones with 2,4,7-trinitrofluorenone (TNF). Retention data for the former solutes were taken from the work of Cooper and co-workers[104, 105] and K' values calculated relative to cis-and trans-decalin; the results are given in Table 10.7. Donor vertical ionization potentials are also presented when available because charge transfer interactions are thought to be a function of I_v^d. Lower I_v^d values in group A

Table 10.7 Ionization Potentials[106, 107] and Relative Formation Constants[103] for Substituted Aromatic Amines

Compound	Structure	I_v^d (eV)	K'_{cis}	K'_{trans}
A. Anilines				
1. Aniline		7.90	33.02	27.18
2. *o*-Toluidine		7.75	35.16	28.93
3. *m*-Toluidine		7.75	37.34	30.73
4. *p*-Toluidine		7.65	40.10	33.00
5. 2,4-Xylidine		...	40.44	33.28
B. *N,N*-Dimethylanilines				
1. *N,N*-Dimethylaniline		...	16.54	13.61
2. *N,N*-Dimethyl-*o*-toluidine		7.37	3.685	3.033
3. *N,N*-Dimethyl-*m*-toluidine		7.35	...	...
4. *N,N*-Dimethyl-*p*-toluidine		7.33	18.86	15.52
5. *N,N*-Dimethyl-3,5-xylidine		7.25	...	...
6. *N,N*-Dimethyl-2,6-xylidine		7.22	2.764	2.275

Table 10.7　(*Continued*)

Compound	Structure	I_v^d (eV)	K'_{cis}	K'_{trans}
7. N,N-Dimethyl-2,4-xylidine	H_3C—⬡—$N(CH_3)_2$ ⎮ CH_3	7.17	3.540	2.913
C. Increased N-Substitution				
1. Aniline	⬡—NH_2	7.90	33.02	27.18
2. N-Methylaniline	⬡—$NH(CH_3)$	7.60	26.52	21.83
3. N,N-Dimethylaniline	⬡—$N(CH_3)_2$	...	16.54	13.61

seem to correlate roughly with higher K' data, although 2,4-xylidine has a lower K' value than would be expected on the basis of increased methyl substitution (i.e., lower I_v^d). In contrast, no correlation is seen in groups B or C. Especially interesting is the pair N,N-dimethyl-o-toluidine/N,N-dimethylaniline. The K' data are the reverse of what was expected on the basis of the trend shown in group A. McRae and Goodman[108] have shown that ortho substitution in N,N-dimethylanilines causes out-of-plane deformations at the nitrogen. Such an effect may well cause steric hindrance to charge transfer and result in the lower (than expected) K' values observed for Nos. A2, A4, B2, B6, B7, C2, and C3.

In attempts to verify this effect the K' values of several cis- and trans-β-ionyl pairs were determined (twisting about the central single bond had previously been measured unambiguously by NMR[109-111]). The results were totally inconsistent with the deformation angles (30–32° for the trans compounds and 34–39° for the cis materials) which may be due to the fact that twisting is so severe in these systems that conjugation is destroyed (each double bond then acting as an independent ethylene unit). Thus solution phenomena other than complexation must be responsible for the K' values in these cases.

10.7　IONIZATION POTENTIAL AND ELECTRON AFFINITY

Laub and Pecsok[112] have determined vertical ionization potentials by GLC: charge transfer K_1 values for several dienes with known I_v^d values were evaluated with TNF in di-n-butyl phthalate (DNBP); K_1 was then plotted versus I_v^d and other unknown diene ionization potentials found from this plot and the K_1 data for each new solute. The plots were curved and there was considerable scatter about the nonlinear least squares fitted lines. In this case, however, the "standard" and "unknown" dienes were closely related structurally and solution interactions other than charge transfer were probably of minimal consequence as reflected in a comparison of the GLC and photoelectron spectral (PES) values[113] (the latter were determined subsequent to the chromatographic work). The data given in Table 10.8 show that the GLC values appear to be accurate to ± 0.1 eV. Interesting was the fact that steric hindrance to charge transfer appeared to affect markedly the GLC I_v^d values shown in Table 10.9. There is a clear correlation between the bulkiness of substituent groups and the I_v^d error and Laub and Pecsok have suggested further investigations with highly branched dienes.[113, 114] [It remains to be seen what place ionization potentials and steric hindrance will hold in light of the solution model of Laub and Purnell.[115]]

Electron affinities may also be determined by the Laub-Pecsok method:[113] K_1 values would first be measured for several solutes with various acceptors of known E_v^a in a common solvent; K_1 would then be plotted versus E_v^a and electron affinities of other acceptors found from the solute K_1 values with those stationary phase additives and the $K_1 - E_v^a$ plot; for example, pyridazinediones[116, 117] should be amenable to this technique. In a separate method Wentworth and Becker[118] employed an electron capture detector (the mechanism of which is now thought to be understood[119-131]) to measure the constant K in the expression

$$K = \frac{C_{A^-}}{C_A C_{e^-}} = \frac{f_{A^-}}{f_A f_{e^-}} \exp\left(-\frac{E^a}{kT}\right) \tag{10.26}$$

Table 10.8   Comparison of GLC- and PES-Determined Diene I_v^d Values[113]

Diene	I_v^d (eV)	
	GLC	PES
cis-1,3-Pentadiene	8.65	...
trans-1,3-Pentadiene	...	8.61
2-Ethyl-1,3-butadiene	8.76	8.79
2-Methyl-1,3-pentadiene	8.53	...
3-Methyl-1,3-pentadiene	8.51	8.40
4-Methyl-1,3-pentadiene	8.49	8.45

**Table 10.9 Apparent Steric Hindrance[113]
to Charge Transfer Reflected by $\Delta(I_v^d)$
$(= I_v^d, \text{GLC} - I_v^d, \text{PES})$**

Diene	I_v^d (eV) GLC	PES	$I\Delta_v^d$ (eV)
	8.70	8.53	0.17
	8.75	8.51	0.24
	8.81	8.47	0.34
	8.71	8.14	0.57
	8.85	9.31	−0.46

where C_{A^-}, C_A, and C_{e^-} are the concentrations of the negatively charged (electron-capturing) solute, neutral solute, and free electrons in the detector cell; f_i represents the statistical thermodynamic partition function[132] of the ith species, E^a is the electron affinity (presumably the adiabatic value), and k is Boltzmann's constant. Wentworth and Becker[118] assumed that the ratio f_{A^-}/f_A is constant and equal to two (because the negative ion is doubly degenerate) and found that K for anthracene was 3.5×10^6 1/mole at 420°K. They then estimated f_{e^-} to be 6.27×10^{-2} and calculated a value of 0.42 eV for E^a from eq. 10.26 (compared with a theoretical estimation[133] of 0.61 eV).

10.8 MISCELLANEOUS

A variety of studies in addition to those described above has been reported. Grob and McGonigle[134] examined 3d π-electron densities of metal chloride adsorbents by gas chromatography. Fischer and King[135] used a piezoelectric crystal detector to measure the oxidation stability of elastomers. Huebner[136] developed relative "polarity" scales for surfactants by GC. Solvation numbers have been determined chromatographically by Meloan and co-workers,[137-138] hydrophilic/lipophilic balance numbers by Becker and Birkmeier,[139] parachors by Wurst,[140] relative permeability coefficients of medicinally important materials by Varsano and Gilbert,[141] critical volumes with a thermal conductivity detector by Barry and co-workers,[142-144] the sizing of aerosols with a flame ionization detector by Ohline,[145] and polymer self-reorganization rates by Schep and De Clerk,[146] all of which is adequate testimony to the versatility of gas chromatography when applied to physicochemical studies.

10.9 REFERENCES

1. K. Friedrich and K. Stammbach, *J. Chromatogr.*, **16**, 22 (1964).

2. H. Hachenberg and A. P. Schmidt. *Gas Chromatographic Headspace Analysis*, Heydon, London, 1977.

3. F. T. Eggertsen, E. E. Siebert, and F. H. Stross, *Anal. Chem.*, **41**, 1175 (1969).

4. A. J. Groszek, *J. Inst. Petrol.*, **48**, 325 (1962).

5. H. Mackle, R. G. Mayrick, and J. J. Rooney, *Trans. Faraday Soc.*, **56**, 115 (1960).

6. A. Gianetto and M. Panetti, *Ann. Chim.*, **50**, 1721 (1960).

7. M. Panetti and G. Musso, *Ann. Chim.*, **52**, 472 (1962).

8. P. Benedek and L. Müller, in *Gas Chromatography*, H. P. Angele and H. G. Struppe, Eds., Akademie Verlag, Berlin, 1963, p. 139.

9. R. C. Duty and W. R. Mayberry, *J. Gas Chromatogr.*, **4**, 115 (1966).

10. J. H. Griffiths and C. S. G. Phillips, *J. Chem. Soc.*, 3446 (1954).

11. S. A. Voykevich, M. M. Shchedrina, N. P. Solovieva, and T. A. Rudolfi, *Masleb.-Zhir. Prom.*, **37**, 27 (1971).

12. A. Rose and V. N. Schrodt, *J. Chem. Eng. Data*, **8**, 9 (1963).

13. N. Dimov, M. Mukhtarova, and D. Shopov, *Neftekhim.*, **15**, 621 (1975).

14. R. A. Keller and H. Freiser, in *Gas Chromatography 1960*, R. P. W. Scott, Ed., Butterworths, London, 1960, p. 301.

15. S. T. Sie, J. P. A. Bleumer, and G. W. A. Rijnders, *Sep. Sci.*, **1**, 41 (1966).

16. D. White and C. T. Cowan, *Trans. Faraday Soc.*, **54**, 557 (1958).

17. H. Mackle and R. G. Mayrick, *Trans. Faraday Soc.*, **58**, 33 (1962).

18. A. Matukuma in *Gas Chromatography 1968*, C. L. A. Harbourn, Ed., Institute of Petroleum, London, 1969, p. 55.

19. A. P. I. Research Project 44, *Selected Values of Properties of Hydrocarbons and Related Compounds*, Carnegie Press, Pittsburgh, 1956.

20. J. J. Walraven and A. W. Ladon, in *Gas Chromatography 1970*, R. Stock, Ed., Institute of Petroleum, London, 1971, p. 358.

21. F. Baumann, A. E. Straus, and J. F. Johnson, *J. Chromatogr.*, **20**, 1 (1965).

22. H. F. Martin, J. L. Driscoll, and B. J. Gudzinowicz, *Anal. Chem.*, **35**, 1901 (1963).

23. D. W. Grant and G. A. Vaughn, *J. Appl. Chem.*, **6**, 145 (1956).

24. F. Trouton, *Phil. Mag.*, **18**(5), 54 (1884).

25. A. W. Francis, *Chem. Rev.*, **42**, 107 (1948).

26. A. C. Olson, *Ind. Eng. Chem.*, **52**, 833 (1960).

27. S. W. Ferris, *Handbook of Hydrocarbons*, Academic, New York, 1955.

28. D. E. Willis, *J. Chromatogr.*, **30**, 86 (1967).

29. L. Sojak, J. Krupcik, K. Tesarik, and J. Janak, *J. Chromatogr.*, **71**, 243 (1972).

30. L. Sojak, J. Hrivnak, A. Simkovicova, and J. Janak, *J. Chromatogr.*, **71**, 243 (1972).

31. L. Sojak and L. Hrivnak, *J. Chromatogr. Sci.*, **10**, 701 (1972).

32. L. Sojak, J. Krupcik, and J. Rijks, *Chromatographia*, **7**, 26 (1974).

33. M. Gesheva, A. Kolachkovsky, and Y. Norseyev, *J. Chromatogr.*, **60**, 414 (1971).

34. N. T. Sultanov and L. G. Austamova, *J. Chromatogr.*, **115**, 553 (1975).

35. A. J. P. Martin and A. T. James, *Biochem. J.*, **63**, 138 (1956).

36. A. T. James and A. J. P. Martin, *Biochem J.*, **63**, 144 (1956).

37. K. E. Murray, *Aust. J. Appl. Sci.*, **10**, 156 (1959).

38. E. A. Johnson, D. G. Childs, and G. H. Beaven, *J. Chromatogr.*, **4**, 429 (1960).

39. D. Hennenberg and G. Schomburg, in *Gas Chromatography 1962*, M. van Swaay, Ed., Butterworths, London, 1962, p. 191.

40. C. L. Guillemin and F. Auricourt, *J. Gas Chromatogr.*, **1**(10), 24 (1963); **2**, 156 (1964).

41. C. L. Guillemin, F. Auricourt, and P. Blaise, *J. Gas Chromatogr.*, **4**, 338 (1966).

42. A. G. Nerheim, *Anal. Chem.*, **35**, 1640 (1963); U.S. Patent 3, 082, 618 (1963); U.S. Patent 3, 091, 113 (1963).

43. A. G. Nerheim and J. H. Ruston, U.S. Patent 3, 082, 619 (1963).

44. I. A. Revelskii, R. I. Borodulina, and G. M. Sovakova, *Pet. Chem. USSR*, **4**, 296 (1965); *Neftekhim.*, **4**, 804 (1964).

45. J. T. Walsh and D. M. Rosie, *J. Gas Chromatogr.*, **5**, 232 (1967).

46. J. T. Walsh, K. J., McCarthy, and C. Merritt, *J. Gas Chromatogr.*, **6**, 416 (1968).

47. J. T. Walsh, R. E. Kramer, and C. Merritt, *J. Chromatogr. Sci.*, **7**, 348 (1969).

48. S. C. Bevan, T. A. Gough, and S. Thorburn, *J. Chromatogr.*, **44**, 241 (1969).

49. E. C. Creitz, *J. Chromatogr. Sci.*, **7**, 137 (1969).

50. A. Adam, K. Focke, and J. Harangozo, *Meres Auto.*, **18**(3), 81 (1970).

51. A. A. Datskevich, V. A. Rotin, B. P. Okhotnikov, V. A. Pirogov, and L. I. Rozanova, *Tr. Vses Nauchn.-Issled. Geol. Neft. Inst.*, **64**, 173 (1970).

52. A. G. Vitenburg, A. K. Pospelova, and V. V. Loff, *Neftekhim.*, **12**, 623 (1972).

53. J. Vermont and C. L. Guillemin, *Anal. Chem.*, **45**, 775 (1973).

54. R. S. Swingle, *J. Chromatogr. Sci.*, **12**, 1 (1974).

55. D. T. Heggie and W. S. Reeburgh, *J. Chromatogr. Sci.*, **12**, 7 (1974).

56. E. Kiran and J. K. Gillham, *Anal. Chem.*, **47**, 983 (1975).

57. S. Claesson, *Arkiv. Kemi. Mineral. Geol.*, **23A**, 1 (1946).

58. A. Liberti, L. Conti, and V. Crescenzi, *Nature*, **178**, 1067 (1956); *Atti Accad. Naz. Lincei Rend.*, **20**, 623 (1956).

59. C. S. G. Phillips and P. L. Timms, *J. Chromatogr.*, **5**, 131 (1961).

60. F. W. Karasek and R. J. Laub, *Anal. Chem.*, **46**, 1349 (1974).

61. J. S. Parsons, *Anal. Chem.*, **36**, 1849 (1964).

62. D. G. Paul and G. E. Umbreit, *Res./Dev.*, **21**(5), 18 (1970).

63. C. E. Bennett, L. W. DiCave, Jr., D. G. Paul, J. A. Wegener, and L. J. Levase, *Am. Lab.*, May 1971.

64. A. C. Lanser, J. O. Ernst, W. F. Kwolek, and H. J. Dutton, *Anal. Chem.*, **45**, 2344 (1973).

65. D. E. Martire and J. H. Purnell, *Trans. Faraday Soc.*, **62**, 710 (1966).

66. D. E. Martire, *Anal. Chem.*, **46**, 626 (1974).

67. R. A. Hively, *Anal. Chem.*, **35**, 1921 (1963).

68. E. L. Eliel, *Elements of Stereochemistry*, Wiley, New York, 1969.

69. F. D. Gunstone, *Basic Stereochemistry*, English Universities Press, London, 1974.

70. J. J. Walraven, A. W. Ladon, and A. I. M. Keulemans, *Chromatographia*, **1**, 195 (1968).

71. L. M. McDonough and D. A. George, *J. Chromatogr. Sci.*, **8**, 158 (1970).

72. Z. A. Radyuk, G. Y. Kabo, and D. N. Andreyevskii, *Neftekhim.*, **12**, 679 (1972).

73. F. Vernon, *J. Chromatogr.*, **87**, 29 (1973).

74. J. K. Haker, *J. Chromatogr. Sci.*, **11**, 144 (1973).

75. D. A. Leathard and B. C. Shurlock, *Identification Techniques in Gas Chromatography*, Wiley-Interscience, London, 1970, p. 51.

76. J. W. Cornforth, R. H. Cornforth, and K. K. Mathew, *J. Chem. Soc.*, 112 (1959).

77. A. Pelter, C. Subrahmanyam, R. J. Laub, K. J. Gould, and C. R. Harrison, *Tetrahedron Lett.*, 1633 (1975).

78. A. Pelter, T. W. Bentley, C. R. Harrison, C. Subrahmanyam, and R. J. Laub, *J. Chem. Soc. Perkin Trans. 1*, 2419 (1976).

79. B. L. Karger, R. L. Stern, H. C. Rose, and W. Keane, in *Gas Chromatography 1966*, A. B. Littlewood, Ed., Institute of Petroleum, London, 1967, p. 240.

80. E. Gil-Av and D. Nurok, *Proc. Chem. Soc.*, 146 (1962).

81. R. Charles-Sigler, G. Fischer, and E. Gil-Av, *Israel J. Chem.*, **1**, 234 (1963).

82. E. Gil-Av, R. Charles-Sigler, G. Fischer, and D. Nurok *J. Gas Chromatogr.*, **4**, 51 (1966).

83. E. Gil-Av, B. Feibush, and R. Charles-Sigler, *Tetrahedron Lett.*, 1009 (1966).

84. E. Gil-Av, G. Feibush, and R. Charles-Sigler, in *Gas Chromatography 1966*, A. B. Littlewood, Ed., Institute of Petroleum, London, 1967, p. 277.

85. E. Gil-Av and B. Feibush, *Tetrahedron Lett.*, 3345 (1967).

86. B. Feibush and E. Gil-Av, *Tetrahedron*, **26**, 1361 (1970).

87. J. A. Corbin, J. E. Rhoad, and L. B. Rogers, *Anal. Chem.*, **43**, 327 (1971).

88. J. A. Corbin and L. B. Rogers, *Anal. Chem.*, **42**, 974 (1970).

89. S. Nakaparksin, P. Birrell, J. Oro, and E. Gil-Av, *J. Chromatogr. Sci.*, **8**, 177 (1970).

90. W. Koenig, W. Parr, H. A. Lichtenstein, E. Bayer, and J. Oro, *J. Chromatogr. Sci.*, **8**, 183 (1970).

91. W. Parr, J. Pleterski, C. Yang, and E. Bayer, *J. Chromatogr. Sci.*, **9**, 141 (1971).

92. W. Parr, C. Yang, E. Bayer, and E. Gil-Av, *J. Chromatogr. Sci.*, **8**, 591 (1970).

93. W. Parr, C. Yang, J. Pleterski, and E. Bayer, *J. Chromatogr.*, **50**, 510 (1970).

94. W. Parr and P. Howard, *Chromatographia*, **4**, 162 (1971).

95. L. Grohmann and W. Parr, *Chromatographia*, **5**, 8 (1972).

96. W. Parr and P. Howard, *J. Chromatogr.*, **71**, 193 (1972).

97. W. Parr and P. Howard, *Anal. Chem.*, **45**, 711 (1973).

98. R. Brazell, W. Parr, and A. Zlatkis, *Chromatographia*, **9**, 57 (1976).

99. W. A. König, *Chromatographia*, **9**, 72 (1976).

100. B. Halpern, in *Handbook of Derivatives for Chromatography*, K. Blau and G. S. King, Eds., Heydon, New York, 1977, Chapter 11.

101. J. W. Westley and B. Halpern, in *Gas Chromatography 1968*, C. L. A. Harbourn, Ed., Institute of Petroleum, London, 1969, p. 119.

102. R. K. Cahn, C. K. Ingold, and V. Prelog, *Experentia*, **12**, 81 (1956).

103. R. J. Laub, V. Ramamurthy, and R. L. Pecsok, *Anal. Chem.*, **46**, 1659 (1974).

104. A. R. Cooper, C. W. P. Crowne, and P. G. Farrell, *Trans. Faraday Soc.*, **62**, 2725 (1966).

105. A. R. Cooper, C. W. P. Crowne, and P. G. Farrell, *Trans. Faraday Soc.*, **63**, 447 (1967).

106. G. Briegleb and J. Czekalla, *Z. Elecktrochem.*, **63**, 6 (1959).

107. P. G. Farrell and J. Newton, *J. Phys. Chem.*, **69**, 3506 (1965).

108. E. G. McRae and L. Goodman, *J. Chem. Phys.*, **29**, 334 (1958).

109. V. Ramamurthy, Y. Butt, C. Yang, P. Yang, and R. S. H. Liu, *J. Org. Chem.*, **38**, 1247 (1973).

110. B. Honig, B. Hudson, S. D. Sykes, and M. Karplus, *Proc. Nat. Acad. Sci.*, **68**, 1289 (1971).

111. V. Ramamurthy, T. T. Bopp, and R. S. H. Liu, *Tetrahedron Lett.*, 3915 (1972).

112. R. J. Laub and R. L. Pecsok, *Anal. Chem.*, **46**, 1214 (1974).

113. R. J. Laub and R. L. Pecsok, *Chromatogr. Rev.*, **19**, 47 (1975).

114. W. B. Forbes, R. Shilton, and A. Balasubramanian, *J. Org. Chem.*, **29**, 3527 (1964).

115. R. J. Laub and J. H. Purnell, *J. Am. Chem. Soc.*, **98**, 35 (1976).

116. R. H. Mizzoni and P. E. Spoerri, *J. Am. Chem. Soc.*, **76**, 2201 (1954).

117. K. Eichenberger, A. Staehelin, and J. Druey, *Helv. Chim. Acta*, **37**, 837 (1954).

118. W. E. Wentworth and R. S. Becker, *J. Am. Chem. Soc.*, **84**, 4263 (1962).

119. J. E. Lovelock, *Nature*, **189**, 729 (1961).

120. J. E. Lovelock, A. Zlatkis, and R. S. Becker, *Nature*, **193**, 540 (1962).

121. J. E. Lovelock and S. R. Lipsky, *J. Am. Chem. Soc.*, **82**, 431 (1960).

122. P. Devaux and G. Guiochon, *Bull. Soc. Chim. Fr.*, 1404 (1966); *J. Gas Chromatogr.*, **5**, 341 (1967).

123. W. E. Wentworth and E. Chen, *J. Gas Chromatogr.*, **5**, 170 (1967).

124. T. Fujinaga and Y. Ogino, *Bull. Chem. Soc. Jap.*, **40**, 434 (1967).

125. J. E. Lovelock, in *Gas Chromatography 1968*, C. L. A. Harbourn, Ed., Institute of Petroleum, London, 1969, p. 95; *J. Chromatogr.*, **99**, 3 (1974).

126. J. E. Lovelock, K. W. Charlton, and P. G. Simmonds, *Anal. Chem.*, **41**, 1048 (1969).

127. A. Zlatkis and D. Fenimore, *Usp. Khromatogr.*, 236 (1972).

128. W. A. Aue and S. Kapila, *J. Chromatogr. Sci.*, **11**, 255 (1973).

129. C. A. Burgett, *Res./Dev.*, **25**(11), 28 (1974).

130. E. D. Pellizzari, *J. Chromatogr.*, **98**, 323 (1974).

131. C. R. Hastings, T. R. Ryan, and W. A. Aue, *Anal. Chem.*, **47**, 1169 (1975).

132. T. L. Hill, *Introduction to Statistical Thermodynamics*, Addison-Wesley, Reading, Mass., 1960, Ch. 10.

133. J. R. Hoyland and L. Goodman, *J. Chem. Phys.*, **36**, 21 (1962).

134. R. L. Grob and E. J. McGonigle, *J. Chromatogr.*, **59**, 13 (1971).

135. W. F. Fischer and W. H. King, Jr., *Anal. Chem.*, **39**, 1265 (1967).

136. V. R. Huebner, *Anal. Chem.*, **34**, 488 (1962).

137. D. Gaede and C. E. Meloan, *Anal. Chem.*, **43**, 1515 (1971).

138. D. Noel and C. E. Meloan, *Sep. Sci.*, **7**, 389 (1972).

139. P. Becker and R. L. Birkmeier, *J. Am. Oil Chem. Soc.*, **41**, 169 (1964).

140. M. Wurst, *Mikrochim. Acta*, 379 (1966).

141. J. Varsano and S. Gilbert, *J. Pharm. Sci.*, **62**, 87, 92 (1973).

142. E. F. Barry, R. S. Fischer, and D. M. Rosie, *Anal. Chem.*, **44**, 1559 (1972).

143. E. F. Barry and D. M. Rosie, *J. Chromatogr.*, **59**, 269 (1971); **63**, 203 (1971).

144. E. F. Barry, R. Trakimas, and D. M. Rosie, *J. Chromatogr.*, **73**, 226 (1972).

145. R. W. Ohline, *Anal. Chem.*, **37**, 93 (1965).

146. R. A. Schep and K. De Clerk, *J. Chromatogr. Sci.*, **10**, 530 (1972).

Index